International Review of Cytology

A Survey of Cell Biology

Cytology | Cell Biology

VOLUME 136

International Review of Cytology
Cytology

A Survey of
Cell Biology

Edited by

Kwang W. Jeon
Department of Zoology
The University of Tennessee
Knoxville, Tennessee

Martin Friedlander
Jules Stein Eye Institute
UCLA School of Medicine
Los Angeles, California

VOLUME 136

Academic Press, Inc.
Harcourt Brace Jovanovich, Publishers
San Diego   New York   Boston   London   Sydney   Tokyo   Toronto

Academic Press, Inc.
1250 Sixth Avenue, San Diego, California 92101-4311

United Kingdom Edition published by
Academic Press Limited
24–28 Oval Road, London NW1 7DX

Library of Congress Catalog Number: 52-5203

International Standard Book Number: 0-12-364536-0

PRINTED IN THE UNITED STATES OF AMERICA
92 93 94 95 96 97   BB   9 8 7 6 5 4 3 2 1

CONTENTS

Functional Components of Microtubule-Organizing Centers

Mary Kimble and Ryoko Kuriyama

Molecular and Cellular Basis of Formation, Hardening, and Breakdown of the Egg Envelope in Fish

K. Yamagami, T. S. Hamazaki, S. Yasumasu, K. Masuda, and I. Iuchi

Ganglia within the Gut, Heart, Urinary Bladder, and Airways: Studies in Tissue Culture

M. J. Saffrey, C. J. S. Hassall, T. G. J. Allen, and G. Burnstock

The Nuclear Envelope of the Yeast *Saccharomyces cerevisiae*

Eduard C. Hurt, Ann Mutvei, and Maria Carmo-Fonseca

The Specialized Junctions of the Lens

G. A. Zampighi, S. A. Simon, and J. E. Hall

Colloidal Gold and Its Application in Cell Biology

Marc Horisberger

Tracheary Element Formation as a Model System of Cell Differentiation

Hiroo Fukuda

CONTRIBUTORS

Numbers in parentheses indicate the pages on which the authors' contributions begin.

T. G. J. Allen (93), *Department of Pharmacology, University College London, London WC1E 6BT, England*

G. Burnstock (93), *Department of Anatomy and Developmental Biology, University College London, London WC1E 6BT, England*

Maria Carmo-Fonseca (145), *European Molecular Biology Laboratory, D-6900 Heidelberg, Germany*

Hiroo Fukuda (289), *Biological Institute, Tohoko University, Sendai 980, Japan*

J.E. Hall (185), *Department of Physiology and Biophysics, University of California, Irvine, Irvine, California 92717*

T. S. Hamazaki (51), *Life Science Institute, Sophia University, Tokyo 102, Japan*

C. J. S. Hassall (93), *Department of Anatomy and Developmental Biology, University College London, London WC1E 6BT, England*

Marc Horisberger (227), *Nestec Ltd., CH-1800 Vevey, Switzerland*

Eduard C. Hurt (145), *European Molecular Biology Laboratory, D-6900 Heidelberg, Germany*

I. Iuchi (51), *Life Science Institute, Sophia University, Tokyo 102, Japan*

Mary Kimble (1), *Department of Cell Biology and Neuroanatomy, University of Minnesota, Minneapolis, Minnesota 55455*

Ryoko Kuriyama (1), *Department of Cell Biology and Neuroanatomy, University of Minnesota, Minneapolis, Minnesota 55455*

K. Masuda (51), *Life Science Institute, Sophia University, Tokyo 102, Japan*

Ann Mutvei (145), *European Molecular Biology Laboratory, D-6900 Heidelberg, Germany*

M. J. Saffrey (93), *Department of Anatomy and Developmental Biology, University College London, London WC1E 6BT, England*

S. A. Simon (185), *Department of Neurosciences, Duke University Medical Center, Durham, North Carolina 27710*

K. Yamagami (51), *Life Science Institute, Sophia University, Tokyo 102, Japan*

S. Yasumasu (51), *Life Science Institute, Sophia University, Tokyo 102, Japan*

G. A. Zampighi (185), *Department of Anatomy and Cell Biology, UCLA School of Medicine, Los Angeles, California 92004*

Functional Components of Microtubule-Organizing Centers

Mary Kimble and Ryoko Kuriyama
Department of Cell Biology and Neuroanatomy, University of Minnesota,
Minneapolis, Minnesota 55455

I. Introduction and Background

Long before techniques were developed which allowed single microtubules to be visualized, the early cell biologists described a cellular structure which was associated with the poles of the mitotic spindle apparatus. Van Beneden and Boveri independently recognized that the spindle pole persisted throughout the cell cycle and suggested that this structure, ultimately named the centrosome, was the true division center of the cell (Fulton, 1971; Mazia, 1987). With the advent of electron microscopy and aldehyde fixatives, it became apparent that microtubules are nearly ubiquitous components of eukaryotic cells and that the microtubules are organized around specific cellular structures (Porter, 1966). These structures have come to be referred to collectively as microtubule-organizing centers (MTOCs) (Pickett-Heaps, 1969). At least three types of MTOCs are found in all cell types: (1) a centrosome, or equivalent structure, which organizes the cytoplasmic microtubule array of interphase cells; (2) the spindle poles, which organize the microtubules of the spindle apparatus; and (3) kinetochores, which are specialized regions at which the chromosomes attach to the microtubules of the spindle apparatus. Centrosomes and spindle poles function both to nucleate microtubules and to determine the three-dimensional arrangement of microtubules in the cell. Kinetochores, on the other hand, do not nucleate microtubules *in vivo* but affect microtubule distribution in the spindle apparatus by binding to and apparently stabilizing microtubules. Electron microscopic studies have shown that the structure of MTOCs varies widely in different cell types. Some cells have very precise structures, such as the trilaminar spindle pole bodies (spbs) of *Saccharomyces cerevisiae*, and the basal bodies with their associated structures, which in many protists are responsible for organization of the cytoplasmic microtubules. The centrosome (the primary MTOC) of most animal cells is less well defined, being

1

composed of a cloud of electron-dense material which may or may not be associated with a pair of centrioles. Plants take the prize for the least well characterized MTOCs. Although plant cells do assemble organized arrays of microtubules, structures analogous to the spbs or centrioles are not generally associated with these MTOCs. The MTOCs of plants are, however, usually associated with membranous structures, such as the nuclear envelope, the plasma membrane, and membrane-bound vesicles (Wick, 1985a). Recent studies have shown that plants share some MTOC antigens with animals and fungi (Clayton *et al.*, 1985; Wick, 1985b; Vandré *et al.*, 1986), suggesting that basic functional components of MTOCs have been conserved throughout eukaryotic evolution. (For detailed discussions of MTOC structures, see reviews by Raff, 1979; Heath, 1981; McIntosh, 1983; Brinkley, 1985; Wick, 1985a; Mazia, 1987; Vorobjev and Nadezhdina, 1987.)

During the past 20 years, much progress has been made in understanding the behavior and roles of microtubule-containing structures through the cell cycle. In recent years, a number of laboratories have initiated studies designed to probe the composition of MTOCs, with the goal of identifying the essential components and their respective functions. These studies are the subject of this review. Before beginning a discussion of the individual components that have been identified, it is useful to outline briefly what has been learned about the dynamics of MTOCs during the cell cycle. Most of the studies thus far have focused on the centrosome of animal cells. The brief description given below is based on studies involving either tissue culture cells or embryos of different species. It should be kept in mind that not all of the details apply to both of these systems, nor will these details necessarily apply to other systems. We have not attempted to provide a complete review of the literature that is summarized below, as this literature has been reviewed elsewhere (in addition to the reviews listed above, see McIntosh and Koonce, 1989; Mitchison and Sawin, 1990).

A. Centrosome Behavior during the Cell Cycle

During interphase, a single centrosome serves as the focal point for most of the microtubules of the cytoskeleton. Although individual microtubules have been shown to be dynamic, undergoing alternating periods of growth and disassembly, the overall array of microtubules is maintained throughout the interphase stage. The centrosome is composed of a cloud of electron-dense material which, in animal cells, is usually associated with a pair of centrioles. Because of this association with the centrioles, the electron-dense material is usually referred to as the pericentriolar material (PCM). Both *in vivo* and *in vitro* analyses of microtubule assembly have shown that the microtubules of the interphase cytoskeleton and the mitotic spindle apparatus nucleate from the PCM rather than from the centrioles (Gould and Borisy, 1977; Rieder and Borisy, 1982; Vorobjev and

Chentsov, 1982). The role of the centrioles in the centrosome is not known. Although centrioles are usually seen in animal cells, there are many examples of animal cells that lack centrioles; thus they are not absolutely required for microtubule organization. Regardless of their role in the centrosome the centrioles do serve as a useful marker for the stage of the cell cycle, as behavior of the centrioles is tightly regulated during the cell cycle (Kuriyama and Borisy, 1981a). During late G1–S phase the centrioles duplicate. In tissue culture cells the duplication of the centrioles appears to require a signal from the nucleus, as when cells are enucleated during the G1 phase, duplication does not occur (Kuriyama and Borisy, 1981a). Duplication of the centrioles does not, however, require DNA synthesis, as duplication can occur in cells in which DNA replication has been blocked with DNA synthesis inhibitors (Rattner and Phillips, 1973; Kuriyama and Borisy, 1981a; Sluder and Lewis, 1987; Raff and Glover, 1988; Sawin and Mitchison, 1991a).

At the onset of mitosis, the G2/M transition, the cytoskeletal microtubules disassemble, the duplicated centrosomes separate, and the spindle microtubules begin to assemble between the separating centrosomes. Immunofluorescence studies have shown that a number of mitosis-specific components become associated with the centrosome (Chaly et al., 1984; McCarty et al., 1984; Pettijohn et al., 1984; Newmeyer and Ohlsson-Wilhelm, 1985; Senécal et al., 1985; Sager et al., 1986; Kellogg et al., 1989; Maekawa et al., 1991; Sellitto et al., 1992), while other components are posttranslationally modified (Davis et al., 1983; Vandré et al., 1984) at this time. Changes are also seen in the organization of the PCM at the G2/M transition. During interphase the PCM is characterized by the presence of clumps of material that have been referred to as PCM satellites. At the G2/M transition the PCM satellites begin to disappear and are replaced by an expanding halo of fibrous-appearing material (Robbins et al., 1968; Rieder and Borisy, 1982; Vorobjev and Chentsov, 1982; Baron and Salisbury, 1988). In vitro studies have shown that the nucleating capacity of centrosomes increases dramatically at the onset of mitosis (Snyder and McIntosh, 1975; Kuriyama and Borisy, 1981b). It has been suggested that the PCM satellites represent condensed foci of microtubule nucleating material which decondense during mitosis, and result in the availability of an increased number of microtubule nucleating sites (Rieder and Borisy, 1982). As the spindle apparatus assembles between the separating centrosomes (spindle poles), some of the microtubules radiating from the poles come into contact with the condensing chromosomes. Those which contact kinetochores are captured and stabilized, and become involved in congression of the chromosomes to the metaphase plate (Nicklas et al., 1979; Mitchison and Kirschner, 1985; Ghosh and Paweletz, 1987; Sawin and Mitchison, 1991a; Rieder, 1991). Studies using biotin or fluorescently tagged tubulin have recently shown that, during prometaphase congression, elongation of the kinetochore microtubules occurs primarily at the kinetochore, while disassembly occurs at the spindle poles (Mitchison et al., 1986; Sawin and Mitchison, 1991b).

The onset of anaphase is marked by the separation of the sister centromeres and movement of the chromosomes to opposite spindle poles (anaphase A). At the same time, the distance between the spindle poles increases (anaphase B). Anaphase A movement appears to involve two distinct processes: (1) disassembly of the chromosome to pole microtubules, with most of the disassembly occurring at the kinetochore (Mitchison *et al.*, 1986; Gorbsky *et al.*, 1988), and (2) movement of kinetochores along microtubules (Nicklas, 1989; Hyman and Mitchison, 1991). Meanwhile, microtubule assembly (elongation) continues in the spindle midzone, the area between the separating sets of chromosomes, and is in part responsible for the continued separation of the spindle poles (anaphase B movement) (reviewed in McIntosh and Koonce, 1989). During anaphase, a decrease in the microtubule nucleating activity of the poles has been documented (Snyder *et al.*, 1982; Kuriyama, 1984), which coincides with the reappearance of the PCM satellites (Robbins *et al.*, 1968; Rieder and Borisy, 1982) and dephosphorylation of centrosomal components (Vandré and Borisy, 1989). As the cells progress through anaphase and telophase, the spindle microtubules are gradually disassembled.

In addition to the structural changes in the PCM material seen in cultured mammalian cells, changes in the distribution of the PCM material have also been documented in sea urchin (Paweletz *et al.*, 1984; Mazia, 1987; Schatten *et al.*, 1987; for review, see Leslie, 1990) and *Drosophila* embryos (Callaini and Riparbelli, 1990). During interphase through metaphase the centrosomes and poles have a spherical appearance. During anaphase the PCM appears to spread, and by telophase it forms a plate or cap at each pole. With the entrance of the cell back into interphase, the PCM again takes on a spherical distribution (Paweletz *et al.*, 1984; Mazia, 1987; Schatten *et al.*, 1987; Callaini and Riparbelli, 1990). Changes in the distribution of the PCM during the cell cycle have also been reported in PtK cells (Joswig and Petzelt, 1990), indicating that the ability to alter the distribution of the PCM is not unique to rapidly dividing embryonic cells.

As can be seen from the foregoing discussion, centrosomes are dynamic cell organelles which follow a strict pattern of changes during the cell cycle. The changes that are seen include (1) duplication of the centrioles during G1–S phase; (2) changes in the overall three-dimensional organization of the centrosomal material, from spherical to platelike to spherical, as the cell progresses through mitosis and reenters interphase; (3) changes in the organization of components within the centrosome, in particular the reorganization of the PCM satellites to form a fibrous halo at the onset of mitosis and reformation of the satellites at the end of mitosis; (4) increase in the microtubule nucleating capacity at the onset of mitosis. The increase in nucleating capacity could involve (a) changes in composition, (b) posttranslational modification of centrosomal components, and/or (c) changes in the organization of the nucleating materials, thereby making additional nucleation sites available.

B. Functional Requirements of MTOCs

Porter (1966) noted that, in order for microtubules to influence cell shape, their distribution within the cell must be regulated, with one end of the microtubules anchored while the other end is free to grow. Porter also postulated that (1) microtubule-initiating (nucleating) sites exist within the cytoplasm, (2) that the nucleating sites form a complex, and (3) that initiation and distribution of microtubules are determined by the spatial and temporal program of the complex (paraphrased from Borisy and Gould, 1977). Or, to phrase it more simply, the fundamental properties of MTOCs likely will include nucleation, orientation, and anchoring of microtubules. Observations of MTOC function, during the intervening years, support this model. As is discussed below, these studies also indicate that the MTOCs affect not only the frequency with which microtubules nucleate, but also determine the structure of the assembled microtubules.

1. Microtubule Nucleation

Clearly, centrosomes and spindle poles must in some way provide an environment which favors microtubule nucleation. Studies of the assembly properties of tubulin, the primary structural component of microtubules, have shown that purified tubulin does not spontaneously assemble into microtubules *in vitro* at physiological concentrations. In general, microtubule assembly *in vitro* requires concentrations of tubulin that are substantially higher than are seen in cells, or nonphysiological ionic conditions (Himes *et al.*, 1977; Mitchison and Kirschner, 1984). Addition of isolated centrosomes to these protein preparations allows microtubule assembly to occur at physiological protein concentrations (Mitchison and Kirschner, 1984). Karsenti *et al.* (1984) demonstrated that, in enucleated cells (cytoplasts), the number and arrangement of microtubules depended on whether the centrosome was retained. Cytoplasts which retained the centrosome assemble a normal-appearing array of cytoplasmic microtubules. Acentrosomal cytoplasts, those which had lost their centrosome, assembled few microtubules. The microtubules that assembled in the acentrosomal cytoplasts were randomly distributed through the cytoplasm, rather than being organized in an astral-like array. This difference was not due to a difference in the amount of tubulin dimers available for assembly, as treatment of acentrosomal cytoplasts with taxol resulted in the assembly of large bundles of microtubules. It was suggested that the centrosome affects the number of microtubules assembled by capping the minus (proximal) end, which has a greater critical concentration for assembly. Under these conditions, the extent of polymerization would be determined by the lower critical concentration required for assembly at the plus (distal) end and would result in an increase in total polymer. In the acentrosomal cytoplasts, the microtubules that assemble result from spontaneous polymerization, and the final amount of polymer that is formed represents a balance between the assembly and disassembly rates at the two ends of the microtubule (Karsenti *et al.*, 1984). If the increase in polymer mass were only a function of the capping of the microtubules, one could just as well achieve the increased amount of polymer by making a few, very long microtubules rather than many shorter microtubules. Thus, these results also indicate that the centrosome provides nucleating sites for microtubule assembly, since the number of microtubules assembled is greater in centrosomal cytoplasts than in acentrosomal cytoplasts.

Studies of microtubule assembly *in vitro* suggest that nucleation of microtubule assembly by centrosomes also influences the structure of microtubules. When microtubules are assembled *in vitro* from either microtubule proteins (Scheele *et al.*, 1982) or purified tubulin (Evans *et al.*, 1985), the majority of the microtubules are composed of 14 or 15 protofilaments. Microtubules nucleated *in vitro* from centrosomes, however, are composed primarily of 13 protofilaments (Evans *et al.*, 1985), as are the vast majority of microtubules assembled *in vivo* (Tilney *et al.*, 1973).

2. Organization of Microtubule Arrays

Borisy and Gould (1977) suggested that the organization of microtubule arrays is most likely a function of the inherent structure of the microtubule nucleating elements (MTNEs) and the manner in which these elements are linked together to form a MTOC. Mazia (1987) added flexibility to this model by proposing that the microtubule nucleating elements were attached to a linear element, and that the distribution of nucleating elements, and thus microtubules, was a function of the three-dimensional organization of the linear element. We would suggest that the centrosome may involve an additional level of organization.

A model for how we imagine that the components of the centrosome are organized is given in Fig. 1A–C. The model involves three levels of organization: (1) Elements that nucleate single microtubules (MTNEs) and which, as suggested by Borisy and Gould (1977), confer directionality on the microtubule (Fig. 1A). (2) At the second level of organization, MTNEs are held together in clusters (Fig. 1B), which we will refer to as microtubule nucleating centers (MTNCs). The organization of MTNEs into MTNCs could be accomplished by interconnections between the individual elements or by attachment of the MTNEs to a structural component of some type. (3) The third level of organization is the centrosome or MTOC, which is composed of multiple MTNCs organized within a matrix, which provides the overall integrity of the centrosome (Fig. 1C). The centrosome matrix could be flexible to allow for the changes in the distribution of the PCM that have been documented in sea urchin (Paweletz *et al.*, 1984; Mazia, 1987; Schatten *et al.*, 1987) and *Drosophila* (Callaini and Riparbelli, 1990) embryos, and in PtK cells (Joswig and Petzelt, 1990), while maintaining the integrity of the centrosome. Flexibility is also necessary within the MTNCs to allow for changes in the

A

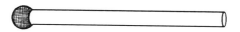

B

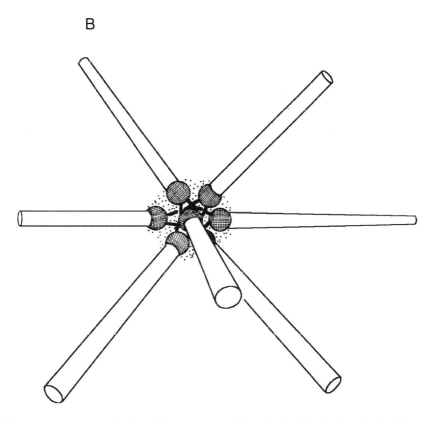

FIG. 1   A model for organization of the centrosome. (A) A single microtubule nucleating element
(MTNE) and microtubule. (B) Single MTNEs are organized into clusters, or microtubule nucleating
centers (MTNCs), held together by structural components. (C) Multiple MTNCs are organized within
a matrix, which provides the overall integrity of the centrosome. (D) A whole mount electron micrograph of an isolated CHO centrosome, onto which microtubules have been assembled. Note that some
of the microtubules form clusters (indicated by the arrowheads), located a short distance from the
centriole. (D is taken from Kuriyama, 1984.)

C

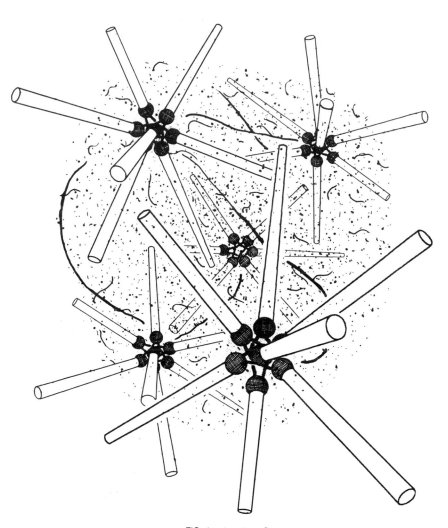

FIG. 1   *(continued)*

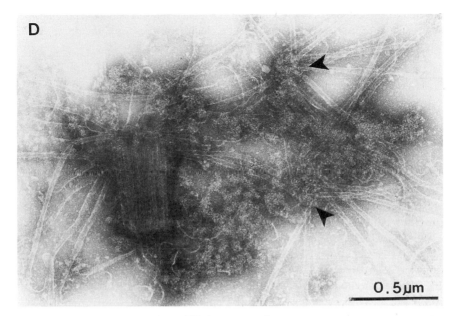

FIG. 1 *(continued)*

orientation of microtubules, and perhaps in the distribution of the MTNEs within the clusters.

The suggestion that the centrosome has a subunit type of organization is based on the following observations. When cells that have been blocked in mitosis, by microtubule-disrupting drugs, are allowed to recover, multipolar spindles frequently form. EM analysis has shown that PCM is associated with each of these spindle poles (Sellitto and Kuriyama, 1988a), indicating that during drug treatment the PCM can become dispersed. Recent data, using anti-centrosome antibodies to observe the distribution of PCM components in drug-treated cells, indicate that the partially dispersed PCM forms numerous discrete foci (clusters of nucleating elements) in the cytoplasm (Sager *et al.*, 1986; Maekawa *et al.*, 1991; Maekawa and Kuriyama, 1991). These observations suggest that the PCM may be composed of packets or building blocks of nucleating materials (i.e., the MTNCs), represented by the foci seen with immunofluorescence. When the cells are blocked at mitosis by treatment with microtubule-disrupting drugs, the MTNCs either disperse, or, if newly synthesized, are unable to aggregate. During recovery from the drug treatment the MTNCs aggregate and eventually form centrosomes or spindle poles. Individual MTNCs could be organized around a structural component, while the organization of the MTNCs into a centrosome could involve association with a structural matrix. The organization of MTNEs into clusters, which are in turn organized into the centrosome is also suggested by studies of isolated centrosomes. As is shown in Fig. 1D, when microtubules are polymerized onto isolated centrosomes, then viewed by electron microscopy, the microtubules often form clusters, located a short distance from the centriole.

3. Microtubule Anchoring

Although the suggestion that microtubules are anchored at one end implies a stable association, numerous observations have suggested that this anchoring or attachment could be a dynamic process. During mitosis microtubule disassembly has been shown to occur at both the kinetochore and the spindle poles (Mitchison *et al.,* 1986; Sawin and Mitchison, 1991b). However, there is no evidence that connections between the MTOCs and the microtubules are lost during disassembly, indicating a mechanism for maintaining the connection between the microtubule and MTOCs even as tubulin subunits are lost from the end of the microtubule. During *in vitro* studies of microtubule assembly from isolated centrosomes, it has been observed that occasionally microtubules detach and move away from the centrosome (Belmont *et al.,* 1990). Studies of the disassembly of the interphase microtubule arrays during prophase in sea urchin embryos (Harris *et al.,* 1980) and in *Dictyostelium* (Kitanishi-Yumura and Fukui, 1987) indicate that release of microtubules from the centrosome occurs *in vivo* as well, and suggests that maintenance of the microtubule-MTOC connection is an active process. Ault and Nicklas (1989) have reported that, during reorientation of kinetochores, which had been detached from the spindle by micromanipulation and allowed to attach to the opposite spindle pole, kinetochore-to-pole microtubules sometimes detach from the pole. When this happens, the microtubules splay out from the kinetochore and eventually disassemble. It was suggested that both the mechanical attachment of microtubules to the poles and kinetochores and the force production required for bipolar orientation could be accomplished by microtubule motor proteins (Ault and Nicklas, 1989).

4. Kinetochore Specific Functions

In addition to the microtubule-anchoring components that one would expect to see associated with kinetochores, it is likely that other components are required for chromosome movement along the microtubules. That chromosomes actively move along microtubules is suggested by numerous observations of chromosome behavior during mitosis (see Rieder, 1991, for review). Perhaps the most convincing evidence that kinetochores actively move along microtubules is the study reported by Nicklas (1989). Nicklas showed that, when kinetochore-to-pole microtubules were cut and the pole removed from the spindle apparatus, chromosomes continued to move away from the metaphase plane, along the remnants of the kinetochore-to-pole microtubules. It has also been shown that, *in vitro,* microtubules will slide across isolated kinetochores (Hyman and Mitchison, 1991).

That is, isolated kinetochores retain a microtubule motor function, which, if present *in vivo,* could function to move kinetochores along microtubules rather than the kinetochores passively moving in response to changes in microtubule length.

The kinetochore represents a discrete region of the chromosome to which microtubules attach. DNA sequences specific to kinetochores were first identified in *S. cerevisiae* by Carbon and co-workers (Bloom *et al.,* 1984; Clarke and Carbon, 1985). The kinetochore or centromere DNAs of *S. cerevisiae* have a relatively simple structure. They are composed of a single 250-bp sequence of DNA. This 250-bp sequence can be further subdivided into three distinct elements, all of which are required for centromere function (Fitzgerald-Hayes *et al.,* 1982). The kinetochores of *Schizosaccharomyces pombe* are larger and more complex, being composed of 50–100 kb of DNA containing several classes of repetitive sequences, while the kinetochores of higher eukaryotes extend for millions of bases along the chromosome and are composed of both heterochromatin and repetitive DNA sequences (for reviews, see Bloom and Yeh, 1989; Brinkley, 1990). A recent model for kinetochore structure has suggested that the kinetochore may be composed of multiple copies of the centromere-specific DNA interspersed with linker DNA regions (Zinkowski *et al.,* 1991). Association of kinetochore proteins with the centromeric DNA forms subunits which are then assembled into mature kinetochores through coiling of the DNA, thereby positioning the subunits along one face of the condensing chromosome (Zinkowski *et al.,* 1991). Thus, any model for kinetochore structure must include components that are able to bind to the centromeric DNA specifically. The subunit model proposed by Zinkowski *et al.* (1991) would also suggest that some kinetochore components might be required for establishing and maintaining the coiled structure of the DNA.

II. Approaches to the Identification of MTOC Components

Identification of MTOC components that are required for function has long been a goal of cell biologists. The application of biochemical and immunological techniques to this problem has allowed us to begin identifying proteins that copurify with either MTOCs or stabilized microtubules as well as those that colocalize to MTOCs in fixed cells. Most of these proteins are known only at the level of a band on a gel and/or a pattern of fluorescence in cells. Clearly, some of these proteins have been conserved over considerable evolutionary distance. Analyses of mutations which disrupt mitosis or meiosis are also beginning to contribute to our knowledge of what gene products are required for MTOC function and suggest mechanisms by which they act. In this section and the next we attempt to summarize the work that has been done thus far, placing special emphasis on components for which a possible function has been suggested.

A. Biochemical and Immunological Dissection of MTOCs

1. Methods for Identification of MTOC Components

Projects directed toward the biochemical and immunological dissection of MTOCs have been undertaken in a number of laboratories. The earliest studies involved analysis of basal body/centriole components and are not considered here. The association of centrioles with centrosomes is not universal even within animals cells, and is rarely seen in plants. Even in those cell types where centrioles are present, they are not directly involved in microtubule nucleation. Whether they have a role in the organization of the PCM and microtubule arrays is not known. Instead, we will focus on studies that have been undertaken to identify components of centrosomes, spindle pole bodies, and kinetochores. Basically, four different approaches have been taken:

1. MTOCs are isolated and tested for retention of the ability to nucleate microtubules *in vitro* (Bornens *et al.*, 1987; Komesli *et al.*, 1989). The protein components of the preparations are then separated by sodium dodecyl sulfate–polyarylamide gel electrophoresis (SDS-PAGE), with or without further fractionation, and antibodies are raised against the SDS-denatured, purified antigens. This approach has been used in the analysis of isolated mitotic spindles (Toriyama *et al.*, 1988) and basal body/striated rootlet complexes of the alga *Polytomella* (Stearns and Brown, 1979; Atchison and Brown, 1986).

2. MTOCs are isolated and tested as above, but the isolated structures are then used directly as antigens to generate monoclonal antibodies. Hybridoma lines are screened by testing for immunofluorescent staining of either the original antigen preparation or of the relevant structure(s) in fixed cells. The antibodies can then be used in immunoblotting experiments to identify the antigen. Mitotic spindles and/or centrosomes from mammalian cell lines, sea urchins, and *Dictyostelium* (Ring *et al.*, 1980; Kuriyama and Borisy, 1985; Keryer *et al.*, 1989; Joswig and Petzelt, 1990; Sellitto *et al.*, 1992), spindle pole bodies of *S. cerevisiae* (Rout and Kilmartin, 1990), and a kinetochore-enriched fraction from HeLa cells (Tousson *et al.*, 1991) have all been studied in this way. Similarly, monoclonal antibodies which recognize kinetochore components have been recovered by immunizing mice with isolated mitotic chromosome scaffolds (Cooke *et al.*, 1987; Compton *et al.*, 1991).

3. The third approach is based on techniques that were developed for the identification and separation of microtubule-associated proteins. In this procedure, total cellular proteins are isolated and those that bind to microtubules are recovered by passing the sample over a column containing taxol-stabilized microtubules. The microtubule-binding proteins can then be selectively eluted from the column by varying concentrations of salts and/or nucleotide triphosphates, and the partially purified proteins analyzed by SDS-PAGE. Antibodies are then raised against either individual purified proteins or against particular column fractions.

So far, this approach has only been used for identification of *Drosophila* embryo microtubule proteins (Kellogg *et al.*, 1989).

4. Autoimmune sera which specifically recognize either centrosomes or kinetochores have been identified in several laboratories. Preimmune rabbit sera have been described that specifically recognize various types of MTOCs (Connolly and Kalnins, 1978; Sauron *et al.*, 1984). However, the richest source of antisera that specifically recognize MTOCs has been sera obtained from patients who have the CREST autoimmune syndrome (Moroi *et al.*, 1980; McCarty *et al.*, 1981, 1984; Tuffanelli *et al.*, 1983; Senécal *et al.*, 1985). Serendipity has, as usual, also played a role in the identification of MTOC components. A number of potentially interesting MTOC and spindle proteins have been identified by researchers who were looking for something else, such as nuclear matrix proteins (Lydersen and Pettijohn, 1980; Chaly *et al.*, 1984; Newmeyer and Ohlsson-Wilhelm, 1985).

2. Testing MTOC Components for Function

Two basic strategies have been used to test whether antigens identified as components of MTOCs are required for the assembly or functioning of microtubular structures.

1. *In vitro* inhibition test: Cells grown on coverslips are gently lysed *in situ* under conditions that cause disassembly of the endogenous microtubules; the cells are then incubated with an antibody that recognizes the antigen of interest. After washing off unbound antibody, the cells are incubated with purified tubulin, and processed for microscopy to determine whether microtubule assembly is affected by binding of the antibody to the antigen. This procedure asks whether a particular component is required for nucleation of microtubules, but it would be hard to elicit information as to whether the antigen in question might be required for other MTOC functions.

2. *In vivo* functional analysis by microinjection: Antibodies are introduced into living cells and progress through the cell cycle is monitored. This procedure has the advantage over the first method that many aspects of MTOC function can be monitored. However, it requires cell types into which the antibody can be readily introduced without severely damaging the cell, and in which the effects of antibody injection can be easily assayed.

Since antibodies bind to a limited portion of the protein and may not recognize regions of the protein that are important for its function, a negative result obtained by these procedures may be meaningless. This is especially true when testing monoclonal antibodies, which recognize only a single epitope of the antigen. Second, the experiments frequently involve the use of whole immune sera or immunoglobulin fractions purified from serum or ascites fluid. One cannot rule out the possibility that the injected material includes antibodies that are not detectable by immunofluorescence or immunoblotting but are contributing to the observed effects. Regardless of their weaknesses, these techniques can be useful for deciding which of the many components identified through immunofluorescence should be given priority in subsequent studies.

B. Identification of Mutations Which Affect MTOC Function

Recent studies of mutations in *S. cerevisiae, S. pombe, Aspergillus nidulans,* and *D. melanogaster* have also contributed to our understanding of components that are required for MTOC function. One advantage to taking a genetic approach to the study of specific cell organelles is, by using the appropriate controls, one can be reasonably certain that the effects of a single gene product are being studied. This is not to say that analysis of mutations always gives unambiguous results. However, combining genetic analysis with biochemical and molecular studies can permit the investigator to suggest specific functions for the gene product. Fortunately, in all of these species, relatively straightforward, albeit somewhat laborious, techniques are available for the identification and cloning of genes of interest.

Most of the mutations that have been identified in the yeast and fungal species were isolated in screens designed to identify mutations that affect the function of microtubular structures. These mutations include (1) extragenic suppressors of tubulin gene mutations (Weil *et al.,* 1986), (2) mutations that affect heterokaryon formation (Polaina and Conde, 1982; Berlin *et al.,* 1990), (3) mutations that disrupt the cell cycle (Hirano *et al.,* 1986; Enos and Morris, 1990), and (4) mutations that cause diploidization of haploid cells (Baum *et al.,* 1986, 1988). Many of the mutations identified in *Drosophila* were isolated from screens that were designed to look for maternal effect lethal mutations that disrupt early embryogenesis. In *Drosophila,* as in most metazoans, the early rapid division stage of embryogenesis depends on products that are loaded into the egg during oogenesis. Once the cellular blastoderm is formed, most cells undergo only a few additional divisions (Foe and Alberts, 1983). Evidence suggests that although high levels of zygotic gene expression commence at about the time of cellularization, some of the maternally loaded gene products persist throughout embryogenesis (Glover, 1989). Animals homozygous for recessive mutations in genes required for the early mitotic divisions can be recovered from matings of heterozygous parents, since the heterozygous female parent can supply enough wild-type gene product to allow embryogenesis to proceed normally. A second class of *Drosophila* mutations that is being studied are those where the homozygous individuals die at the larval/pupal transition. During larval development, most cell division is confined to the imaginal discs, small pockets of tissue that will give rise to most of the adult structures and to specific stem cells within the brain (the larval neuroblasts). Flies homozygous for mutations in genes required for mitosis could

develop to the larval stage on maternally provided product, but would not be able
to metamorphose due to failure of division of the adult precursor tissues (Gatti and
Baker, 1989; Wilson and Fuller, 1990).

III. MTOC Components

Immunological and genetic analyses indicate that MTOCs are complex structures.
A large number of antibodies (summarized in Table I–III) have been shown to
recognize antigens that colocalize with MTOCs. The immunofluorescent studies
have also shown that the distribution of these antigens is very complex. Some
antigens are associated with MTOCs throughout the cell cycle, while others are
stage specific. Some are associated with more than one type of MTOC, while
others associate not only with MTOCs but with other cellular components as well
(Kellogg et al., 1989; Sellitto et al., 1992). Based on their distribution, as determined by immunofluorescence, MTOC components can be divided into three
broad categories: (1) those that are associated only with MTOCs, (2) those that
associate with both nuclei and MTOCs, and (3) those that associate with MTOCs
and other cellular structures. The antibodies and their antigens from each of these
categories are summarized in Tables I, II, and III, respectively.

Although all of the antibodies listed in Tables I–III have been shown to stain
MTOCs, only a few have been tested to determine whether they can inhibit MTOC
function. It should be kept in mind that not all of these antigens are necessarily
required for MTOC function. Some components could associate with MTOCs,
especially with the centrosome/spindle poles, to ensure equal distribution to the
daughter cells during mitosis, while others might be involved in centrosome
functions not directly related to microtubule nucleation or organization. The
antigens whose antibodies have been shown to inhibit MTOC functions can be
divided into three classes: (1) those involved in kinetochore function, (2) those
whose antibodies inhibit microtubule nucleation, and (3) those whose antibodies
affect the distribution of microtubule arrays.

Mutations which appear to affect MTOC function are summarized in Table IV.
Based on their phenotypes, the mutations can be divided into five classes, depending on whether they appear to affect (1) kinetochore function, (2) microtubule
nucleation, (3) centrosome or spb duplication, (4) centrosome or spb separation,
or (5) centrosome organization. Unfortunately, the products encoded by many of
these genes have not been identified nor have their distributions in the cell been
determined. Thus, in some cases, it is possible that, although mutations in the gene
appear to affect MTOC function, the product of the gene may not be associated
with MTOCs in vivo. One should also keep in mind, when discussing mutations,
that the final phenotype of the mutation may not directly indicate the function of

TABLE I
MTOC-Specific Proteins/Antigens

Antibody	Antigen	Protein	Species + for Ab staining	Functional test[a]	Refs.[b]
I. Kinetochore specific					
EK	Human autoimmune	80 kDa CENP-B	Vertebrates, invertebrates, and plants	(+) *in vivo*	1
ACA	Human autoimmune	17, 80, 140, and 50 kDa (CENP-A to D)	Mammals	(+) *in vivo*	2
cen1, cen2	Human autoimmune	20, 23, and 34 kDa	Ptk$_1$ and mammalian cell lines	(+) *in vitro*	3
II. Centrosome/spindle pole specific					
5051	Human autoimmune	200 kDa (mammals), 59 kDa (yeast)	Ubiquitous	(+) *in yeast*	4–8
0013	Rabbit preimmune	Multiple bands 140–250 kDa	Primates	(−) *in vitro*	9–11
CHO2	CHO centrosomes	66, 225, and 260 kDa	CHO cells	(+) *in vitro*	12
Anti-230K	Polyplastron contractile protein	62–64 kDa doublet	Mammalian cell lines	(+) *in vitro* and *in vivo*	13
Anti-centrin	*Tetraselmis* striated rootlets	20 kDa	Algae, protozoans, and vertebrate tissue culture cells		14–16
Anti-caltractin	20 kDa *Chlamydomonas* centrin homolog		*Chlamydomonas*		16, 17
CTR-532	Human centrosomes	170 kDa	Mammals, *Paramecium*, and *Tetrahymena*		18
Anti-estradiol	Preimmune or autoimmune?		Mammals		19
MPM-13	Mitotic HeLa cell extracts	43 kDa	*Tetrahymena*, mammals		20
S1-4	*Drosophila* microtubule-associated proteins (mt proteins)	222 kDa	*Drosophila*		21
S5-45	*Drosophila* (mt. proteins)	52 kDa	*Drosophila*		21
21D9 group	Yeast spbs	90 kDa	*S. cerevisiae*		22

Name	Source/antigen	kDa	Description	Ref.
SU4	Sea urchin spindles	20, 80, 180 and 190 kDa	Sea urchin embryos	23
AX1	D. discoideum nucleus/NAB	66 and 140 kDa	Dictyostelium, interphase specific	12
POPA	Human autoimmune	100/115 kDa	Mammals, mitosis specific	24
ATP-12	Drosophila mt proteins	77 kDa	Drosophila, mitosis specific	21
S1-1		335 kDa		
S1-6		194 kDa		
S1-20		102 kDa		
S1-24		89 kDa		
S1-25		86 kDa		
HuSP-1	Human autoimmune	80 kDa	Human and sea urchin, mitosis specific	25

III. Centrosome and kinetochore specific

Name	Source/antigen	kDa	Description	Ref.
MPM-1, MPM-2	HeLa mitotic cell extract	Multiple phosphorylated proteins	Ubiquitous/mitosis specific	26–30 MPM-2: (+) in vitro
2D3	Human autoimmune	210 kDa	Primate specific	31
S5-39	Drosophila mt proteins	59 kDa	Drosophila	21
9H8 group	Human mitotic chromosome scaffolds	275 kDa	Human	32
3H1	Human mitotic chromosome scaffolds	205 kDa	Human	32

[a] In vitro antibody inhibition of microtubule nucleation from centrosomes of cells lysed in situ. In vivo antibody inhibition of mitosis by injection into cells or embryos.

[b] List of references: (1) Simerly et al., 1990; (2) Bernat et al., 1990; (3) Cox et al., 1983; (4) Callarco-Gillam et al., 1983; (5) Clayton et al., 1985; (6) Wick, 1985b; (7) Bastmeyer et al., 1986; (8) Snyder and Davis, 1988; (9) Sauron et al., 1984; (10) Gosti-Testu et al., 1986; (11) Gosti et al., 1987; (12) Sellitto et al., 1992; (13) Moudjou et al., 1989; (14) Baron and Salisbury, 1988; (15) Baron et al., 1991; (16) J. Salisbury, personal communication; (17) Huang et al., 1988a; (18) Keryer et al., 1989; (19) Nenci and Marchetti, 1978; (20) Rao et al., 1989; (21) Kellogg et al., 1989; (22) Rout and Kilmartin, 1990; (23) Kuriyama and Borisy, 1985; (24) Sager et al., 1986; (25) Leslie et al., 1991; (26) Davis et al., 1983; (27) Vandré et al., 1984, 1986; (28) Engle et al., 1988; (29) Wordeman et al., 1989; (30) Centonze and Borisy, 1990; (31) Tousson et al., 1991; (32) Compton et al., 1991.

TABLE II
Components Seen in Nucleus and Centrosome

Antibody	Antigen	Protein	Species and structures + for Ab staining[a]	Functional test[b]	Refs.[c]
CHO1	CHO spindles	95 and 105 kDa	CHO, also stains midbody	(+) *in vivo*	1, 2
CHO3	CHO spindles	Multiple bands 225–350 kDa, 225 kDa in sea urchin	CHO, *Dictyostelium* and sea urchin, also stains midbody and cross-reacts with MAP1	(−) *in vitro*	3
51 kDa	Sea urchin spindles	51 kDa, immunologically related to EF1α	Sea urchin	(+) *in vivo*	4–6
SU5	Sea urchin spindles	50 kDa, ~70% identity with EF1α	Sea urchin		7, 8
Anti-cdc2	Yeast fusion protein	34 kDa	Yeast to humans, also faint cytoplasmic and midzone staining		9–12
Anti-cyclin B	Yeast fusion protein	63 kDa	Yeast		9, 13
Anti-cyclin B	*Drosophila* fusion protein	65 kDa	*Drosophila*, accumulates in G2 with peak at metaphase		14, 15
Anti-cyclin A	*Drosophila* fusion protein	60 kDa	*Drosophila*, accumulates in G2 with peak at prophase		14, 15
J17	Rat chromatin	220–230 kDa	Mammals, marsupials, and amphibians, MTOC in mitosis only		16
NuMA	HeLa nuclear matrix	250 kDa	Mammalian cell lines, MTOC in mitosis only		17
2E4	Partially pure NuMA	250 kDa	Mammalian cell lines, MTOC in mitosis only		18
SP-H	Human autoimmune	210–230 kDa	Mammalian cell lines, MTOC in mitosis only		19
2D3	Kinetochore fraction	180 and 210 kDa	Primate specific, also stains midbody		20

18

Name	Source	kDa	Description	Ref
AX5	*D. discoideum* nucleus/NAB	Multiple bands 64–200 kDa	*Dictyostelium*, MTOC in mitosis only	3
Bx63	*Drosophila* embryo nuclei	66 and 185 kDa	*Drosophila* specific, faint nuclear staining	21
F72	Affinity purified 185 kDa + associated peptides	185 kDa	*Drosophila* specific, faint nuclear staining	21
AX4	*D. discoideum* nucleus/NAB	Multiple bands 59–280 kDa	*Dictyostelium*	3
AX6	*D. discoideum* nucleus/NAB	73 and 125 kDa	*Dictyostelium*	3
S1-8	*Drosophila* microtubule (mt) proteins	175 kDa	*Drosophila*, faint staining of spindle and midbody	22
S1-18	*Drosophila* mt proteins	110 kDa	As above but with faint staining of interzone mts	22
S5-32	*Drosophila* mt proteins	68 kDa	*Drosophila*, MTOC anaphase to telophase and early cycle 14	22
MAP1	Bovine brain MAPs	MAPs 1A and B	HeLa, PtK_2, and human skin fibroblasts	23
3G3	Human mitotic chromosomes	205 kDa	Human, kinetochores and midbody	24
2D3 group	Human mitotic chromosomes	205 kDa	Human, midbody	24
1F1 group	Human mitotic chromosomes	205 kDa	Human	24
CTR 393	Human centrosomes	205 kDa	Human and dinoflagellates, MTOC in mitosis only	25

[a] Only structures other than centrosomes and nuclei, which stain with the antibodies, are listed.

[b] *In vitro* antibody inhibition of microtubule nucleation from centrosomes of cells lysed *in situ*. *In vivo* antibody inhibition of mitosis by injection into cells or embryos.

[c] List of references: (1) Sellito and Kuriyama, 1988b; (2) Nislow *et al.*, 1990; (3) Sellito *et al.*, 1992; (4) Toriyama *et al.*, 1988; (5) Sakai *et al.*, 1989; (6) Ohta *et al.*, 1990; (7) Kuriyama and Borisy, 1985; (8) Kuriyama *et al.*, 1990; (9) Alfa *et al.*, 1990; (10) Bailly *et al.*, 1989; (11) Riabowol *et al.*, 1989; (12) Rattner *et al.*, 1990; (13) Booher *et al.*, 1989; (14) Whitfield *et al.*, 1990; (15) Lehner and O'Farrell, 1990; (16) Newmeyer and Ohlsson-Wilhelm, 1985; (17) Lydersen and Pettijohn, 1980; (18) Pettijohn *et al.*, 1984; (19) Maekawa *et al.*, 1991; (20) Tousson *et al.*, 1991; (21) Whitfield *et al.*, 1988; (22) Kellogg *et al.*, 1989; (23) Sato *et al.*, 1983; (24) Compton *et al.*, 1991; (25) Perret *et al.*, 1991.

19

TABLE III

Components Seen in MTOCs and Other Organelles/Structures

Antibody	Antigen	Protein	Species and structures + for Ab staining[a]	Functional test[b]	Refs.[c]
ATP-2	*Drosophila* mt proteins	205 kDa	*Drosophila*, faint mt staining		1
ATP-5	*Drosophila* mt proteins	147 kDa	*Drosophila*, faint mt staining after cycle 14		1
S1-28	*Drosophila* mt proteins	79 kDa	*Drosophila*, weak spindle staining		1
S5-47	*Drosophila* mt proteins	50 kDa	*Drosophila*, weak spindle staining		1
34E12	Yeast spbs	80 kDa	*Saccharomyces cerevisiae*, punctate spindle		2
CR	Human sera preimmune?	112 kDa	*Pales* spermatocytes, centrosome, and midbody		3
154	Human mitotic chromosomes	250 kDa	Human, kinetochore and midbody		4
3D3	Mitotic chromosome scaffolds	135 and 155 kDa	Chicken, kinetochore and chromosome arms through metaphase, midbody in anaphase and telophase		5
CLiPs	Human autoimmune	50 kDa	Indian muntjac, kinetochore and chromosome arms		6
Anti-kinesin	Bovine brain kinesin		Bovine, squid, and *Drosophila* kinesin, membranous components		7
CHO5	CHO spindles		Mammalian cells, Golgi	(−) *in vitro*	8
CTR 210	Human centrosomes	72 kDa	Human and dinoflagellates, golgi		9

Antibody	Antigen/source	Bands	Staining pattern	Ref.
Anti-dynein	Chick brain cytoplasmic dynein		Chick embryo, fibroblasts and CHO, punctate in cytoplasm in interphase. Spindle poles, kinetochores, and some spindle fibers during mitosis	10
Anti-dynein	*Dictyostelium* cytoplasmic dynein		*Dictyostelium*, variable nuclear and punctate cytoplasmic staining	11
R-1	*Drosophila* embryo cytoplasmic dynein		*Drosophila*, HeLa and PtK staining as above in CHO and chick	12
C-1	79-kDa protein that copurifies with *Drosophila* dynein		*Drosophila*, HeLa and PtK staining as above in CHO and chick	12
AX7	*D. discoideum* nucleus/NAB	Multiple bands 34–255 kDa	*Dictyostelium*, cytoplasmic granules in interphase, spindle poles during mitosis	8
CHO 6	CHO spindles	135, 160, and 215 kDa	CHO, stress fibers (–) *in vitro*	8

[a] Only structures, other than MTOCs, which stain with the antibodies are listed. MTOC staining is centrosomal unless otherwise indicated.

[b] *In vitro* antibody inhibition of microtubule nucleation from centrosomes of cells lysed *in situ*. *In vivo* antibody inhibition of mitosis by injection into cells or embryos.

[c] List of references: (1) Kellogg *et al.*, 1989; (2) Rout and Kilmartin, 1990; (3) Bastmeyer and Russell, 1987; (4) Compton *et al.*, 1991; (5) Cooke *et al.*, 1987; (6) Rattner *et al.*, 1988; (7) Neighbors *et al.*, 1988; (8) Sellitto *et al.*, 1992; (9) Perret *et al.*, 1991; (10) Steuer *et al.*, 1990; (11) Koonce and McIntosh, 1990; (12) Pfarr *et al.*, 1990.

21

TABLE IV

Mutations Affecting MTOC and Spindle Function

Mutation	Species/protein	Phenotype	Refs.[a]
ncd	Drosophila, kinesin-like	High rates of meiotic chromosome nondisjunction and chromosome loss, disorganized spindle poles	1–4
nod	Drosophila, kinesin-like	Nondisjunction of nonexchange bivalents, nonexchange chromosomes do not remain associated with karyosphere	5, 6
cut7	Schizosaccharomyces pombe, kinesin-like	Cells blocked in mitosis, spbs duplicate but fail to separate	7
bimC	Aspergillus nidulans, kinesin-like	Same as cut7	8
kar3	Saccharomyces cerevisiae, kinesin-like	Nuclear fusion and spindle elongation blocked	9
spo15	S. cerevisiae, dynamin-like	spbs in meiosis duplicate but fail to separate	10
mipA	A. nidulans, (γ-tubulin)	Microtubule nucleation reduced or absent	11, 12
esp1	S. cerevisiae	Multiple spbs without nuclear division—negative regulator of spb duplication	13
cut1	S. pombe, ESP1-like	Multiple spbs without nuclear division	14
cut2	S. pombe	Multiple spbs without nuclear division	14
cdc31	S. cerevisiae	spbs increase in size but do not duplicate	15
mps1	S. cerevisiae	spbs fail to duplicate, an abnormally long half-bridge structure is assembled	16
mps2	S. cerevisiae	spbs duplicate but fail to insert into the nuclear membrane, lack the inner plaque and nuclear microtubules	16
mes1	S. pombe	spbs fail to duplicate at meiosis II and spore formation is blocked	17
mgr	Drosophila	Centrosomes fail to duplicate in larval neuroblasts	18
asp	Drosophila	Same as mgr, also disorganized spindles during embryogenesis	19, 20
kar1	S. cerevisiae	Defects in spbs, abnormally long cytoplasmic mts, aberrant chromosome disjunction. Overexpression or underexpression of wild-type gene causes failure of spb duplication	21
kem1	S. cerevisiae	Enhance the kar1 phenotype, hypersensitive to benomyl, high rates of chromosome loss, sporulation minus, spbs fail to duplicate and/or separate	22
spa1	S. cerevisiae	Abnormally shaped spindles, high rates of nondisjunction, and nuclear migration blocked	23
polo	Drosophila	Dispersed centrosomal material and multipolar spindles	24, 25

TABLE IV *(continued)*

Mutation	Species protein	Phenotype	Refs.[a]
aurora and *thule*	*Drosophila*	Dispersed centrosomal material during late syncytial cleavage divisions	25
brls	*Drosophila*	Barrel shaped and multipolar spindles seen during embryogenesis and spermatogenesis	26
dv	Maize	Meiotic spindles remain broad, rather than focusing at the poles; results in multinucleate cells	27
1(3)7m62 and *1(1)d.deg11*	*Drosophila*	Chromosomes fail to segregate, larval neuroblasts polyploid with multipolar spindles	25, 28

[a]List of references: (1) Lewis and Gencarella, 1952; (2) Kimble and Church, 1983; (3) Endow *et al.*, 1990; (4) McDonald and Goldstein, 1990; (5) Zhang *et al.*, 1990; (6) Hawley *et al.*, 1990; (7) Hagan and Yanagida, 1990; (8) Enos and Morris, 1990; (9) Meluh and Rose, 1990; (10) Yeh *et al.*, 1991; (11) Oakley and Oakley, 1989; (12) Oakley *et al.*, 1990; (13) Baum *et al.*, 1988; (14) Uzawa *et al.*, 1990; (15) Baum *et al.*, 1986, (16) Winey *et al.*, 1991, (17) Shimoda *et al.*, 1985, (18) Gonzales *et al.*, 1988, (19) Ripoll *et al.*, 1985; (20) Gonzales *et al.*, 1990; (21) Rose and Fink, 1987; (22) Kim *et al.*, 1991; (23) Snyder and Davis, 1988; (24) Sunkel and Glover, 1988; (25) Glover, 1989; (26) Wilson and Fuller, 1990; (27) Staiger and Cande, 1990; (28) Gatti and Baker, 1989.

the gene product. For example, a mutation which disrupts the organization of the MTOC might be described as blocking microtubule nucleation, if the disruption prevented the recruitment of nucleating elements into the MTOC.

Although many components have been shown to associate with MTOCs, we have restricted the following discussion to those components that have been implicated in MTOC function through either antibody inhibition tests or analysis of mutations. We have organized the material according to *apparent* function, using the five divisions listed above. That is, components implicated in (1) kinetochore function (Section III,A), (2) microtubule nucleation (Section III,B), (3) organization of centrosomes or microtubule arrays (Section III,D), (4) centrosome or spb duplication (Section III,E), and (5) centrosome or spb separation (Section III,F). A brief discussion of the protein kinase systems that have been shown to associate with microtubular structures and are thought to be involved in regulation of the assembly and functioning of these structures is also included (Section III,C).

A. Components Implicated in Kinetochore Function

Components that may be required for kinetochore function have been identified through both immunological and genetic analyses. Among the antibodies that have been shown to recognize kinetochore components (Table I), four autoantibodies (EK, ACA, cen1, and cen2) have been shown to affect kinetochore function either *in vitro* or *in vivo.*

In early studies of kinetochores, Cox *et al.* (1983) showed that the cen1 and cen2 antibodies blocked microtubule nucleation from kinetochores in lysed cells. Kinetochores do not appear to nucleate microtubules *in vivo* (Rieder, 1991). However, it has been shown that microtubule elongation does occur at the kinetochore (Sawin and Mitchison, 1991b). It may be that the cen1 and cen2 antibodies blocked the addition of tubulin subunits onto residual fragments of microtubules that were attached to the kinetochores.

The EK and ACA antibodies were tested *in vivo* by injecting either whole serum into mouse oocytes (Simerly *et al.,* 1990) or protein A purified IgGs into tissue culture cells (Bernat *et al.,* 1990). The EK serum, which recognizes the 80-kDa CENP-B protein on immunoblots, blocks chromosome congression but does not interfere with microtubule attachment or chromosome segregation during anaphase. The CENP-B protein has been shown to bind to kinetochore-specific DNA sequences (Masumoto *et al.,* 1989) and may interact with tubulin (Balczon and Brinkley, 1987). Simerly *et al.* (1990) suggested that the CENP-B protein could be involved in either microtubule assembly at the kinetochore or in sliding of the kinetochore along the microtubules. The purified ACA IgGs recognize four kinetochore-specific proteins, CENP-A, B, C, and D (Bernat *et al.,* 1990). Injection of these purified IgGs into tissue culture cells disrupts microtubule attachment to kinetochores, chromosome congression, and chromosome segregation, depending on the stage of the cell cycle at the time of injection. The authors suggest that most of their results can be explained by defects in the assembly or maturation of the kinetochores such that either microtubules do not attach to the kinetochores or the attachment is defective. The CENP-A protein has been shown to be a centromere-specific histone (Palmer *et al.,* 1987), and thus would certainly be expected to function in the organization of the centromeric DNA. Bernat *et al.* (1990) do not, however, present any evidence for morphological defects in the kinetochores in the treated cells; thus, their results do not rule out the possibility that the effects seen are due to interference with particular kinetochore functions rather than with kinetochore assembly. For example, the block to chromosome segregation could be due to the failure of the sister centromeres to separate, to the failure of microtubule disassembly at the kinetochore, or to the inability of the kinetochores to move along the microtubules.

Mutational analysis has identified three genes in *Drosophila (nod+, l(3)7m62+,* and *l(1)d.deg11+),* whose products may be required for kinetochore function. Mutations in the *nod+* locus cause increased rates of chromosome nondisjunction and loss during meiosis. Nondisjunction and loss in these mutations primarily affect chromosomes that have not undergone recombination, and thus would normally segregate via a secondary disjunctional pathway, the distributive segregation system (Carpenter, 1973; Zhang and Hawley, 1990). Sequence analysis of the nod^+ gene indicates that the nod^+ protein includes a domain that shows sequence similarity to the motor domain of kinesin, a plus end directed microtubule motor protein, however, motor function has not yet been demonstrated for the nod^+ protein (Zhang *et al.*, 1990).

In order to understand how a putative microtubule motor protein might affect a specific subset of chromosomes, it is useful to look at the process of spindle

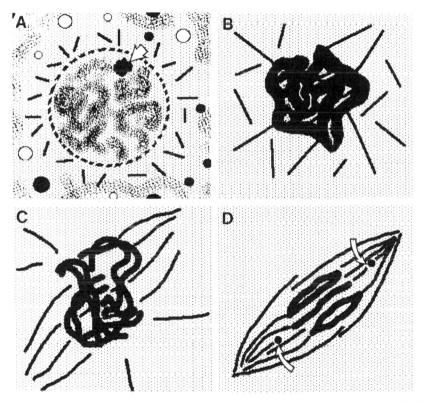

FIG. 2 Diagramatic representation of meiotic spindle assembly in *Drosophila* oocytes. (A) The nucleus at the time of nuclear envelope breakdown is shown. The chromosomes form a highly condensed mass, the karyosome (indicated by the arrow), and numerous microtubules are seen in the cytoplasm surrounding the nucleus. (B) Following nuclear envelope breakdown, microtubules are captured by the chromosome mass. (C) The microtubules begin to reorient, ultimately forming a bipolar spindle (D), with the exchange bivalents positioned at the metaphase plate, while nonexchange chromosomes (arrows) assume a position midway between the metaphase plate and the poles. [Based on the description by Theurkauf and Hawley (1992) except that the chromosomes are shown in a more dispersed configuration, which is seen when the oocytes receive a hypotonic shock.]

assembly in *Drosophila* oocytes (shown in Fig. 2). Using immunofluorescence and confocal microscopy, Theurkauf and Hawley (1992) have shown that the meiotic spindles in *Drosophila* females assemble in an unusual manner. During late prophase of meiosis I, the chromosomes form a highly condensed mass within the nucleus, called the karyosome (arrow in Fig. 2A). At this stage, no centrosomes are present in the oocyte but numerous microtubules are seen in the cytoplasm surrounding the nucleus. Following nuclear envelope breakdown, microtubules are captured by the chromosome mass (Fig. 2B), and ultimately are organized to form a bipolar spindle (Fig. 2C and D). During spindle assembly, nonexchange chromosomes (arrows in Fig. 2D) precociously separate from the karyosome, and take up positions midway between the poles and the metaphase plate (Puro and Nokkala, 1977; Theurkauf and Hawley, 1991). In females homozygous for *nod* mutations, the nonexchange chromosomes drift away from the karyosome during the earliest stages of spindle assembly (Theurkauf and Hawley, 1991). Once spindle assembly is complete, the chromosomes may reassociate with the spindle, but they do so at random, resulting in the high rates of nondisjunction and loss observed during meiosis in *nod/nod* females. Also, nonexchange chromosomes which reassociate with the spindle are usually located closer to the spindle poles than is normal. These observations suggest that the *nod*⁺ protein is required to hold the nonexchange chromosomes in the karyosome during the early stages of spindle assembly, and to maintain the intermediate position of these chromosomes between the metaphase plate and the spindle poles (Theurkauf and Hawley, 1991).

Nicklas and co-workers have shown that, when chromosomes first capture microtubules during spindle assembly, the chromosomes often move rapidly toward the pole (the minus end of the microtubules) to which they first become attached. Once the sister or homologous kinetochore has attached to microtubules nucleated from the opposite pole, the forces on the two kinetochores eventually become balanced and the chromosomes are positioned at the metaphase plate (see Rieder, 1991, for review). Since the nonexchange chromosomes do not remain paired (Fig. 2D), the balancing force could be provided by a plus-end-directed motor acting on the same kinetochore. In the absence of the plus-end-directed motor, the chromosomes would simply move to the minus ends of the microtubules. During the early stages of spindle assembly in *Drosophila* oocytes, the microtubules radiate in all directions (Fig. 2B and C); thus, movement of chromosomes along the microtubules, toward the minus end, could result in the chromosomes being transported a considerable distance from the developing spindle apparatus. This simple model would explain some aspects of the *nod* phenotype. However, Theurkauf and Hawley (1991) have also shown that the precise behavior of different nonexchange chromosomes in *nod/nod* females is a function of their relative sizes; thus *nod*⁺ function may be more complex than suggested by this model.

The association of microtubule motor proteins with kinetochores has long been suspected, based on the studies of chromosome movement in living cells, which suggested that kinetochores actively move along microtubules rather than passively responding to changes in microtubule length (reviewed in Rieder, 1991). Recently, Steuer *et al.* (1990) and Pfarr *et al.* (1990) reported that antibodies raised against cytoplasmic dynein stain kinetochores, while Hyman and Mitchison (1991) have reported the detection of both minus-end- and plus-end-directed microtubule motor functions associated with isolated mammalian kinetochores.

Two other mutations in *Drosophila* which may be involved in kinetochore function are the *l(3)7m62* and *l(1)d.deg11* mutations. The phenotypes of these mutations have only been described in larval neuroblasts where most mitotic cells are seen to be polyploid and assemble multipolar spindles (Gatti and Baker, 1989; Glover, 1989). This phenotype suggests that spindle assembly occurs normally but that the sister chromosomes are unable to separate. A defect in kinetochore function could result in such a phenotype.

B. Nucleation of Microtubules Appears to Require Several Distinct Proteins

Microtubule nucleation by centrosomes and spbs appears to require several different proteins, including a distinct type of tubulin (γ-tubulin), Ca^{2+}-binding proteins, and an unknown protein recognized by the CHO2 antibody.

γ-Tubulin was originally identified through a screen for extragenic suppressors of mutations in the *benA* β-tubulin gene of *A. nidulans*, which yielded the *mipA* mutation (Weil *et al.*, 1986). The *mipA* gene was cloned by chromosome walking and testing different regions of the walk for the ability to rescue the *mipA* phenotype. Sequence analysis of the complementary DNA predicted a protein that is equally related to both α- and β-tubulin (29–32 and 32–35% identity, respectively) and approximately as diverged from either as α and β are from each other (36–42%) (Oakley and Oakley, 1989). Disruption of the *mipA* gene results in cells which are blocked in nuclear division and which assemble few to no microtubules, suggesting that γ-tubulin may be required for microtubule nucleation. Recently, the isolation of genes for γ-tubulin from *Homo sapiens, Drosophila* (Zheng *et al.*, 1991), *Xenopus, S. pombe*, maize, and diatoms (Stearns *et al.*, 1991) has been reported. Immunofluorescent analysis has shown that, in each of these species, γ-tubulin is not assembled into microtubules, but rather is localized to the spbs or centrosomes, suggesting the conservation throughout eukaryotic evolution of a tubulin that may be specifically involved in microtubule nucleation. The involvement of centrosomal tubulin in microtubule nucleation had previously been suggested by Pepper and Brinkley (1979), who showed that a monospecific rabbit tubulin antibody blocked nucleation of microtubules from centrosomes in lysed

PtK cells. This result was seen even after extensive washing, and thus was probably not due to binding of the antibody to the exogenous tubulin, thereby blocking polymerization. Whether this antibody was able to bind to γ-tubulin is not known.

CHO2 (Table I) is a monoclonal antibody which was raised against isolated spindles from CHO cells (Sellitto *et al.*, 1992). When CHO cells are treated to depolymerize their microtubules, lysed, exposed to the CHO2 antibody, and then incubated with exogenous tubulin, very few microtubules are nucleated from the centrosome. Control cells, treated in the same way but incubated with non-immune or other immune sera are able to nucleate large numbers of microtubules. These results suggest that the CHO2 antigen is involved in microtubule nucleation (Sellitto *et al.*, 1992). As shown in Table I, the CHO2 antibody recognizes three distinct proteins on immunoblots (66, 225, and 260 kDa). Which of these antigens are actually involved in centrosome function will require additional studies.

There is evidence suggesting that Ca^{2+}-binding proteins may be involved in microtubule nucleation. Moudjou *et al.* (1991) have probed centrosomes in human cell lines using antibodies raised against a 230-kDa contractile protein from *Polyplastron* cytoskeletons. On immunoblots of proteins from HeLa and KE37 cell lines, this antibody recognizes a 62–64 kDa doublet, whose behavior during gel electrophoresis is dependent on Ca^{2+}. On immunoblots the protein is also recognized by antibodies raised against centrin, a 20-kDa, Ca^{2+}-binding, contractile protein isolated from striated rootlets of *Tetraselmis*. Moudjou *et al.* (1991) observed that the anti-230K antibody was able to block microtubule nucleation from the centrosome both *in vitro* and *in vivo*. These results would suggest that Ca^{2+}-binding proteins may be involved in microtubule nucleation. Other investigators have also reported the presence of Ca^{2+}-binding, contractile proteins in the centrosome, but have suggested other functions for the proteins. Baron *et al.* (1991) have suggested that the Ca^{2+}-binding, centrosomal proteins in marsupial and mammalian cell lines, identified by antibodies raised against *Tetraselmis* centrin (Table I), may have contractile properties and be involved in the changes in the distribution of the PCM that occur during the cell cycle. Specifically, they suggest that these proteins could be involved in the interconversion of the PCM from satellites to fibrous halo and back to satellites. Huang *et al.* (1988a) analyzed a 20-kDa, Ca^{2+}-binding, contractile protein from *Chlamydomonas*. The protein is a component of the striated rootlets in interphase cells and of the spindle poles during mitosis. DNA sequence analysis indicates that this protein is related to calmodulins, with the greatest sequence similarity to the product of the *CDC31* gene of *S. cerevisiae* (Huang *et al.*, 1988b). Mutations in the *CDC31* gene result in cells where the spbs double in size but fail to split into two separate structures; thus, a single large spb is formed which assembles a monopolar spindle (Baum *et al.*, 1986, 1988). This suggests a role for Ca^{2+}-binding proteins in the proper duplication of the spb.

C. Posttranslational Phosphorylation May Regulate Many Aspects of MTOC Function

The onset of mitosis is accompanied by increased level of phosphorylation, brought about by an increase in the activity of the M-phase promoting factor (MPF). MPF, which is composed of the p34^{cdc2} protein kinase and cyclin B, triggers a cascade of phosphorylation events that regulate the cells' progress through mitosis (reviewed in Lee and Nurse, 1988; Lohka, 1989; Lewin, 1990; Pines and Hunter, 1990). p34^{cdc2} kinase was initially identified through a mutation in *S. pombe* which could block the cell cycle at either START, a point after which the cell is committed to dividing, or at the G2/M transition (Nurse and Bissett, 1981). The activity of p34^{cdc2} kinase is regulated in part by association with a number of other proteins including cyclins, proteins that undergo cyclic accumulation and destruction during the cell cycle, with the peak of accumulation at metaphase followed by rapid degradation at anaphase (for reviews, see Lee and Nurse, 1988; Lohka, 1989; Lewin, 1990; Pines and Hunter, 1990). Recently, the genes for p34^{cdc2}, the cyclins, and cdc25, a protein required for activation of p34^{cdc2}, have been isolated from other species, such as human (Lee and Nurse, 1987) and *Drosophila* (Jimenez *et al.*, 1990), by selecting for sequences that can rescue the relevent mutations in *S. pombe*, indicating that these proteins have been functionally conserved throughout eukaryotic evolution.

Immunofluorescent localization of p34^{cdc2} kinase (Table II) in *S. pombe* has shown that the protein is located within the nucleus during interphase. With the onset of mitosis the nuclear staining is lost, and the protein becomes associated with the spbs. Staining of the spbs persists until late anaphase to telophase, at which time it disappears (Alfa *et al.*, 1990). A similar pattern of distribution is seen in mammalian tissue culture cells (Bailly *et al.*, 1989; Riabowol *et al.*, 1989; Rattner *et al.*, 1990). During interphase the protein is dispersed throughout the nucleus and the cytoplasm. During G2 the protein begins to accumulate in the centrosomal region and is associated with the spindle poles, kinetochores, and kinetochore fibers during metaphase. With the onset of chromosome separation at anaphase the staining is redistributed to the midzone of the spindle apparatus (Bailly *et al.*, 1989; Riabowol *et al.*, 1989; Rattner *et al.*, 1990). Similar patterns of distribution have been seen in *S. pombe* for the cdc13 protein (cyclin B) (Alfa *et al.*, 1990), and in HeLa cells for the p13^{suc1} protein (Bailly *et al.*, 1989), two of the proteins that are involved in the regulation of p34^{cdc2} kinase.

Numerous substrates for p34^{cdc2} kinase have been identified, including histone H1, the lamins, a tyrosine kinase (pp60^{c-src}), and various transcription factors (for review, see Moreno and Nurse, 1990; Pines and Hunter, 1990). MTOC components are also among the cellular proteins that are hyperphosphorylated at the G2/M transition (Vandré *et al.*, 1984), and recent studies indicate that a 225-kDa centrosomal protein, which is recognized by the CHO3 monoclonal antibody

(Kuriyama, 1989; Table II), may be a substrate for p34^{cdc2} kinase (R. Kuriyama, J. Ruderman, M. Dorée, and T. Maekawa, unpublished data).

The role of phosphorylation in the regulation of MTOC function is not known, but recent studies suggest that changes in phosphorylation may affect the activity of proteins involved in microtubule nucleation and movement of kinetochores along microtubules. MPM-2 (Table I), a monoclonal antibody which recognizes phosphorylated epitopes on several proteins that are specifically phosphorylated at the onset of mitosis (Davis et al., 1983; Vandré et al., 1984), has been shown to block microtubule assembly onto centrosomes in an in vitro inhibition assay (Centonze and Borisy, 1990). It was also shown that treatment of lysed cells with alkaline phosphatase could inhibit microtubule nucleation assembly from centrosomes (Centonze and Borisy, 1990). Mitotic centrosomes are capable of nucleating up to five times as many microtubules as interphase centrosomes (Kuriyama and Borisy, 1981b). Verde et al. (1990) demonstrated that the capacity of isolated interphase centrosomes to nucleate microtubules increased on incubation in Xenopus egg extracts in which p34^{cdc2} kinase activity is high. These results, together with the results reported by Centonze and Borisy (1990), suggest that changes in phosphorylation are important in regulating microtubule nucleating activity.

Kinetochore components are also phosphorylated at the G2/M transition, as shown by staining with antibodies specific to phosphoproteins (Vandré et al., 1984; Kuriyama, 1989; Wordeman et al., 1991). Hyman and Mitchison (1991) have shown that preincubation of isolated kinetochores with ATP-γ-S, an ATP analog which can be used by kinases to phosphorylate proteins irreversibly, caused a reversal in the direction in which microtubules moved, relative to the kinetochores, suggesting that changes in phosphorylation may regulate the activity of the motor proteins that are localized to kinetochores.

Cyclic-AMP (cAMP)-dependent kinases may also play a role in regulation of microtubular structures. The type II regulatory subunit (RII) of cAMP-dependent kinase has been shown to colocalize with centrosomes, and with the spindle apparatus during mitosis (Browne et al., 1980; Nigg et al., 1985; De Camilli et al., 1986). The localization of this protein is due to association of the RII subunit with the microtubule-associated protein MAP2, and is thought to regulate the effects of MAP2 on microtubule bundling and stability through changes in phosphorylation (Vallee et al., 1984; Lohmann et al., 1984). Whether cAMP-dependent kinases also have a role in MTOC function is not known.

D. Assembly and Organization of the Centrosome Appears to Involve Many Components, Including Microtubule Motors and RNA-Binding Proteins

It has been suggested that the morphology of microtubule arrays depends on the three-dimensional distribution of the microtubule nucleating elements (MTNEs) within the centrosome or spindle poles (Borisy and Gould, 1977; Mazia, 1987;

also see Fig. 1). A number of proteins or genes have been identified which, when tested for function either *in vivo* or *in vitro,* appear to affect the assembly or organization of the centrosome/spindle poles and thus alter the morphology of the microtubule arrays. Some have been studied using both molecular and genetic techniques (the SPA1 and *ncd⁺* gene products of *S. cerevisiae* and *D. melanogaster,* respectively), while others have been characterized primarily by immunological methods (the 51-kDa EF1α-like protein of sea urchins). In addition, eight genes have been identified in which mutational analysis suggests a role for the gene product in centrosome/spb or spindle organization. They are the *KAR1* and *MPS2* genes of *S. cerevisiae,* the *asp⁺, polo⁺, aurora⁺, thule⁺,* and *brls⁺* genes of *Drosophila,* and the *dv⁺* gene of maize. These components will be discussed in order from the best to the least well characterized.

The *ncd⁺* gene of *D. melanogaster* was originally identified through the claret-nondisjunctional (*ca^{nd}*) mutation. *ca^{nd}* causes a change in the eye color, from bright red to ruby red, and affects chromosome behavior during meiosis in females, and during the early cleavage divisions in the egg. As suggested by its name, this mutation causes high rates of chromosome nondisjunction during meiosis I. In addition, high rates of chromosome loss are seen during meiosis II and the early mitotic divisions in the egg (Lewis and Gencarella, 1952). Chromosome loss in the egg occurs primarily at the first mitotic division and is preferential for the maternally derived chromosomes (Hinton and McEarchern, 1963; Davis, 1969). Cytological analysis of meiosis in *ca^{nd}* females showed a number of different defects, including oocytes at first meiosis which assembled more than one spindle, and abnormally shaped bipolar or monopolar spindles (Kimble and Church, 1983). The early mitotic divisions, in approximately 70% of the eggs, were characterized by the presence of what appeared to be multipolar spindles and/or nuclei of variable sizes and shapes. In addition, the nuclei usually formed clusters, rather than becoming dispersed in the ooplasm (Kimble and Church, 1983).

Screening for additional mutations at the *ca^{nd}* locus has shown that the eye color defect and the meiotic/mitotic defects are due to two distinct genes, *ca⁺* and *ncd⁺,* respectively (Sequeira *et al.,* 1989; Yamamoto *et al.,* 1989). The *ncd⁺* gene has recently been cloned and sequenced by two different groups (Endow *et al.,* 1990; McDonald and Goldstein, 1990). The *ncd⁺* protein shows sequence similarity to the kinesin heavy chain; however, the arrangements of functional domains within the two molecules are different. In kinesin heavy chains, the motor domain has been shown to be located at the amino-terminal end of the protein, while in the *ncd⁺* protein it is located in the carboxy-terminal region. Another difference that has been seen between *ncd⁺* and kinesin is that kinesins are plus-end-directed motor proteins, while *ncd⁺* has been shown *in vitro* to be a minus-end-directed motor (Walker *et al.,* 1990; McDonald *et al.,* 1990). *In vitro* studies have also shown that the *ncd⁺* protein causes bundling of microtubules (McDonald *et al.,* 1990).

Several different theories have been put forward to explain how the *ca^{nd}* mutation affects chromosome behavior. Davis (1969) suggested that the ca^{nd} mutation might result in formation of defective spindle fibers. Bivalents which attached to these defective fibers would fail to separate and thus would be drawn together to one pole, resulting in nondisjunction. Chromosome loss during meiosis II and mitosis would occur when chromatids, attached to defective fibers, separated from their sister chromosomes but failed to move toward the pole. Based on cytological data, Kimble and Church (1983) suggested that the primary defect in ca^{nd} was in the organization of the spindle poles, resulting in the assembly of defective spindles. These authors also suggested that a defect in the spindle pole/centrosome could account for the failure of the mitotic nuclei to disperse through the ooplasm, since dispersion of the mitotic nuclei had been shown to require an intact cytoskeleton (Zalokar and Erk, 1976). Baker and Hall (1976) and Endow et al. (1990) have suggested that a kinetochore defect could better explain the nonrandom chromosome loss seen during meiosis II and mitosis, and suggested that interaction of defective kinetochores with spindle fibers could account for the spindle abnormalities reported by Davis (1969) and by Kimble and Church (1983). It is difficult, however, to imagine how a kinetochore defect could affect dispersion of the cleavage stage nuclei.

The functions that have been demonstrated for the ncd^+ protein in vitro, and recent data on the assembly of spindles in Drosophila eggs, suggest that some elements of both the kinetochore and spindle pole models may be correct. As described in Section III,A, and diagrammed in Fig. 2, the meiotic spindles in Drosophila females are assembled in an unusual manner. Prior to nuclear envelope breakdown, numerous microtubules are present in the cytoplasm surrounding the nucleus (Fig. 2A). On nuclear envelope breakdown, these microtubules are first captured by the chromosomes (Fig. 2B), and then organized into a bipolar spindle (Fig. 2C and D) (Theurkauf and Hawley, 1992). Once microtubules are captured by the chromosomes, the ncd^+ protein, by virtue of its minus-end-directed motor, could move along the microtubules, bundling adjacent microtubules. This would result in the convergence of the microtubules, thereby defining the poles of the spindle. This model implies that the ncd^+ protein might be associated with the chromatin, though not necessarily with the kinetochores, prior to nuclear envelope breakdown. One could imagine that the ncd^+ protein might have a low affinity for chromatin, and a higher affinity for microtubules. Following nuclear envelope breakdown, ncd^+ molecules which come into contact with microtubules would preferentially bind to the microtubules and function in the assembly of the spindle apparatus. ncd^+ molecules which failed to contact microtubules would remain associated with the chromosomes and could be involved in spindle assembly during subsequent divisions.

The mode of spindle assembly described above has been seen for the first meiotic spindle, and is probably used for assembly of the second meiotic spindle as well. In Drosophila embryos the maternal and paternal pronuclei do not fuse prior to the first mitotic division, although they do take up positions adjacent to one another in the cytoplasm (Huettner, 1924). At the first mitotic division, two spindles are assembled, one organized around the maternal chromosomes, and the second organized around the paternal chromosomes. The two spindles lie close together in the cytoplasm such that at anaphase the maternal and paternal chromosome complements mingle, forming the zygotic nuclei (Huettner, 1924). Thus, it is possible that the maternal half of the spindle may be assembled in a manner similar to that seen during the meiotic divisions. This model would account for the high rates of chromosome misbehavior during meiosis, and the preferential loss of maternal chromosomes during the early mitotic divisions seen in the ca^{nd} and ncd mutations. Although the genetic studies had suggested that the ncd^+ protein functions primarily during meiosis and the first cleavage division, cytological studies suggested that the effects of mutations in the gene persisted well into the mitotic divisions, as evidenced by the presence of large numbers of abnormal nuclei in embryos from ca^{nd} females (Kimble and Church, 1983). Komma et al. (1991) have recently shown that the ncd^+ mRNA is present in the egg through mid-embryogenesis (8–10 hr postfertilization). Assuming that the protein is also present during this time, it would suggest that the ncd^+ product may play a role in the functioning of the mitotic spindles throughout embryogenesis.

Although the mechanism of assembly of the meiotic spindle in *Drosophila* is unusual, there are several other reports which suggest that motor proteins might be involved in the organization of centrosomes or spindle poles (Weisenberg et al., 1986; Urrutia et al., 1989). Verde et al. (1991) showed that assembly of taxol-induced asters in *Xenopus* egg extracts involved first a bundling of microtubules followed by reorganization of the microtubules into a radial aster. Formation of the asters could be inhibited by vanadate, a drug which is known to selectively inhibit dynein, a minus-end-directed motor protein. This assembly mechanism might also explain the observations by Dietz and co-workers that, in *Pales* spermatocytes, bipolar spindles having one normal pole and one aster-free pole assemble when one centrosome is mechanically separated from the nucleus during diakinesis (Bastmeyer et al., 1986; Steffen et al., 1986).

The 59-kDa *SPA1* protein of *S. cerevisiae* appears to be involved in both spindle assembly and chromosome segregation. The *SPA1* gene was isolated from an expression library using the 5051 antiserum (Snyder and Davis, 1988). Mutations in the *SPA1* gene, created by gene disruption, result in assembly of morphologically abnormal spindles, failure of nuclear migration, and a 7- to 20-fold increase in chromosome nondisjunction, depending on the marker chromosome being followed (Snyder and Davis, 1988). Since microtubules have been shown to be required for chromosome segregation and nuclear migration in this species, the primary role of the *SPA1* protein may be in the assembly or organization of microtubules. Although the 5051 antiserum has been shown to recognize centrosome/spindle pole components in many species (Table I), and antibodies raised against *SPA1* fusion protein stain MTOCs in other species (Snyder and Davis, 1988), it is not clear whether the *SPA1* protein is related to the antigens recognized by the 5051 serum in other species. A preliminary report of the identification of the 5051 antigen in mice and rats indicates that in mammalian species the antigen is a 220-kDa protein (Doxsey *et al.,* 1990), and DNA hybridization analysis indicates that the *SPA1* clone is not related to clones for the 220-kDa mammalian protein (S. Doxsey, personal communication). On the other hand, injection of antibody raised against the 220-kDa protein into mouse or *Xenopus* eggs disrupted the orientation of microtubules and the organization of the meiotic spindles, suggesting that, like the *SPA1* protein, the 220-kDa 5051 antigen may be involved in centrosome organization (Doxsey *et al.,* 1990; S. Doxsey, personal communication).

The 51-kDa EF1α-like protein (Table II) was originally identified as a component of the spindle poles in the sea urchin *Hemicentrotus* (Toriyama *et al.,* 1988) and later as a microtubule-interacting protein (Ohta *et al.,* 1988). It has been shown to be related to elongation factor 1α (EF1α) of yeast (Ohta *et al.,* 1990). When antibodies against the 51-kDa protein are injected into sea urchin eggs before prophase, spindle assembly is suppressed. When the antibodies are injected during prophase–prometaphase, astral microtubule assembly is unaffected, but spindle formation is abnormal. That is, the assembled spindles are shorter and wider than normal, and the total number of nonastral spindle microtubules appears to be decreased. The authors suggest that the anti-51-kDa antibodies may inhibit the assembly of nucleating sites into the centrosome (Sakai *et al.,* 1989). The 51-kDa protein may also be involved in the three-dimensional organization of the microtubule nucleating materials. Interestingly, a 50-kDa EF1α-like protein (SU5; Table II) has also been identified in the MTOCs of a second sea urchin species, *Strongylocentrotus purpuratus* (Kuriyama and Borisy, 1985; Kuriyama *et al.,* 1990). Whether these proteins actually function as elongation factors *in vivo* as well as having a role in spindle assembly is not known. This may represent another case of evolution recruiting proteins to new functions through gene duplication and divergence, or of gene products acquiring new functions while retaining the old (Kuriyama *et al.,* 1990).

Mutations in the *KAR1* and *MPS2* genes of *S. cerevisiae* result in morphological defects in spbs, assembly of abnormally long cytoplasmic microtubules, and aberrant chromosome disjunction. Both overexpression and underexpression of the *KAR1* gene block spb duplication (Conde and Fink, 1976; Rose and Fink, 1987). Rose and Fink (1987) suggested that overproduction of the *KAR1* protein may block spb assembly due to the excess protein forming inappropriate aggregates either with itself or with other spb components, and thus the protein(s) is not available for assembly into the spb. Cells carrying mutations in the *MPS2* gene assemble monopolar spindles (Winey *et al.,* 1991). Electron microscopic analysis of these cells shows that each cell contains two spbs. One appears normal and is responsible for assembly of the monopolar spindle. The second spb does not appear to insert into the nuclear membrane, lacks the inner plaque and has microtubules associated only with its cytoplasmic face (Winey *et al.,* 1991).

Among the five *Drosophila* mutations which appear to affect centrosome organization, *asp* is characterized by the assembly of disorganized and/or monopolar spindles. Also, during embryogenesis, free centrosomes are occasionally observed, and the amount of material associated with different centrosomes seems to vary, suggesting that the centrosome material is not equally distributed (Ripoll *et al.*, 1985; Gonzales *et al.*, 1990). The primary defect among the *polo, aurora,* and *thule* mutations appears to be a loss in integrity of the centrosomes/spindle poles such that the centrosome material, as detected by staining with a *Drosophila*-specific, anti-centrosome antibody, Bx63 (Table II), is more dispersed than normal. The *polo* mutation causes the most severe defects. In early embryos, larval neuroblasts, and spermatocytes, multipolar spindles are frequently seen. When *polo* embryos are probed with the Bx63 antibody, no staining is seen in very early embryos. During the later stages of syncytial cleavage punctate staining can be detected; however, the label is not always associated with microtubule arrays (Sunkel and Glover, 1988). Similarly, mutations in the *brls*+ gene have been reported to affect the organization of spindle poles, resulting in assembly of spindles having a barrel-shaped appearance, or in multipolar spindles (Wilson and Fuller, 1990). In *aurora* and *thule* embryos, the centrosomes, during the early syncytial cleavages, appear normal, while in later stages the spindle pole material spreads out rather than remaining focused as is seen in wild-type embryos. These embryos can develop to the larval stage, and analysis of larval neuroblasts often shows mitotic cells with multipolar spindles (Glover, 1989). Despite the occurrence of monopolar spindles in *asp* animals, these results suggest that the primary function of the *asp*+, *polo*+, *aurora*+, *thule*+, and *brls*+ gene products may be in the organization of the centrosomal material.

The *dv*+ gene of maize has been reported to be involved in spindle assembly during meiosis (Staiger and Cande, 1990). Assembly of meiotic spindles in maize shows some similarity to meiotic spindle assembly in *Drosophila* females. That is, microtubules surround the nucleus during prophase. Following nuclear envelope breakdown the microtubules reorganize into the spindle. When first organized, the poles of the spindle are quite broad, but become tightly focused by metaphase. In plants homozygous for the *dv* mutation, microtubule assembly appears normal but the spindle poles remain broad, and in some cases appear to split, suggesting that the *dv*+ gene product is required for focusing of the spindle poles (Staiger and Cande, 1990). The suggested function of the *dv*+ gene product leads one to wonder whether this gene might represent the maize homolog of the *ncd*+ gene.

E. Duplication of spbs and Centrosomes May Involve Both Negative and Positive Regulators

While duplication and morphological changes in MTOCs have been the subject of intense morphological analysis, little is known about how these changes are regulated. Analyses of mutations in *S. cerevisiae* (*ESP1, CDC31,* and *MPS1*), *S. pombe* (*CUT1, CUT2,* and *MES1*), and *Drosophila* (*mgr*) are beginning to identify the processes involved.

The *ESP1* gene product has been suggested to function as a negative regulator of spb duplication, as cells carrying mutations in this gene accumulate multiple spbs but do not divide (Baum *et al.,* 1988). The *CUT1* and *CUT2* gene products have been suggested to serve a similar function in *S. pombe.* This is based on the fact that, when cell septum formation is blocked in cells carrying mutations in these genes, multiple spbs accumulate but nuclear division does not occur. The predicted amino acid sequence for the *CUT1* protein is 35% identical to the *ESP1* protein, suggesting that the two proteins could have similar functions. The predicted sequence of the *CUT2* gene product, on the other hand, does not show significant similarity to any sequences in the data base (Uzawa *et al.,* 1990). Genetic analysis suggests that the *CUT1* and *CUT2* products interact, as it has not been possible to isolate cells carrying mutations in both genes, while overexpression of the wild-type *CUT1* gene can rescue mutations in the *CUT2* gene, suggesting that the gene products have overlapping functions (Uzawa *et al.,* 1990).

Mutations which appear to block spb or centrosome duplication include *cdc31, mes1, mps1* and *mgr.* As described previously (Section III,B), in cells which carry mutations in *CDC31* the spbs do not duplicate, rather, a single abnormally large spb is observed (Baum *et al.,* 1986). The protein product of the *CDC31* gene shows significant homology to Ca^{2+}-binding proteins, especially with the calmodulins, and is similar to centrin, the 20-kDa, Ca^{2+}-binding protein characterized by Salisbury (1989) and Huang *et al.* (1988b). The phenotype of the *mes1* mutation is similar to that seen in the *cdc31* mutations, except that the mutation only affects separation of the spbs during meiosis II (Shimoda *et al.,* 1985). Cells carrying mutations in *MPS1* fail to duplicate their spbs, but do assemble an abnormally long half-bridge structure (Winey *et al.,* 1991). Mutations in all three genes, *CDC31, MES1* and *MPS1,* result in the assembly of monopolar spindles.

The *mgr* mutation of *Drosophila* is a recessive lethal mutation in which the homozygous individuals die, primarily, during pupal development. When brains from *mgr* larvae are analyzed for mitotic defects, the most common phenotype seen is polyploid cells that appear to be blocked in metaphase with all of the chromosomes organized around a half spindle (Gonzales *et al.,* 1988). Analysis of spermatogenesis in *mgr* pupae indicates that meiotic spindles fail to assemble as well, although the mitotic divisions of the stem cells and the spermatogonial cells appear to occur normally, suggesting that the product of the *mgr*+ gene may not be required in all cell types (Gonzales *et al.,* 1988). In those cells where the *mgr*+ gene product is required, it may be required for centrosome duplication.

F. More Motors

In all of the species in which mutations affecting MTOC function have been described (*D. melanogaster, S. cerevisiae, S. pombe,* and *A. nidulans),* one or more of these mutations has been shown to affect genes that encode proteins having sequence homology to microtubule motor proteins (Endow *et al.,* 1990; McDonald and Goldstein, 1990; Zhang *et al.,* 1990; Meluh and Rose, 1990; Hagan and Yanagida, 1990; Enos and Morris, 1990; Rothman *et al.,* 1990). In *Drosophila* two kinesin-like proteins, the proteins encoded by the *nod*[+] and *ncd*[+] genes, have been suggested to be involved in MTOC functions (Sections III,A and III,D, respectively). Four genes, which encode motorlike proteins, have been identified in fungal species. The proteins derived from three of the genes (*CUT7, KAR3,* and *BIMC*) are similar to kinesin, while a fourth gene (*SPO15*) is most closely related to a second type of motor protein, dynamin.

Among the fungal kinesin-like genes, the defects associated with mutations include the inability of the duplicated spindle pole bodies to separate and assemble a spindle apparatus (*cut7* and *bimC*) (Hagan and Yanagida, 1990; Enos and Morris, 1990) or blockage of spindle elongation and nuclear fusion (*kar3*) (Meluh and Rose, 1990). All three defects could be explained by a block in translocation of structures relative to microtubules.

Mutations in the *SPO15* gene of *S. cerevisiae* have a phenotype similar to that seen for the *cut7* and *bimC* mutations. That is, the duplicated spbs are unable to separate and assemble a spindle. However, *spo15* mutations only affect meiosis. The DNA sequence of the *SPO15* gene shows that it is identical to the previously described *VPS1* gene which has been shown to be involved in vacuolar sorting (Rothman *et al.,* 1990). The predicted protein product of the *SPO15/VPS1* gene has 45% overall identity to a recently described microtubule-associated protein, dynamin (Yeh *et al.,* 1991), which has been shown to cause bundling of microtubules and appears to have some motor function (Shpetner and Vallee, 1989).

G. Do Centrosomes and spbs Include Nucleic Acids?

No review of the functional components of MTOCs would be complete without at least a brief discussion of nucleic acids (for more complete reviews, see Fulton, 1971; Pepper and Brinkley, 1980; Vorobjev and Nadezhdina, 1987). Most of the studies designed to search for nucleic acids in MTOCs have focused on the centriole, and have produced a large and controversial body of literature with few unchallenged results. For readers interested in the literature concerning nucleic acids in the centrioles and basal bodies, we direct their attention to the reviews listed above, and to papers by Hall *et al.* (1989), Johnson and Rosenbaum (1990), Kuroiwa *et al.* (1990), and Johnson and Dutcher (1991). As in previous sections of this manuscript, we do not concern ourselves with the centriole proper, but rather focus on the questions of whether nucleic acids might be a component of the PCM, and what role, if any, nucleic acids might play in centrosome function.

McGill et al. (1976), while looking at the effects of ethidium bromide, a dye which binds to nucleic acids, on centriole morphology and spindle assembly in tissue culture cells, showed that ethidium bromide treatment resulted in a high frequency of cells that contained multipolar spindles. The frequency of multipolar spindles increased with the concentration of the ethidium bromide, with the length of exposure, and with time after treatment, reaching a peak at two to three generations after treatment. Electron microscopic analysis of treated cells showed that, in addition to effects on centriole morphology, the treatment resulted in changes in the appearance of the amorphous material surrounding the centrioles. Specifically, the material surrounding the centrioles appeared more dense than in untreated cells (McGill et al., 1976). Unfortunately, EM reconstruction of whole cells was not undertaken in this study; thus it is not known whether this dense material was associated with all poles of the multipolar spindles, or only with those containing centrioles. In light of data suggesting that the formation of multipolar spindles in cells treated with microtubule-disrupting drugs results from dispersion of the PCM during the drug treatment (Sellitto and Kuriyama, 1988a), it seems reasonable to suggest that perhaps the formation of multipolar spindles in the ethidium bromide-treated cells also resulted from disruption of the three-dimensional organization of the PCM.

Berns et al. (1977) and Peterson and Berns (1978) reported the results of studies which also suggest that the centrosome may contain a nucleic acid component. In both studies PtK_2 cells were treated with nucleic acid-specific dyes and the centrosome regions irradiated using a laser microbeam. When irradiated cells were followed through the cell cycle, it was seen that centrosome separation and spindle assembly could occur. The chromosomes formed a metaphase-like array in the center of the spindle, and cytokinesis was initiated; however, chromosome segregation did not occur. Ultrastructural analysis of the irradiated centrosomes indicated that the primary site of damage was the PCM, as very little PCM could be detected in association with the centrioles, and what material was present appeared to be abnormally organized. Also, very few microtubule profiles were identified in association with the irradiated centrosome (Berns et al., 1977). Data obtained using different dyes which were specific for DNA or RNA indicated that the centrosomal nucleic acid is probably an RNA (Peterson and Berns, 1978).

Additional evidence that RNA may be a component of the centrosome comes from a study by Rieder (1979), who showed, using histochemical techniques, that both kinetochores and the fibrous material (presumably the PCM) in newt lung cells are rich in ribonucleoproteins.

A number of reports have appeared in which in vitro procedures were used to address the question of whether nucleic acids are components of centrosomes. The procedures involve the incubation of cells lysed in situ (Pepper and Brinkley,

1980; Snyder, 1980), or incubation of isolated MTOCs from cultured mammalian cells (Ring *et al.*, 1980; Kuriyama, 1984), yeast (Byers *et al.*, 1978; Hyams and Borisy, 1978), *Dictyostelium* (Kuriyama *et al.*, 1982), and *Physarum* (Roobol *et al.*, 1982) with different RNases. Subsequently, the cells or isolated organelles are incubated with exogenous tubulin to determine whether they retain the ability to nucleate microtubules. Two of the reports indicated that RNase digestion of lysed PtK cells destroyed the ability of the centrosome to nucleate microtubules (Pepper and Brinkley, 1980; Snyder, 1980). However, in the studies in which isolated structures were incubated with the enzymes, the digestion did not have any effect on microtubule nucleation (Byers *et al.*, 1978; Hyams and Borisy, 1978; Ring *et al.*, 1980; Kuriyama *et al.*, 1982; Roobol *et al.*, 1982; Kuriyama, 1984). The differences between these results could be due to differences in the basic procedures used, or they could reflect cell-specific differences.

The results discussed above show that at least the distribution of the PCM can be altered by chemicals that bind nucleic acids, suggesting that nucleic acid, in particular RNA, may be a component of the PCM. Clearly, more studies will be needed before we can say for sure whether RNA is a component of PCM, whether it is required for centrosome function, and what role(s) it may play in the centrosome.

IV. Summary

In the introduction of this review (Section I), we outlined the proposed functional requirements of centrosomes and proposed a model for the organization of elements within the centrosome (Section I,B). We can now begin to suggest where within this complex structure some of the components that have been shown to be required for centrosome function might act. For example, γ-tubulin, the CHO2 antigen, and a 62–64 kDa Ca^{2+}-binding protein (Section III,B) have all been implicated in microtubule nucleation, and thus may be components of the MTNEs.

Microtubule motors are clearly important in the assembly and functioning of MTOCs. Motor proteins have been implicated in kinetochore function (Section III,A), in the organization of the centrosome (Section III,D), and in the separation of spindle poles during spindle assembly (Section III,F). Despite the relatively large number of reports concerning the role of motor proteins in spindle assembly and functioning, a recent paper by Endow and Hatsumi (1991) suggests that we may be seeing only the tip of the iceberg. These authors used PCR techniques to amplify kinesin-related sequences from *Drosophila* cDNA libraries. The amplified DNAs were then hybridized to larval polytene chromosomes to determine the number of genes with homology to the amplified products. In addition to the locations that correspond to the previously identified kinesin heavy-chain, *ncd+* and *nod+* genes, hybridization was detected at 32 distinct positions along the polytene chromosome map. One of the sites of hybridization appears to cor-

TABLE V

Antibodies Which Recognize IF-Like Proteins and MTOCs

Antibody	Antigen	Protein	Species and structures + for Ab staining[a]	Functional test[b]	Refs.[c]
3D2 group	Yeast spbs	110 kDa	*Saccharomyces cerevisiae*	(−) *in vitro*	1
CHO4	CHO spindles	275 kDa	CHO, intermediate filaments		2
Anti-NSP1	Yeast nuclei, insoluble fraction	100 kDa	Yeast, spbs and nuclear membrane	Overexpression blocks cell division and disrupts cell morphology	3
F5.1	PtK	32 kDa	PtK cells		4
GP1	Centrosomin A (32-kDa PtK protein)		PtK cells		5
AH-6	*Drosophila* intermediate filaments		Sea urchin		6
Anti-tektins	Sea urchin tektin proteins A, B, or C		Mammalian cells, centrosome or centriole with variable staining of spindle fibers and midbody		7
CTR2611	Human centrosomes	74 kDa	Mammalian cells, *Xenopus* and dinoflagellates intermediate filaments		8, 9
A46.2	Mouse centrosomes		Mouse, microtubules and intermediate filaments		10
D73.3	Mouse centrosomes	200 kDa	Mouse, microtubules and intermediate filaments		10

[a] Only structures, other than MTOCs, which stain with the antibodies are listed. MTOC staining is centrosomal unless otherwise indicated.

[b] *In vitro* antibody inhibition of microtubule nucleation from centrosomes of cells lysed *in situ*. *In vivo* antibody inhibition of mitosis by injection into cells or embryos.

[c] List of references: (1) Rout and Kilmartin, 1990; J. V. Kilmartin, personal communication; (2) Sellitto et al., 1992; (3) Hurt, 1988, 1989; (4) Joswig and Petzelt, 1990; (5) Joswig et al., 1991; (6) Schatten et al., 1988; (7) Steffen and Linck, 1989; (8) Buendia et al., 1990; (9) Perret et al., 1991; (10) Ring et al., 1980.

respond to the location of a gene (paternal loss) whose product is predicted to function in male meiosis in a manner analogous to that of *ncd⁺* (Endow and Hatsumi, 1991), suggesting that others of these genes may also be involved in spindle assembly and function.

The assembly of either MTNCs or MTOCs may also require structural elements. In fact, we would suggest that intermediate filament-like (IF-like) proteins are likely to be components of one or possibly both of these structures. Although we have not previously discussed IF-like proteins in this review, there is a great deal of evidence to suggest that IF-like proteins are components of centrosomes and spbs. Several antibodies have been described which recognize both intermediate filaments and centrosomes (Table V). Some of these antibodies were raised against IF-like proteins (Schatten *et al.*, 1988; Steffen and Linck, 1989), while others were raised against isolated MTOCs (Ring *et al.*, 1980; Buendia *et al.*, 1990; Joswig and Petzelt, 1990; Joswig *et al.*, 1991; Rout and Kilmartin, 1990; Sellitto *et al.*, 1992) or the insoluble fraction of yeast nuclei (Hurt, 1988). The best evidence that IF-like proteins are components of MTOCs is that the 110-kDa protein that localizes to yeast spbs is an IF-like protein (J. V. Kilmartin, personal communication). Assuming that IF-like proteins are components of MTNCs, one could imagine that changes in the distribution of the MTNEs within individual centers could be regulated by changes in interactions between individual IF-like molecules. For example, structural interactions between IF-like molecules could be altered or disrupted at the onset of mitosis and reestablished during telophase or interphase. This behavior is very similar to the behavior of nuclear lamins (another group of the intermediate filament protein super family), which disassemble during prophase, allowing nuclear envelope breakdown, then reassemble during telophase and early interphase (Lohka and Maller, 1985). The regulation of assembly and disassembly of lamin- (Miake-Lye and Kirschner, 1985; Ottaviano and Gerace, 1985) and vimentin-type intermediate filaments (Inagaki *et al.*, 1987) has been shown to involve changes in phosphorylation. Like these other IF systems, changes in centrosomal intermediate filaments could be triggered by phosphorylation, which clearly plays a significant role in changes in centrosome function (Section III,C).

Molecules that are candidates for centrosome matrix components include (1) the proteins recognized by the 5051 autoimmune serum (i.e., the *SPA1* protein of *S. cerevisiae* and the 220-kDa protein that has been identified in mammalian tissue culture cells) and (2) the 51-kDa EF1α-like protein identified in sea urchin embryos (Section III,D). As yet, we have no clues as to the precise role of the 5051 antigens or the EF1α-like protein in centrosomes and spbs. The presence in the centrosome of a protein that would be expected to bind to RNA (i.e. the EF1α-like protein) is intriguing in light of the evidence that RNA might be involved in the three-dimensional organization of the pericentriolar material (Section III,G).

Antibody inhibition of kinetochore-specific proteins supports the suggestion that kinetochore components play an active role in microtubule capture and microtubule elongation at the kinetochore during prometaphase congression. These studies also indicate that kinetochore components may function in sister chromatid segregation. Kinetochore components could be required for holding sister chromatids together, and/or for disassembly of microtubules at the kinetochore during anaphase. It seems likely that both functions will ultimately be demonstrated for kinetochore components. That microtubule motors are important in kinetochore function is indicated by several lines of evidence, including (1) immunofluorescence localization of cytoplasmic dynein to the kinetochore (Steuer et al., 1990; Pfarr et al., 1990), (2) identification of the kinesin-like protein encoded by the nod⁺ gene of Drosophila which appears to be involved in kinetochore function (Hawley et al., 1990), (3) in vitro studies of microtubule interaction with kinetochores which have provided evidence for both minus-end- and plus-end-directed microtubule motor activities associated with kinetochores (Hyman and Mitchison, 1991).

V. Concluding Remarks

Although much remains to be done, we have begun to make progress in the identification of cellular components that are associated with and required for the functioning of centrosomes, spindle poles, and kinetochores. The first of these structures that are likely to be understood in detail are the spbs of S. cerevisiae. The spbs in this species have a very well defined morphology, and the locations of several components within this structure have already been determined (Rout and Kilmartin, 1990; also see Alfa and Hyams, 1990, for review). Given the relative ease of using a combined genetic and molecular approach to study gene function in S. cerevisiae, we predict that the molecular organization of spbs and the functions of the various components will be established in this species within the next few years. Although obtaining a thorough understanding of the centrosome will undoubtedly take somewhat longer, the prospects for progress in this area are also good. Studies involving the analysis of mutations which appear to disrupt centrosome function, followed by molecular analysis of the gene products, and studies using the antibody inhibition tests have already begun to provide information regarding the assembly and functioning of centrosomes. Although the techniques for molecular and genetic analysis of MTOCs are somewhat more laborious and less precise in higher eukaryotes, the techniques are available and improvements are appearing almost on a daily basis. In addition, the identification of basic MTOC components and their functions in yeast will likely provide additional new tools for probing the less well defined centrosomes of higher eukaryotes.

Once the basic components of MTOCs have been defined, we can begin to try to understand the variations. Although we have focused on the centrosome as the primary MTOC of animal cells, in some cell types microtubules emanate from other sites in the cell as well (Spiegelman *et al.,* 1979). Are microtubules actually nucleated at these sites, and, if so, are these other nucleating sites biochemically equivalent to the centrosome?

Likewise, during development, changes in the organization of centrosomes and microtubules or in the morphology of microtubule nucleating sites often accompany final differentiation. For example, during myogenesis the single cell myoblasts fuse to form multinucleate myotubes. During this process the microtubules are reorganized to form arrays running parallel to the long axis of the myotube (Warren, 1974). It has been shown that the centrioles associated with the centrosomes of the myoblasts disappear during this process, while microtubule nucleating material becomes associated with the nuclear surface (Tassin *et al.,* 1985). By what mechanism does this reorganization occur? Changes in MTOCs during final differentiation also occur during development of the wing in *D. melanogaster* (Tucker *et al.,* 1986). In the epithelial cells that make up the wing blade, microtubules are not organized around centrosomes during the pupal stage of development, but rather nucleate from electron-dense structures associated with the apical surface of the cells. During the early stages of pupal development, the microtubules nucleated from these sites are composed of 13 protofilaments, while at later stages most of the microtubules are composed of 15 protofilaments (Tucker *et al.,* 1986). Are there perhaps differences in the components of the MTOCs between those in undifferentiated cells and in determined or differentiated cell types?

During fertilization, cells are created which contain the potential for assembling two distinct centrosomes. In some species, such as *Spisula solidissima,* the maternal centrosome is activated following fertilization and assembles the meiotic spindles, while the paternal centrosome is inactive (Kuriyama *et al.,* 1986). During the second meiotic division the paternal centrosome is activated and assembles an aster. Following the second meiotic division the maternal centrosome is inactivated, presumably through degradation, and the paternal centrosome is utilized for assembly of the mitotic spindles (Kuriyama *et al.,* 1986). What is the mechanism by which two centrosomes, within a common cytoplasm, are differentially regulated? This leads to a second question, What determines which of the centrosomes will be utilized for assembly of the mitotic divisions in the egg? As indicated above, in marine invertebrates the spindle poles for the first mitotic division are usually organized around the paternally derived centrioles. In other species, such as mammals, the paternally contributed centrioles are not incorporated into the mitotic spindle poles (Schatten *et al.,* 1986). A third strategy for assembly of the first mitotic spindle is seen in *Drosophila,* where both the maternal and paternal chromosome complements assemble separate, but closely apposed spindles for the first mitotic division (Huettner, 1924); presumably, the maternal spindle does not include centrioles while the paternal does. What determines which strategy will be used?

Identification of MTOC components will also open new avenues to evaluating whether centrosomes have functions other than nucleating and organizing microtubules, and what these functions might be. The next 10–20 years promise to be a very exciting period in cytoskeletal research.

Acknowledgments

We thank Drs. W. E. Theurkauf and R. S. Hawley for providing us with a copy of their manuscript prior to publication, and Drs. J. V. Kilmartin, S. Doxsey, and J. L. Salisbury for allowing us to include their unpublished results in this article. We also thank Drs. M. E. Porter and T. S. Hayes for their careful reading of the manuscript and helpful suggestions for revision, Drs. A. L. Khodjakov and T. Maekawa for their suggestions in the preparation of the diagrams shown in Figs. 1 and 2, and Mr. Jerry Sedgewick for reproduction of these figures. The work described here from the authors' laboratory was supported by NIH research grant GM41350 to RK. MK was supported on an NIH fellowship, GM13062.

References

Alfa, C. E., and Hyams, J. S. (1990). *Nature (London)* **348,** 484.
Alfa, C. E., Ducommun, B., Beach, D., and Hyams, J. S. (1990). *Nature (London)* **347,** 680–682.
Atchison, M. E., and Brown, D. L. (1986). *Cell Motil. Cytoskel.* **6,** 122–127.
Ault, J. G., and Nicklas, R. B. (1989). *Chromosoma* **98,** 33–39.
Bailly, E., Doree, M., Nurse, P., and Bornens, M. (1989). *EMBO J.* **8,** 3985–3995.
Baker, B. S., and Hall, J. C. (1976). *In* "The Genetics and Biology of Drosophila" (M. Ashburner and E. Novitski, eds.), Vol. 1a, pp. 351–434. Academic Press, New York.
Balczon, R. D., and Brinkley, B. R. (1987). *J. Cell Biol.* **105,** 855–862.
Baron, A. T., and Salisbury, J. L. (1988). *J. Cell Biol.* **107,** 2669–2678.
Baron, A. T., Greenwood, T. M., and Salisbury, J. L. (1991). *Cell Motil. Cytoskel.* **18,** 1–14.
Bastmeyer, M., and Russell, D. G. (1987). *J. Cell Sci.* **87,** 431–438.
Bastmeyer, M., Steffen, W., and Fuge, H. (1986). *Eur. J. Cell Biol.* **42,** 305–310.
Baum, P., Furlong, C., and Byers, B. (1986). *Proc. Natl. Acad. Sci. U.S.A.* **83,** 5512–5516.
Baum, P., Yip, C., Goetsch, L., and Byers, B. (1988). *Mol. Cell. Biol.* **8,** 5386–5397.
Belmont, L. D., Hyman, A. A., Sawin, K. E., and Mitchison, T. J. (1990). *Cell (Cambridge, Mass.)* **62,** 579–589.
Berlin, V., Styles, C. A., and Fink, G. R. (1990). *J. Cell Biol.* **111,** 2573–2586.
Bernat, R. L., Borisy, G. G., Rothfield, N. F., and Earnshaw, W. C. (1990). *J. Cell Biol.* **111,** 1519–1533.
Berns, M. W., Rattner, J. B., Brenner, S., and Meredith, S. (1977). *J. Cell Biol.* **72,** 351–367.
Bloom, K. S., and Yeh, E. (1989). *Curr. Opin. Cell Biol.* **1,** 526–532.
Bloom, K. S., Amaya, E., Carbon, J., Clarke, L., Hill, A., and Yeh, E. (1984). *J. Cell Biol.* **99,** 1559–1568.
Booher, R. N., Alfa, C. E., Hyams, J. S., and Beach, D. H. (1989). *Cell (Cambridge, Mass.)* **58,** 485–497.
Borisy, G. G., and Gould, R. R. (1977). *In* "Mitosis: Facts and Questions" (M. Little, N. Paweletz, C. Petzelt, H. Ponstingle, D. Schroeter, and H.-P. Zimmermann, eds.), pp. 78–87. Springer-Verlag, Berlin.

Bornens, M., Paintrand, M., Berges, J., Marty, M.-C., and Karsenti, E. (1987). *Cell Motil. Cytoskel.* **8,** 238–249.

Brinkley, B. R. (1985). *Annu. Rev. Cell Biol.* **1,** 145–172.

Brinkley, B. R. (1990). *Cell Motil. Cytoskel.* **16,** 104–109.

Browne, C. L., Lockwood, A. H., Su, J.-U., Beavo, J. A., and Steiner, A. L. (1980). *J. Cell Biol.* **87,** 336–345.

Buendia, B., Antony, C., Verde, F., Bornens, M., and Karsenti, E. (1991). *J. Cell Sci.* **97,** 259–271.

Byers, B., Shriver, K., and Goetsch, L. (1978). *J. Cell Sci.* **30,** 331–352.

Callaini, G., and Riparbelli, M. G. (1990). *J. Cell Sci.* **97,** 539–543.

Callarco-Gillam, P. D., Siebert, M. C., Hubble, R., Mitchison, T., and Kirschner, M. (1983). *Cell (Cambridge, Mass.)* **35,** 621–629.

Carpenter, A. T. C. (1973). *Genetics* **73,** 393–428.

Centonze, V. E., and Borisy, G. G. (1990). *J. Cell Sci.* **95,** 405–411.

Chaly, N., Bladon, T., Setterfield, G., Little, J. E., Kaplan, J. G., and Brown, D. L. (1984). *J. Cell Biol.* **99,** 661–671.

Clarke, L., and Carbon, J. (1985). *Annu. Rev. Genet.* **19,** 29–56.

Clayton, L., Black, C. M., and Lloyd, C. W. (1985). *J. Cell Biol.* **101,** 319–324.

Compton, D. A., Yen, T. J., and Cleveland, D. W. (1991). *J. Cell Biol.* **112,** 1083–1097.

Conde, J., and Fink, G. R. (1976). *Proc. Natl. Acad. Sci. U.S.A.* **73,** 3651–3655.

Connolly, J. A., and Kalnins, V. I. (1978). *J. Cell Biol.* **79,** 526–532.

Cooke, C. A., Hect, M. M. S., and Earnshaw, W. C. (1987). *J. Cell Biol.* **105,** 2053–2067.

Cox, J. V., Schenk, E. A., and Olmsted, J. B. (1983). *Cell (Cambridge, Mass.)* **35,** 331–339.

Davis, D. G. (1969). *Genetics* **61,** 577–594.

Davis, F. M., Tsao, T. Y., Fowler, S. K., and Rao, P. N. (1983). *Proc. Natl. Acad. Sci. U.S.A.* **80,** 2926–2930.

De Camilli, P., Moretti, M., Donini, S. D., Walter, U., and Lohmann, S. M. (1986). *J. Cell Biol.* **103,** 189–203.

Doxsey, S., Calarco, P., Seibert, P., Evans, L., Stein, P., and Kirschner, M. (1990). *J. Cell Biol.* **111,** 179a.

Endow, S. A., and Hatsumi, M. (1991). *Proc. Natl. Acad. Sci. U.S.A.* **88,** 4424–4427.

Endow, S. A., Henikoff, S., and Soler-Niedziela, L. (1990). *Nature (London)* **345,** 81–83.

Engle, D. B., Doonan, J. H., and Morris, R. N. (1988). *Cell Motil. Cytoskel.* **10,** 432–437.

Enos, A. P., and Morris, R. N. (1990). *Cell (Cambridge, Mass.)* **60,** 1019–1027.

Evans, L., Mitchison, T., and Kirschner, M. (1985). *J. Cell Biol.* **100,** 1185–1191.

Fitzgerald-Hayes, M., Clarke, L., and Carbon, J. (1982). *Cell (Cambridge, Mass.)* **29,** 235–244.

Foe, V. E., and Alberts, B. M. (1983). *J. Cell Sci.* **61,** 31–70.

Fulton, C. (1971). *In* "Origin and Continuity of Cell Organelles" (J. Reinert and H. Ursprung, eds.), pp. 170–221. Springer-Verlag, New York.

Gatti, M., and Baker, B. S. (1989). *Genes Dev.* **3,** 438–453.

Ghosh, S., and Pawelctz, N. (1987). *Chromosoma* **95,** 136–143.

Glover, D. M. (1989). *J. Cell Sci.* **92,** 137–146.

Gonzales, C., Casal, J., and Ripoll, P. (1988). *J. Cell Sci.* **89,** 39–47.

Gonzales, C., Saunders, R. D. C., Casal, J., Molina, I., Carmena, M., Ripoll, P., and Glover, D. M. (1990). *J. Cell Sci.* **96,** 605–616.

Gorbsky, G. J., Sammak, P. J., and Borisy, G. G. (1988). *J. Cell Biol.* **106,** 1185–1192.

Gosti, F., Marty, M.-C., Courvalin, J. C., Maunoury, R., and Bornens, M. (1987). *Proc. Natl. Acad. Sci. U.S.A.* **84,** 1000–1004.

Gosti-Testu, F., Marty, M.-C., Berges, J., Maunoury, R., and Bornens, M. (1986). *EMBO J.* **5,** 2545–2550.

Gould, R. P., and Borisy, G. G. (1977). *J. Cell Biol.* **73,** 601–615.

Hagan, I., and Yanagida, M. (1990). *Nature (London)* **347,** 563–566.
Hall, J. L., Ramanis, Z., and Luck, D. J. L. (1989). *Cell (Cambridge, Mass.)* **59,** 121–132.
Harris, P., Osborn, M., and Weber, K. (1980). *J. Cell Biol.* **84,** 668–679.
Hawley, S., Theurkauf, W. E., Sullivan, W., Goldstein, L. S. B., Zhang, P., and Knowles, B. A. (1990). *J. Cell Biol.* **111,** 134a.
Heath, I. B. (1981). *Int. Rev. Cytol.* **69,** 191–221.
Himes, R. H., Burton, P. R., and Gaito, J. M. (1977). *J. Biol. Chem.* **252,** 6222–6228.
Hinton, C. W., and McEarchern, W. (1963). *Drosophila Inf. Serv.* **37,** 90.
Hirano, T., Funahashi, S., Uemura, T., and Yanagida, M. (1986). *EMBO J.* **5,** 2973–2979.
Huang, B., Watterson, D. M., Lee, V. D., and Schibler, M. J. (1988a). *J. Cell Biol.* **107,** 121–131.
Huang, B., Mengersen, A., and Lee, V. D. (1988b). *J. Cell Biol.* **107,** 133–140.
Huettner, A. F. (1924). *J. Morphol.* **39,** 249–265.
Hurt, E. C. (1988). *EMBO J.* **7,** 4323–4334.
Hurt, E. C. (1989). *J. Cell Sci., Suppl.* **12,** 243–252.
Hyams, J. S., and Borisy, G. G. (1978). *J. Cell Biol.* **78,** 401–414.
Hyman, A. A., and Mitchison, T. J. (1991). *Nature (London)* **351,** 206–211.
Inagaki, M., Nishi, Y., Nishizawa, K., Matsuyama, M., and Sato, C. (1987). *Nature (London)* **328,** 649–652.
Jimenez, J., Alphey, L., Nurse, P., and Glover, D. M. (1990). *EMBO J.* **9,** 3565–3571.
Johnson, D. E., and Dutcher, S. K. (1991). *J. Cell Biol.* **113,** 339–346.
Johnson, K. A., and Rosenbaum, J. L. (1990). *Cell (Cambridge, Mass.)* **62,** 615–619.
Joswig, G., and Petzelt, C. (1990). *Cell Motil. Cytoskel.* **15,** 181–192.
Joswig, G., Petzelt, C., and Werner, D. (1991). *J. Cell Sci.* **98,** 37–43.
Karsenti, E., Kobayashi, S., Mitchison, T., and Kirschner, M. (1984). *J. Cell Biol.* **98,** 1763–1776.
Kellogg, D. R., Field, C. M., and Alberts, B. M. (1989). *J. Cell Biol.* **109,** 2977–2991.
Keryer, G., De Loubresse, N. G., Bordes, N., and Bornens, M. (1989). *J. Cell Sci.* **93,** 287–298.
Kim, J., Ljungdahl, P. O., and Fink, G. R. (1991). *Genetics* **126,** 799–812.
Kimble, M., and Church, K. (1983). *J. Cell Sci.* **62,** 301–318.
Kitanishi-Yumura, T., and Fukui, Y. (1987). *Cell Motil. Cytoskel.* **8,** 106–117.
Komesli, S., Tournier, F., Paintrand, M., Margolis, R. L., Job, D., and Bornens, M. (1989). *J. Cell Biol.* **109,** 2869–2878.
Komma, D. J., Horne, A. S., and Endow, S. A. (1991). *EMBO J.* **10,** 419–424.
Koonce, M. P., and McIntosh, J. R. (1990). *Cell Motil. Cytoskel.* **15,** 51–62.
Kuriyama, R. (1984). *J. Cell Sci.* **66,** 277–295.
Kuriyama, R. (1989). *Cell Motil. Cytoskel.* **12,** 90–103.
Kuriyama, R., and Borisy, G. G. (1981a). *J. Cell Biol.* **91,** 814–821.
Kuriyama, R., and Borisy, G. G. (1981b). *J. Cell Biol.* **91,** 822–826.
Kuriyama, R., and Borisy, G. G. (1985). *J. Cell Biol.* **101,** 524–530.
Kuriyama, R., Sato, C., Fukui, Y., and Nishibayashi, S. (1982). *Cell Motil. Cytoskel.* **2,** 257–272.
Kuriyama, R., Borisy, G. G., and Masui, Y. (1986). *Dev. Biol.* **114,** 151–160.
Kuriyama, R., Savereide, P., Lefebvre, P., and Dasgupta, S. (1990). *J. Cell Sci.* **95,** 231–236.
Kuroiwa, T., Yorihuzi, T., Yabe, N., Ohta, T., and Uchida, H. (1990). *Protoplasma* **158,** 155–164.
Lee, M. G., and Nurse, P. (1987). *Nature (London)* **327,** 31–35.
Lee, M. G., and Nurse, P. (1988). *Trends Genet.* **4,** 287–290.
Lehner, C. F., and O'Farrell, P. H. (1990). *Cell (Cambridge, Mass.),* **61,** 535–547.
Leslie, R. J. (1990). *Cell Motil. Cytoskel.* **16,** 225–228.
Leslie, R. J., Kohler, E., and Wilson, L. (1991). *Cell Motil. Cytoskel.* **19,** 80–90.
Lewin, B. (1990). *Cell (Cambridge, Mass.)* **61,** 743–752.
Lewis, E. B., and Gencarella, W. (1952). *Genetics* **37,** 600–601.
Lohka, M. J. (1989). *J. Cell Sci.* **92,** 131–135.
Lohka, M. J., and Maller, J. L. (1985). *J. Cell Biol.* **101,** 518–523.

Lohmann, S. M., De Camilli, P., Einig, I., and Walter, U. (1984). *Proc. Natl. Acad. Sci. U.S.A.* **81**, 6723–6727.

Lydersen, B. K., and Pettijohn, D. E. (1980). *Cell (Cambridge, Mass.)* **22**, 489–499.

Maekawa, T., and Kuriyama, R. (1991). *J. Cell Sci.* **100**, 533–540.

Maekawa, T., Leslie, R., and Kuriyama, R. (1991). *Eur. J. Cell Biol.* **54**, 255–267.

Masumoto, H., Masukata, H., Muro, Y., Nozaki, N., and Okazaki, T. (1989). *J. Cell Biol.* **109**, 1963–1973.

Mazia, D. (1987). *Int. Rev. Cytol.* **100**, 49–92.

McCarty, G. A., Valencia, D. W., Fritzler, M. J., and Barada, F. A. (1981). *N. Engl. J. Med.* **305**, 703.

McCarty, G. A., Valencia, D. W., and Fritzler, M. J. (1984). *J. Rheumatol.* **11**, 213–218.

McDonald, H. B., and Goldstein, L. S. B. (1990). *Cell (Cambridge, Mass.)* **61**, 991–1000.

McDonald, H. D., Stewart, R. J., and Goldstein, L. S. B. (1990). *Cell (Cambridge, Mass.)* **63**, 1159–1165.

McGill, M., Highfield, D. P., Monahan, T. M., and Brinkley, B. R. (1976). *J. Ultrastruct. Res.* **57**, 43–53.

McIntosh, J. R. (1983). *Mod. Cell Biol.* **2**, 115–142.

McIntosh, J. R., and Koonce, M. P. (1989). *Science* **246**, 622–628.

Meluh, P. B., and Rose, M. D. (1990). *Cell (Cambridge, Mass.)* **60**, 1029–1041.

Miake-Lye, R., and Kirschner, M. W. (1985). *Cell (Cambridge, Mass.)* **41**, 165–175.

Mitchison, T., and Kirschner, M. (1984). *Nature (London)* **312**, 232–237.

Mitchison, T., and Kirschner, M. (1985). *J. Cell Biol.* **101**, 766–777.

Mitchison, T. J., and Sawin, K. E. (1990). *Cell Motil. Cytoskel.* **16**, 93–98.

Mitchison, T. J., Evans, L., Schulze, E., and Kirschner, M. (1986). *Cell (Cambridge, Mass.)* **45**, 515–527.

Moreno, S., and Nurse, P. (1990). *Cell (Cambridge, Mass.)* **61**, 549–551.

Moroi, Y., Peebles, C., Fritzler, M. J., Steigerwald, J., and Tan, E. M. (1980). *Proc. Natl. Acad. Sci. U.S.A.* **77**, 1627–1631.

Moudjou, M., Paintraud, M., Vigues, B., and Bornens, M. (1991). *J. Cell Biol.* **115**, 129–140.

Neighbors, B. W., Williams, R. C., Jr., and McIntosh, J. R. (1988). *J. Cell Biol.* **106**, 1193–1204.

Nenci, I., and Marchetti, E. (1978). *J. Cell Biol.* **76**, 255–260.

Newmeyer, D. D., and Ohlsson-Wilhelm, B. M. (1985). *Chromosoma* **92**, 297–303.

Nicklas, R. B. (1989). *J. Cell Biol.* **109**, 2245–2255.

Nicklas, R. B., Brinkley, B. R., Pepper, D. A., Kubai, D. F., and Rickards, G. K. (1979). *J. Cell Sci.* **35**, 87–104.

Nigg, E. A., Schäfer, G., Hilz, H., and Eppenberger, H. M. (1985). *Cell (Cambridge, Mass.)* **41**, 1039–1051.

Nislow, C., Sellitto, C., Kuriyama, R., and McIntosh, J. R. (1990). *J. Cell Biol.* **111**, 511–522.

Nurse, P., and Bissett, Y. (1981). *Nature (London)* **292**, 558–560.

Oakley, B. R., Oakley, C. E., Yoon, Y., and Jung, M. K. (1990). *Cell (Cambridge, Mass.)* **61**, 1289–1301.

Oakley, C. E., and Oakley, B. R. (1989). *Nature (London)* **338**, 662–664.

Ohta, K., Toriyama, M., Endo, S., and Sakai, H. (1988). *Cell Motil. Cytoskel.* **10**, 496–505.

Ohta, K., Toriyama, M., Miyazaki, M., Murofushi, H., Hosoda, S., Endo, S., and Sakai, H. (1990). *J. Biol. Chem.* **265**, 3240–3247.

Ottaviano, Y., and Gerace, L. (1985). *J. Biol. Chem.* **260**, 624–632.

Palmer, D. K., O'Day, K., Wener, M. H., Andrews, B. S., and Margolis, R. L. (1987). *J. Cell Biol.* **104**, 805–815.

Paweletz, N., Mazia, D., and Finze, E. M. (1984). *Exp. Cell Res.* **152**, 47–65.

Pepper, D. A., and Brinkley, B. R. (1979). *J. Cell Biol.* **82**, 585–591.

Pepper, D. A., and Brinkley, B. R. (1980). *Cell Motil.* **1**, 1–15.

Perret, E., Albert, M., Bordes, N., Bornens, M., and Soyer-Gobillard, M.-O. (1991). *Biosystems* **25,** 53–65.

Peterson, S. P., and Berns, M. W. (1978). *J. Cell Sci.* **34,** 289–301.

Pettijohn, D. E., Henzl, M., and Price, C. (1984). *J. Cell Sci., Suppl.* **1,** 187–201.

Pfarr, C. M., Coue, M., Grissom, P. M., Hays, T. S., Porter, M. E., and McIntosh, J. R. (1990). *Nature (London)* **345,** 263–265.

Pickett-Heaps, J. D. (1969). *Cytobios* **3,** 257–280.

Pines, J., and Hunter, T. (1990). *New Biol.* **2,** 389–401.

Polaina, J., and Conde, J. (1982). *Mol. Gen. Genet.* **186,** 253–258.

Porter, K. R. (1966). *In* "Principles of Biomolecular Organization" (G. E. W. Wolstenholme and M. O'Connor, eds.), pp. 308–345. Churchill, London.

Puro, J., and Nokkala, S. (1977). *Chromosoma* **63,** 273–286.

Raff, E. C. (1979). *Int. Rev. Cytol.* **59,** 1–96.

Raff, J. W., and Glover, D. M. (1988). *J. Cell Biol.* **107,** 2009–2019.

Rao, P. N., Zhao, J.-Y., Ganju, R. K., and Ashorn, C. L. (1989). *J. Cell Sci.* **93,** 63–69.

Rattner, J. B., and Phillips, S. B. (1973). *J. Cell Biol.* **57,** 359–372.

Rattner, J. B., Kingwell, B. G., and Fritzler, M. J. (1988). *Chromosoma* **96,** 360–367.

Rattner, J. B., Lew, J., and Wang, J. H. (1990). *Cell Motil. Cytoskel.* **17,** 227–235.

Riabowol, K., Draetta, G., Brizuela, L., Vandré, D., and Beach, D. (1989). *Cell (Cambridge, Mass.)* **57,** 393–401.

Rieder, C. L. (1979). *J. Cell Biol.* **80,** 1–9.

Rieder, C. L. (1991). *Curr. Opin. Cell Biol.* **3,** 59–66.

Rieder, C. L., and Borisy, G. G. (1982). *Biol. Cell.* **44,** 117–132.

Ring, D., Hubble, R., Caput, D., and Kirschner, M. (1980). *In* "Microtubules and Microtubule Inhibitors" (M. De Brabander and J. De May, eds.), pp. 297–309. Elsevier/North-Holland Biomedical Press, Amsterdam.

Ripoll, P., Pimpinelli, S., Valdivia, M. M., and Avila, J. (1985). *Cell (Cambridge, Mass.)* **41,** 907–912.

Robbins, E., Jentzsch, G., and Micali, A. (1968). *J. Cell Biol.* **36,** 329–339.

Roobol, A., Havercroft, J. C., and Gull, K. (1982). *J. Cell Sci.* **55,** 365–381.

Rose, M. D., and Fink, G. R. (1987). *Cell (Cambridge, Mass.)* **48,** 1047–1060.

Rothman, J. H., Raymond, C. K., Gilbert, T., O'Hara, P. J., and Stevens, T. H. (1990). *Cell (Cambridge, Mass.)* **61,** 1063–1074.

Rout, M. P., and Kilmartin, J. V. (1990). *J. Cell Biol.* **111,** 1913–1927.

Sager, P. R., Rothfield, N. L., Oliver, J. M., and Berlin, R. D. (1986). *J. Cell Biol.* **103,** 1863–1872.

Sakai, H., Ohta, K., Toriyama, M., and Endo, S. (1989). *Adv. Exp. Med. Biol.* **255,** 471–480.

Salisbury, J. L. (1989). *J. Phycol.* **25,** 201–206.

Sato, C., Nishizawa, K., Nakamura, H., Komagoe, Y., Shimada, K., Ueda, R., and Suzuki, S. (1983). *Cell Struct. Funct.* **8,** 245–254.

Sauron, M. E., Marty, M. C., Maunoury, R., Courvalin, J. C., Maro, B., and Bornens, M. (1984). *J. Submicrosc. Cytol.* **16,** 133–135.

Sawin, K. E., and Mitchison, T. J. (1991a). *J. Cell Biol.* **112,** 925–940.

Sawin, K. E., and Mitchison, T. J. (1991b). *J. Cell Biol.* **112,** 941–954.

Schatten, H., Schatten, G., Mazia, D., Balczon, R., and Simerly, C. (1986). *Proc. Natl. Acad. Sci. U.S.A.* **83,** 105–109.

Schatten, H., Walter, M., Mazia, D., Biessmann, H., Paweletz, N., Coffe, G., and Schatten, G. (1987). *Proc. Natl. Acad. Sci. U.S.A.* **84,** 8488–8492.

Schatten, H., Walter, M., Biessmann, H., and Schatten, G. (1988). *Cell Motil. Cytoskel.* **11,** 248–259.

Scheele, R. B., Bergen, L. G., and Borisy, G. G. (1982). *J. Mol. Biol.* **154,** 485–500.

Sellitto, C., and Kuriyama, R. (1988a). *J. Cell Sci.* **89,** 57–65.

Sellitto, C., and Kuriyama, R. (1988b). *J. Cell Biol.* **106,** 431–439.

Sellitto, C., Kimble, M., and Kuriyama, R. (1992). *Cell Motil. Cytoskel.* (in press).

Senécal, J.-L., Oliver, J. M., and Rothfield, N. (1985). *Arthritis Rheum.* **28**, 889–898.

Sequeira, W., Nelson, C. R., and Szauter, P. (1989). *Genetics* **123**, 511–524.

Shimoda, C., Hirata, A., Kishida, T., and Tanaka, K. (1985). *Mol. Gen. Genet.* **200**, 252–257.

Shpetner, H. S., and Vallee, R. B. (1989). *Cell (Cambridge, Mass.)* **59**, 421–432.

Simerly, C., Balczon, R., Brinkley, B. R., and Schatten, G. (1990). *J. Cell Biol.* **111**, 1491–1504.

Sluder, G., and Lewis, K. (1987). *J. Exp. Zool.* **244**, 89–100.

Snyder, J. A. (1980). *Intl. Rep.* **4**, 859–868.

Snyder, J. A., and McIntosh, J. R. (1975). *J. Cell Biol.* **67**, 744–760.

Snyder, J. A., Hamilton, B. T., and Mullins, J. M. (1982). *Eur. J. Cell Biol.* **27**, 191–199.

Snyder, M., and Davis, R. W. (1988). *Cell (Cambridge, Mass.)* **54**, 743–754.

Spiegelman, B. M., Lopata, M. A., and Kirschner, M. W. (1979). *Cell (Cambridge, Mass.)* **16**, 239–252.

Staiger, C. J., and Cande, W. Z. (1990). *Dev. Biol.* **138**, 231–242.

Stearns, M. E., and Brown, D. L. (1979). *Nature (London)* **76**, 5745–5749.

Stearns, T., Evans, L., and Kirschner, M. (1991). *Cell (Cambridge, Mass.)* **65**, 825–836.

Steffen, W., and Linck, R. W. (1989). *In* "Cell Movement" (F. D. Warner and J. R. McIntosh, eds.), Vol. 2, pp. 67–81. Alan R. Liss, New York.

Steffen, W., Fuge, H., Dietz, R., Bastmeyer, M., and Müller, G. (1986). *J. Cell Biol.* **102**, 1679–1687.

Steuer, E. R., Wordeman, L., Schroer, T. A., and Sheetz, M. P. (1990). *Nature (London)* **345**, 266–268.

Sunkel, C. E., and Glover, D. M. (1988). *J. Cell Sci.* **89**, 25–38.

Tassin, A.-M., Maro, B., and Bornens, M. (1985). *J. Cell Biol.* **100**, 35–46.

Theurkauf, W. E., and Hawley, R. S. (1992). *J. Cell Biol.* **116**, 1167–1180.

Tilney, L. G., Bryan, J., Bush, D. J., Fujiwara, K., Mooseker, M. S., Murphy, D. B., and Snyder, D. H. (1973). *J. Cell Biol.* **59**, 267–275.

Toriyama, M., Ohta, K., Endo, E., and Sakai, H. (1988). *Cell Motil. Cytoskel.* **9**, 117–128.

Tousson, A., Zeng, C., Brinkley, B. R., and Valdivia, M. M. (1991). *J. Cell Biol.* **112**, 427–440.

Tucker, J. B., Milner, M. J., Currie, D. A., Muir, J. W., Forrest, D. A., and Spencer, M.-J. (1986). *Eur. J. Cell Biol.* **41**, 270–289.

Tuttanelli, D. L., McKeon, F., Kleinsmith, D. M., Burnham, T. K., and Kirschner, M. (1983). *Arch. Dermatol.* **119**, 560–566.

Urrutia, R., McNiven, M. A., Albanesi, J. P., Murphy, D. B., and Kachar, B. (1989). *J. Cell Biol.* **109**, 81a.

Uzawa, S., Samejima, I., Hirano, T., Tanaka, K., and Yanagida, M. (1990). *Cell (Cambridge, Mass.)* **62**, 913–925.

Vallee, R. B., Bloom, G. S., and Theurkauf, W. E. (1984). *J. Cell Biol.* **99**, 38s–44s.

Vandré, D. D., and Borisy, G. G. (1989). *J. Cell Sci.* **94**, 245–258.

Vandré, D. D., Davis, F. M., Rao, P. N., and Borisy, G. G. (1984). *Proc. Natl. Acad. Sci. U.S.A.* **81**, 4439–4443.

Vandré, D. D., Davis, F. M., Rao, P. N., and Borisy, G. G. (1986). *Eur. J. Cell Biol.* **41**, 72–81.

Verde, F., Labbé, J. C., Dorée, M., and Karsenti, E. (1990). *Nature (London)* **343**, 233–238.

Verde, F., Berrez, J. M., Antony, C., and Karsenti, E. (1991). *J. Cell Biol.* **112**, 1177–1187.

Vorobjev, I. A., and Chentsov, Y. S. (1982). *J. Cell Biol.* **98**, 938–949.

Vorobjev, I. A., and Nadezhdina, E. S. (1987). *Int. Rev. Cytol.* **106**, 227–293.

Walker, R. A., Salmon, E. D., and Endow, S. A. (1990). *Nature (London)* **347**, 780–782.

Warren, R. H. (1974). *J. Cell Biol.* **63**, 550–566.

Weil, C. F., Oakley, C. E., and Oakley, B. R. (1986). *Mol. Cell. Biol.* **6**, 2963–2968.

Weisenberg, R. C., Allen, R. D., and Inoue, S. (1986). *Proc. Natl. Acad. Sci. U.S.A.* **83**, 1728–1732.

Whitfield, W. G. F., Millar, S. E., Saumweber, H., Frasch, M., and Glover, D. M. (1988). *J. Cell Sci.* **89**, 467–480.

Whitfield, W. G. F., Gonzalez, C., Maldonado-Codina, G., and Glover, D. M. (1990). *EMBO J.* **9**, 2563–2572.

Wick, S. M. (1985a). *Cell Biol. Intl. Rep.* **9,** 357–371.
Wick, S. M. (1985b). *Cytobios* **43,** 285–294.
Wilson, P. G., and Fuller, M. T. (1990). *J. Cell Biol.* **111,** 179a.
Winey, M., Goetsch, L., Baum, P., and Byers, B. (1991). *J. Cell Biol.* **114,** 745–754.
Wordeman, L., Davis, F. M., Rao, P. N., and Cande, W. Z. (1989). *Cell Motil. Cytoskel.* **12,** 33–41.
Wordeman, L., Steuer, E. R., Sheetz, M. P., and Mitchison, T. (1991). *J. Cell Biol.* **114,** 285–294.
Yamamoto, A. H., Komma, D. J., Shaffer, C. D., Pirrotta, V., and Endow, S. A. (1989). *EMBO J.* **8,** 3543–3552.
Yeh, E., Driscoll, R., Coltrera, M., Olins, A., and Bloom, K. (1991). *Nature (London)* **349,** 713–715.
Zalokar, M., and Erk, I. (1976). *J. Microsc. Biol. Cell.* **25,** 97–106.
Zhang, P., and Hawley, R. S. (1990). *Genetics* **125,** 115–127.
Zhang, P., Knowles, B. A., Goldstein, L. S. B., and Hawley, R. S. (1990). *Cell (Cambridge, Mass.)* **62,** 1053–1062.
Zheng, Y., Jung, M. K., and Oakley, B. R. (1991). *Cell (Cambridge, Mass.)* **65,** 817–823.
Zinkowski, R. P., Meyne, J., and Brinkley, B. R. (1991). *J. Cell Biol.* **113,** 1091–1110.

Notes Added in Proof

The *ncd*+ microtubule motor protein has been found to be associated with meiotic spindles (M. Hatsumi and S. A. Endow, in preparation), consistent with the idea that the protein acts to define or form spindle poles in meiosis (this review and M. Hatsumi and S. A. Endow (1992) *J. Cell Sci.* **101**).

The results of *in vitro* and *in vivo* antibody inhibition studies suggest that γ-tubulin functions in microtubule nucleation in mammalian cells, as in *Aspergillus* (Joshi, H. C., Palacios, M. J., McNamara, L., and Cleveland, D. W. (1992) *Nature (London)* **356,** 80–83).

Molecular and Cellular Basis of Formation, Hardening, and Breakdown of the Egg Envelope in Fish

K. Yamagami, T. S. Hamazaki*, S. Yasumasu[†], K. Masuda[‡], and I. Iuchi
Life Science Institute, Sophia University, Chiyoda-ku, Tokyo 102, Japan[1]

I. Introduction

The egg envelope (egg membrane) is an acellular structure(s) enclosing the egg and the embryo of all multicellular animals except sponges and some coelenterates. The number, structure, origin, and biological functions of the egg envelope vary in different animal groups. The number of egg envelopes varies from one to several in different animal species. In the eggs of birds and reptiles, which have the largest number of egg envelopes of all animals in order to protect the eggs from desiccation during the long period of incubation in the air, the egg envelopes from the innermost to the outermost consist of a sheet of vitelline envelope, variable numbers of layers of egg white (albumen), two sheets of shell membrane, and an egg shell of calcified collagen. In contrast, the eggs of sponges and most coelenterates lack egg envelopes (Ludwig, 1874; Waldeyer, 1906; Balinsky, 1965), and therefore some of them move by means of amoeboid movement. These cases of oviparous amniotes and some lower invertebrates seem to be the two extremes, and the eggs of most animal groups are equipped with at least a sheet of egg envelope, the vitelline envelope, which is in direct contact with the plasma membrane of the egg cell. The presence of an additional jellylike layer of egg envelope surrounding the vitelline envelope is also well known in the eggs of many aquatic invertebrates, some fishes, and amphibians. The constitution of egg envelopes seems to be a consequence of a combination of evolutionary trends and adaptational processes, which can be seen in many facets of reproductive and developmental phenomena. The morphology and the chemical properties of egg en-

[1]*Present addresses*: *Department of Biomaterials Science, Faculty of Dentistry, Tokyo Medical and Dental University, 1–5–45 Yushima, Bunkyo-ku, Tokyo 113, Japan †Zoological Institute, Hiroshima University, 1–3–1 Kagamiyama, Higashi-Hiroshima City 724, Japan ‡Department of Antibiotics, National Institute of Health, 2–10–35 Kami-Oosaki, Shinagawa-ku, Tokyo 141, Japan

velope(s) are also very different within the many layers of the egg envelope of an animal and in the corresponding layers of the egg envelopes of different animal groups. However, it can be said, in general, that the common basic morphological feature is a filmy structure composed of gelatinous or fibrous elements and the chemical constituents of most egg envelopes are glycoproteins rather than lipoproteins. These characteristics may be better understood when we realize that the egg envelope is not a membranous structure *inside* the cells but a structure *outside* the cells that is composed of secretions from the cells. There is a problem in determining the cells from which the materials of egg envelopes originate. We discuss this problem later, as it is one of the important themes of the present article.

The egg envelope plays several physiological roles which are more significant than those conceived from its apparent features as an accessory ornament of the egg. Among the roles, the following apply to the egg envelopes of many animal species: (1) attraction of spermatozoa to the egg, (2) activation of spermatozoa, (3) mechanical blocking to polyspermy at fertilization, and (4) protection of the egg or the enclosed developing embryo against various external physical, chemical, and biological hazards including desiccation. According to Miller (1985), sperm attractants are secreted from tissues of female reproductive organs in some marine invertebrates. Induction of important actions such as binding, agglutination, and activation, and the acrosomal reaction of spermatozoa by the jelly substance of echinoderm eggs or the zona pellucida of mammalian eggs are well known (Aketa *et al.*, 1979; Tyler, 1948, 1949; Dan, 1956; Wassarman *et al.*, 1985; Wassarman, 1988). Some peptides obtained from the echinoderm egg jelly, namely, speract, resact, and mosact, are found to induce activation of sperm respiration and/or metabolism (Hansbrough and Garbers, 1981a,b; Suzuki *et al.*, 1981, 1984). Such sperm-activating peptides (SAPs) are probably distributed in the egg envelopes in various animal groups (Suzuki, 1990).

It has long been supposed that, at fertilization, there occur rapid and slow changes blocking polyspermy in the eggs (Just, 1919; Rothschild and Swann, 1952). The rapid change is considered to be closely related to the change in the membrane potential (fertilization potential) of the egg cell itself (Jaffe and Gould, 1985), which has been compared to the signal transduction mechanism occurring in nervous tissues (Hagiwara and Jaffe, 1979; Berridge and Irvine, 1989). The slow change to block polyspermy is, however, mostly due to the change in the egg envelope, especially in the vitelline envelope. A change of the vitelline envelope into a fertilization envelope is best known in sea urchin eggs, and comparable changes are observed in the eggs of many other invertebrates and vertebrates as well. The slow block to polyspermy is due not only to mechanical change of the envelope, as seen in the sea urchin eggs, but also to some physiological changes of the egg envelope, which are associated with its transformation upon activation of the egg. This transformation implies a reduction in the number of sperm receptors and/or a reduction in the inducing activity of the acrosomal reaction of spermatozoa (Greve and Hedrick, 1978; Bleil and Wassarman, 1980, 1983). Thus, the mechanism of slow change to block polyspermy in the egg envelope itself also differs in various animal groups.

It is not the aim of this article, however, to describe and compare the structure, properties, and physiological functions of the egg envelopes of various animal groups. In the present article, our description concentrates on some of the processes from formation to breakdown of fish egg envelope from the viewpoint of cell and developmental biology. There are two different facets of the biology of the egg envelope that can be investigated. One is to analyze directly the egg envelope itself. Thus, we analyze the structure and function of the egg envelope, compare the egg envelopes of various species, and deduce a biological concept of the egg envelope. The other is to exploit the egg envelope as a probe or as a clue that can be used in the analysis of molecular, cellular, and developmental phenomena in living systems. Although the structure of the egg envelope is quite simple as compared with the cell itself, various biological activities with many accompanying molecular and cellular events are needed for the formation of the egg envelope, performance of its biological roles, and its breakdown. Analyses of the mechanisms of these events should be closely related not only to the biology of the egg envelope but also to understanding of cellular and developmental bases of biological phenomena.

In this connection, we shall consider some experimental results concerning molecular and cellular processes relating to formation, hardening, and breakdown of the egg envelope in fish.

II. Terminology of the Fish Egg Envelope

There have been many articles describing the terminology of the egg envelope (egg membrane) of animals, and many of them deal with the fish egg envelope (Ludwig, 1874; Balfour, 1880; Wilson, 1928; Chaudhury, 1956; Anderson, 1967; Laale, 1980; Dumont and Brummett, 1985). This would suggest that the fish egg envelope has been the subject of many investigations and that there are some problems with its terminology. Indeed, there is some confusion with regard to its terminology.

A. Conventional Nomenclature

The egg envelope has also been called the egg membrane. It is different from many kinds of membranous structures constituting cell organelles: the egg envelope is a structure with a kind of extracellular matrix (Somers and Shapiro, 1989), and is much larger in size and dimension than these cellular membranes. This seems to be the main reason why the designation of egg envelope was used instead of egg membrane. In addition, some confusion possibly arose concerning the usage of the term egg membrane, as it was sometimes used for the cell membrane (plasma membrane) of the egg cell (e.g., Barth, 1953).

Nomenclature of the egg envelope in general is closely related to its origin. In a classic treatise on oogenesis in animals, Ludwig (1874) classified the egg envelope into two categories based on its origin. One category is named the primary egg envelope, and is derived from the oocyte itself and/or the follicle cells, and the other is the secondary egg envelope, which originates from the cells of other reproductive organs such as the epithelial cells of the oviduct. Ludwig's terminology was slightly modified by Balfour (1880), who subdivided the primary egg envelopes into those of oocyte origin and follicle cell origin. Following these classic studies, Waldeyer (1906) and Wilson (1928) established a definition which is accepted nowadays as a conventional and common concept of egg envelope formation, i.e., (1) the primary egg envelope is derived from the oocyte, (2) the secondary egg envelope originates from the follicle cells, and (3) the tertiary egg envelope is formed by some cells of other reproductive organs such as oviducts.

There are various designations for the egg envelope of fish. Most fish egg envelopes consist of two or three layers. These layers are different in morphology, ultrastructure, stainability, and therefore, presumably, in chemical properties. The outer one or two layers are thin, and the innermost layer is usually the thickest. Although the designated names may differ from layer to layer, the most commonly employed terms for the *whole* egg envelope are vitelline envelope and chorion. According to Waldeyer (1883; cf. Anderson, 1967), the vitelline envelope is, in general, the follicular products enveloping the egg yolk (vitellus), which is an envelope just adjacent to the oocyte (plasma) membrane; it is considered to be mostly a primary egg envelope. The term chorion is also very often used for the envelopes of unfertilized and fertilized eggs of fish. According to Laale (1980), Lereboullet (1862) was the first to use chorion to refer to the membrane of oocyte origin, but it was later used to designate the membrane of follicle cell origin by Balfour (1885). As Anderson (1967) noted, the chorion may be an augmented egg envelope. Thus, the envelopes of unfertilized eggs, fertilized eggs, and developing embryos of fish may all be labeled chorion.

B. Designations Employed in the Present Article

Besides the designations described above, there are many terms indicating the whole egg envelope or a part (some layers) of it (Laale, 1980; Dumont and Brummet, 1985). Among them, zona radiata and zona pellucida are often employed, but these terms may sometimes cause confusion (see Anderson, 1967). In fish, either whole egg envelope (Flügel, 1967b; Chaudhry, 1956) or its thickest inner layer (Oppen-Berntsen *et al.*, 1990) is referred to as zona radiata, while another term, zona pellucida, is used for either whole egg envelope (Chaudhry,

1956; Tesoriero, 1977a,b, 1978; Hamazaki *et al.*, 1989) or the outer thin layer (Oppen-Berntsen *et al.*, 1990). The zona radiata, or radiate zone, is the name given to the vitelline envelope with radial pore canals pierced through by the microvilli of the oocyte and follicle cells (granulosa cells). The zona pellucida, or pellucid zone, on the other hand, originally referred to a covering structure (vitelline envelope) around oocytes, which is difficult to stain with ordinary dyes (Anderson, 1967). As mentioned above, once the cytoplasmic projections of dual origin have pierced the vitelline envelope (zona pellucida) radially, it is named zona radiata (Waldeyer, 1906; Chaudhury, 1956; Balinsky, 1965; Flügel, 1967b). Thus, as far as designation is concerned, zona pellucida is, in general, replaced by zona radiata. In mammals, however, zona pellucida refers to the vitelline envelope at any stage of formation, especially at the fully grown stage.

When the oocyte grows sufficiently, the microvilli of dual origin are retracted prior to or at the time of ovulation (Kishimoto *et al.*, 1984). It seems that the zona radiata and zona pellucida substantially overlap, but are not identical with each other. The terms zona pellucida and vitelline envelope are thus often used in reference to the same structure (Haddon, 1887; Anderson, 1967), and, if we pay attention only to the substance of the vitelline envelope which is deposited between the microvilli, zona pellucida and zona radiata seem to be identical. However, the use of both zona pellucida and zona radiata as synonyms of whole egg envelope in fish (Hamazaki *et al.*, 1989a) may cause confusion. Moreover, neither vitelline envelope, zona pellucida, nor zona radiata seems to be appropriate to refer to the envelope of the fertilized egg and the developing embryo.

In the present article, we therefore employ the term envelope, such as oocyte envelope or egg envelope, and refer to the distinct concentric strata of the envelope as the layer, e.g., the outer layer of the oocyte envelope, the inner layer of the egg envelope. The term chorion is also employed in the present article for a whole envelope, as this designation can be applied widely to the augmented envelope, from the ovulated eggs (unfertilized eggs) to the prehatching embryos in fish. Moreover, a simple word such as chorion is convenient to combine with a prefix or suffix when making compound words, such as extrachorionic and choriolysis.

III. Formation of the Inner Layer of Egg Envelope in *Oryzias latipes*

As described above, the egg envelope of fish has been considered to be synthesized in oocytes or follicle cells, and it is classified as the primary or secondary egg envelope according to the conventional concept (Ludwig, 1874; Wilson, 1928). However, the experimental results that are described in this section are incompatible with this concept.

A. Presence of a Spawning Female-Specific Substance, an Egg Envelope Glycoprotein-Like Immunoreactive Substance, Outside the Ovary

The envelope (chorion) of fertilized egg of many fishes is a hard and tough structure with strong elasticity and is also insoluble in water. The constituent proteins of the egg envelope are, therefore, inconvenient as immunogens to raise antibodies. However, the hatching enzyme-digests of the hardened egg envelope seemed to be appropriate for immunological analyses of the egg envelope, as the hatching enzyme does not break down the egg envelope completely into free amino acids or small peptides, but, by limited proteolysis (see Section V), produces a mixture of water-soluble, high-molecular-weight glycoproteins (Iuchi and Yamagami, 1976; Yamagami, 1981). Thus, we can isolate two glycoprotein fractions: Fr1 (or F1) which is heterogeneous, consisting of several glycoprotein molecules with similar characteristics (Fr1 family; Iuchi and Yamagami, 1976; Yamagami, 1981), and Fr2 (or F2), which is almost homogeneous by Sephadex column chromatography and can serve as the antigen to raise polyclonal anti-egg envelope glycoprotein antibodies in rabbits.

When either anti-F1 IgG or anti-F2 IgG was reacted with the extracts of various tissues of adult fish in breeding condition, some unexpected results were obtained. As shown in Fig. 1, a distinct cross-reaction of the antibody was observed not only with the ovarian extracts but also with the liver extracts and blood of spawning female fish (Hamazaki *et al.*, 1984). The result indicated that some egg envelope glycoprotein-like immunoreactive substance(s) was present outside as well as in the ovary. The liver and blood of male fish were not reactive. Immunohistochemical examinations using fluorescent isothiocyanate (FITC)-conjugated anti-F1 antibody also revealed that the parenchymal cells of the liver and blood plasma of spawning female fish only were strongly stained; those of male fish were not (Fig. 1). In the ovary, the staining was observed in the follicle cell layers and intercellular spaces as well as in the oocyte envelopes themselves. It should be noted here that, when the female fish ceases spawning after the breeding season, the immunoreactivity of the tissues decreases gradually, resulting finally in no reactivity, like a male fish (Hamazaki *et al.*, 1985). The immunoreactivity disappears first from the liver and last from the ovary. This behavior of the egg envelope glycoprotein-like immunoreactive substance(s) suggested that this substance(s) might be synthesized in the liver and transported to the oocyte envelope in the ovary through blood circulation. At the same time, however, the behavior of this substance made us suspect that we might confuse this substance(s) with vitellogenin, as it is well established that, in oviparous vertebrates, vitellogenin, a precursor of egg yolk protein, is synthesized in the liver of the egg-laying female and transported to the ovary to be incorporated into vitellogenic oocytes (Wallace, 1978, 1985; Bergink *et al.*, 1974). The intrahepatic synthesis of vitellogenin is also known to be induced by estrogen administration, even in male animals.

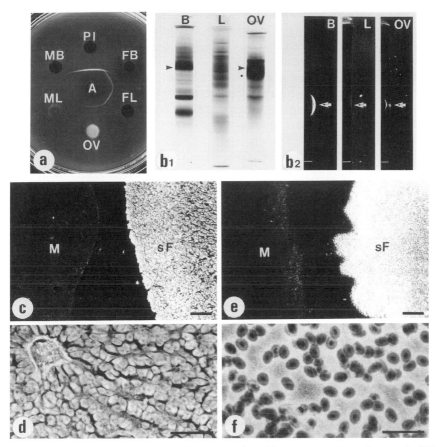

FIG. 1 Presence of a substance which is immunoreactive with anti-egg envelope glycoprotein (F1) antibody and is distinct from vitellogenin in blood plasma and the liver as well as in the ovary of spawning female fish of *Oryzias latipes*. (a) Double diffusion analysis. A, Anti-F1 IgG; P1, egg envelope glycoprotein (F1) employed as the immunogen; MB and ML, blood plasma and liver extract of male fish. FB, FL, and OV, blood plasma and extracts of liver and ovary of spawning female fish. (b₁ and b₂) Immunoelectrophoretograms of blood plasma (B), and extracts of liver (L) and ovary (OV) of spawning female fish. The arrowheads in b1 and the arrows in b2 indicate vitellogenin and immunoprecipitation, respectively. (c f) Immunofluorescent manifestation of the immunoreactive substance in the liver (c, d) and blood plasma (e, f) of spawning female fish. M, Male fish tissue; sF, spawning female fish tissue. Bar, 40 μm. (From Hamazaki *et al.* (1985)

On examination of the liver and ovarian extracts and of blood plasma of spawning female fish using native polyacrylamide gel disc electrophoresis (disc PAGE) combined with immunodiffusion analysis, the relative mobility (R_m) of the immunoreactive substance was found to be clearly different from that of vitellogenin, although this immunoreactive substance could not be identified as a protein band but was noticed only by a precipitation line on one-dimensional (1D-)PAGE (Fig. 1). Ferguson plot analysis showed that the molecular weight of this substance was about 60K, while that of intact vitellogenin of *O. latipes* on PAGE was about 420K (Hamazaki *et al.*, 1985). The clear distinction between this substance and vitellogenin allowed us to name this substance tentatively as spawning female-specific (SF) substance (Hamazaki *et al.*, 1985, 1987a,c).

B. Spawning Female-Specific Substances as Component Glycoproteins of the Inner Layer of Oocyte Envelope

1. Intrahepatic Formation of an SF Substance in Response to Estrogen and Its Accumulation in the Ascitic Fluid of the Estrogenized Fish

Purification of the SF substance from the tissues of the spawning female fish seemed to be almost impossible at first, as the concentration of this substance in the tissue extracts or blood plasma was so low that it could hardly be identified as a protein band on 1D-PAGE. However, a similarity of the behavior of the SF substance to that of vitellogenin led us to expect that intrahepatic synthesis of the SF substance might be stimulated by estrogen, even in male fish, just as vitellogenin synthesis is. A continued feeding of estrogen (17 β-estradiol, E_2)-dusted TetraMin powders to adult male or female fish under nonbreeding conditions, i.e., on a short-day length regime and at temperatures below 20°C, results in accumulation of ascitic fluid (Fig. 2).

During the E_2 treatment, the ultrastructure of the hepatic parenchymal cells was found to change remarkably, especially in male fish. The male fish hepatocytes contain abundant glycogen granules with relatively sparse endoplasmic reticulum, while the cytoplasm of the hepatocytes of the spawning female fish is mostly occupied by a well developed endoplasmic reticulum with a relatively small number of glycogen granules, suggesting active synthesis of proteins such as vitellogenin. The continued administration of E_2 to male fish changed the hep-

FIG. 2 Accumulation of ascitic fluid and changes in the ultrastructure of hepatic cells in the estrogenized fish. (a) Normal spawning female (sF) and male (M) fish; (b) nonbreeding female (F) and a male (M) fish fed with 17 β-estradiol-dusted TetraMin powder (1 mg/g) for 7 or 14 days; (c–f) ultrastructure of hepatocytes of a spawning female (c), a normal male (d), and a male fed with estrogen for 14 days (e, f). N, Nucleus. M, mitochondria; Gc, glycogen granules; Ga, Golgi apparatus; V, blood vessel; EDD, electron-dense droplet. Bar, 2 μm. From Hamazaki *et al.* (1987a).

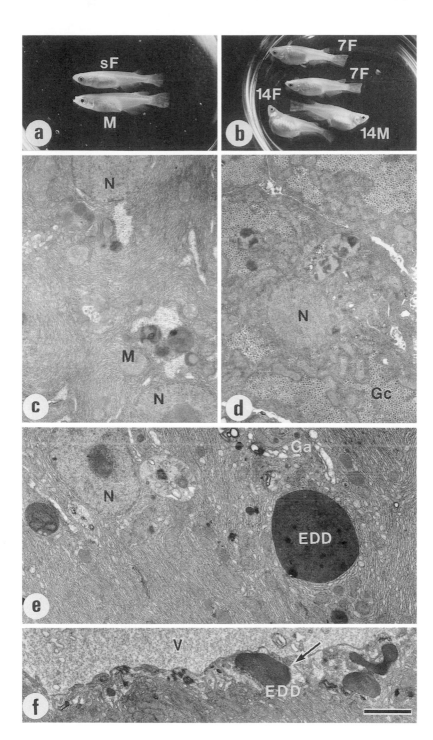

atocytes from male type into those of spawning female type (Fig. 2), indicating an augmentation of protein synthesis in the male hepatocytes. In parallel with the increase in the density of endoplasmic reticulum, some electron-dense droplets were found to have accumulated in the hepatocytes (Fig. 2). These droplets were considered to contain both vitellogenin and the SF substance, the synthesis of which was forced by E_2, even though they were not utilized in the male or the nonbreeding female fish. Accumulation of ascitic fluid occurs as the last and the most remarkable change induced by E_2 administration in both the male and the nonspawning female fish. Examination of the ascitic fluid using PAGE and immunoelectrophoresis revealed that it contained both vitellogenin and the SF substance as the two major component proteins. The latter substance was so abundant here as to be identified as a fairly dense band on PAGE (Fig. 3) (Hamazaki *et al.*, 1987a). According to these experimental results, the purification of the SF substance as well as vitellogenin of *O. latipes* became possible.

2. Similarity of the SF Substance to ZI-3, a Component of the Inner Layer of Oocyte Envelope, and Incorporation of Labeled SF Substance into Oocyte Envelope

The SF substance was isolated from ascitic fluid of the estrogenized fish by combined use of Toyopearl HW50 gel filtration column chromatography and repeated high-performance liquid chromatography (HPLC) with an AX300 DEAE

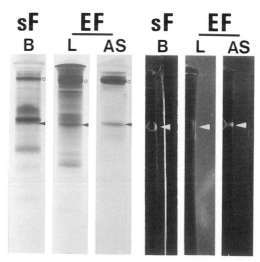

FIG. 3   Immunoelectrophoretic demonstration of accumulation of the SF substance in ascitic fluid of estrogen-fed fish. sF and EF, Normal spawning female and estrogen-fed male fish, respectively; B, blood plasma; L, liver extract; AS, ascitic fluid. Small circles and arrowheads indicate the positions of vitellogenin and the SF substance, respectively. From Hamazaki *et al.* (1987a).

silica column. The apparent molecular weight of the purified SF substance as estimated on SDS-PAGE is about 49K. Considering the above mentioned molecular weight (60K) on Ferguson plot, this substance seems to exist in tissues as a monomeric protein. As shown in Table I, it contains about 3% carbohydrate (Hamazaki *et al.*, 1987c).

The constituent proteins of the oocyte envelope were also electrophoretically examined. When the envelopes were isolated from fully grown ovarian follicles, washed with a medium containing EDTA, and treated with 0.05 *N* NaOH, the inner layer proteins of the envelope were obtained in solution. On SDS-PAGE, the inner layer solution gives rise to three major and several minor protein bands. The major bands are tentatively named ZI-1, ZI-2, and ZI-3, in order of decreasing molecular weight. As shown in Fig. 4, ZI-3, the most abundant of the major components, represented a band of molecular weight of about 49K, almost comparable to the band of the SF substance. Moreover, as shown in Table I, the amino acid composition of both was found to be almost identical. Thus, the SF substance was found to be a glycoprotein which is closely related to ZI-3, and was probably

TABLE I

Amino Acid Composition and Carbohydrate Contents of the SF Substance, ZI-3, and the Component Glycoproteins F1 and F2 of the Inner Layer of Oocyte Envelope[a]

Amino acid	SF substance	ZI-3	F1	F2
Asx	8.7 (9)	8.7 (9)	9.8 (10)	9.6 (10)
Thr	5.8 (6)	5.7 (6)	5.6 (6)	5.6 (6)
Ser	6.6 (7)	7.2 (7)	6.3 (6)	4.8 (5)
Glx	12.2 (12)	12.2 (12)	9.4 (9)	10.1 (10)
Gly	7.4 (7)	7.1 (7)	6.4 (6)	6.3 (6)
Ala	9.2 (9)	8.8 (9)	8.4 (8)	8.7 (9)
Cys/2	1.3 (1)	0.9 (1)	3.2 (1)	3.2 (1)
Val	7.2 (7)	6.6 (7)	8.2 (8)	8.4 (8)
Met	0.8 (1)	0.8 (1)	1.2 (1)	1.2 (1)
Ile	3.4 (3)	3.3 (3)	4.6 (5)	4.6 (5)
Leu	8.0 (8)	7.8 (8)	8.6 (9)	8.6 (9)
Phe	4.9 (5)	4.8 (5)	4.9 (5)	5.1 (5)
Tyr	4.0 (4)	4.4 (4)	5.1 (5)	4.8 (5)
Lys	5.3 (5)	5.2 (5)	3.3 (3)	3.4 (3)
His	2.7 (3)	2.9 (3)	2.7 (3)	2.8 (3)
Arg	3.5 (4)	3.6 (4)	3.8 (4)	3.8 (4)
Pro	8.5 (9)	9.4 (9)	8.1 (8)	8.7 (9)
Trp	0.5 (1)	0.5 (1)	0.3 (0)	0.3 (0)
Total Carbohydrate content (μg/mg protein)	30.4	—	32.3	24.9

[a]From Hamazaki *et al.* (1987c). Round numbers of the amino acid residues (mol%) are shown in parentheses.

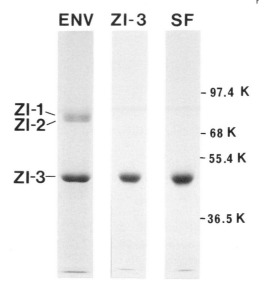

FIG. 4   SDS-PAGE patterns of the isolated SF substance and the component glycoproteins (ZI-1, ZI-2, and ZI-3) of the inner layer of oocyte envelope (ENV). From Hamazaki *et al.* (1987c).

responsible for construction of the inner layer of oocyte envelope as a probable precursor.

In order to verify this probability, the isolated SF substance was labeled with ^{125}I and injected into the abdominal cavity of spawning female fish to examine whether the labeled SF substance is actually incorporated into the inner layer of oocyte envelopes. As shown in Fig. 5, when the SF substance is injected into the abdominal cavity it continues to be incorporated into the ovary, more than 80% of the total radioactivity incorporated into a fish being found in the ovary 48 hr after the injection (Hamazaki *et al.*, 1989a). Autoradiographic examination of the ovary revealed that the labels, which had been found at first in intercellular spaces within the ovary, moved into the envelopes of growing oocytes. Finally, the incorporated labels were found to be localized exclusively in the inner layer of the envelope (Fig. 6). Thus, the SF substance formed in the liver in response to E_2 was almost conclusively proved to be a constituent of the inner layer of oocyte envelope, probably as a precursor of ZI-3 (Hamazaki *et al.*, 1989a).

3. Other SF Substances

Some egg envelope glycoprotein-like immunoreactive substances other than the SF substance have also been found in the liver and ovarian extracts and blood plasma of the spawning female of *O. latipes* by immunoblotting analyses (Murata *et al.*, 1991). They could not be detected obviously by the immunoelectrophoretic

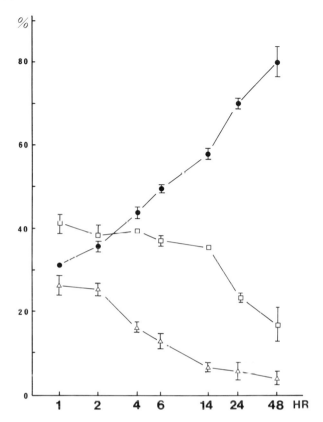

ГIG. 5   Changes in the distribution of the ^{125}I-labeled SF substance injected into the abdominal cavity of spawning female fish. About 7 μCi of the labeled SF substance (7 μCi/μg SF substance) was injected into a spawning female. The percentage refers to the radioactivity of a whole organ per total radioactivity of the fish. ●, Ovary; Δ, liver; □, whole body exclusive of the ovary and liver. The radioactivity was counted at the indicated times after injection. From Hamazaki *et al.* (1989a).

methods employed in the previous studies (Hamazaki *et al.*, 1985, 1987a,c), probably because of their lower diffusibility. On SDS-PAGE, they represent some bands of higher molecular weights (~76K) than that of the SF substance (49K). They include the bands corresponding to ZI-1 and ZI-2, two other major constituent glycoproteins of the inner layer of oocyte envelope. These substances seem to possess epitopes, some of which are common to that of the SF substance and others are specific for themselves. Therefore, we can make antibodies specific for either the SF substance or the new immunoreactive substances by absorbing the original antibody [anti-egg envelope glycoprotein (F1) IgG] with either the new substances or the SF substance (Fig. 7). Exploiting the specific antibodies thus prepared as probes, the behavior of the new substances was examined and found

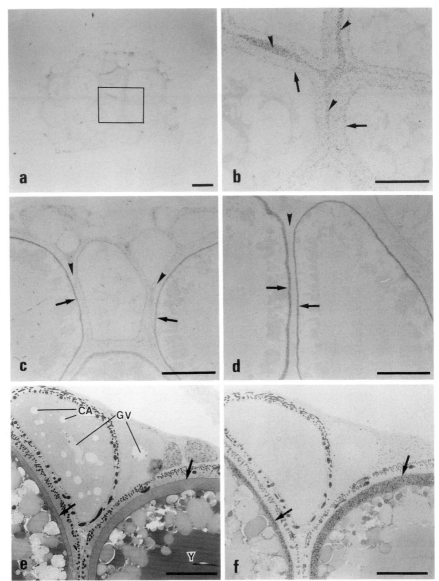

FIG. 6 Autoradiograms indicating a successive change in the intraovarian localization of the ^{125}I-labeled SF substance injected into spawning female fish. (a–d) Autoradiograms of unstained sections prepared 1 hr (a, b), 6 hr (c), and 14 hr (d) after injection. (e and f) A toluidine blue-stained section and its autoradiogram, respectively, prepared 14 hr after injection. This section shows some oocytes at advanced oogenetic stages. Arrowheads and arrows indicate the intercellular space and the oocyte envelope, respectively. Bar, 50 μm. From Hamazaki *et al.* (1989b).

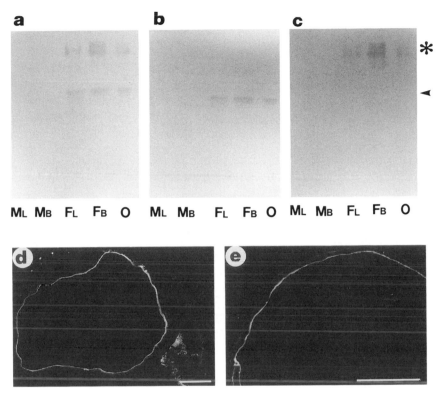

FIG. 7 Immunoblotting and immunohistochemical detection of H-SF substances as well as L-SF substance in blood plasma, liver, and ovary of the spawning female of *Oryzias latipes*. (a–c) Immunoblots of the liver (L) and ovarian (O) extracts and blood plasma (B) of male (M) and spawning female fish (F). After native PAGE, the samples were transblotted to nitrocellulose membranes and stained by the original antibody (anti-F1 IgG) (a), the antibody specifically reactive with the SF (L-SF) substance (b), and the antibody specifically reactive with the new (H-SF) substances (c). (d and e) Sections of the growing oocyte envelope stained by the L-SF-specific antibody (d) and the H-SF-specific antibody (e). Bar, 40 μm. From Murata *et al.* (1991).

to be very similar to that of the SF substance; i.e., they are localized to the inner layer of the oocyte envelope and their intrahepatic formation is also stimulated by E_2 (Murata *et al.*, 1991). On the E_2 stimulation, their intrahepatic synthesis is initiated and continues concurrently with that of the SF substance. Like the SF substance in the previous reports (Hamazaki *et al.*, 1987a), they are also found to be accumulated in the ascitic fluid of the E_2-treated male fish. From these results, the new immunoreactive substances can also be regarded as spawning female-specific (SF) substances, as they are formed only in the spawning females of this species under natural conditions. To avoid confusion, they may be named high-molecular-weight SF (H-SF) substances, while the SF substance of the previous reports is named low-molecular-weight SF (L-SF) substance (Murata *et al.*, 1991). All of them, now collectively named SF substances, are considered to be precursory substances (proteins) of the major constituent proteins of the inner layer of oocyte (egg) envelope, and are formed in the liver in response to estrogen. Partial purification of H-SF substances from the ascitic fluid, separating them from concomitant vitellogenin(s), has been performed recently by gel filtration column chromatography using a Sephacryl column (K. Murata *et al.*, unpublished).

C. Relation between the Conventional Concept of Oocyte (Egg) Envelope Formation and the Present Results

The experimental results described above lead us to propose, though with prudence, the idea that all major constituents of the inner layer of oocyte (egg) envelope of this fish are probably formed neither in the oocyte nor in the follicle cell but in the mother's liver in response to estrogen.

As mentioned before, however, the fish oocyte envelope, especially its inner layer, is considered to be a so-called primary egg envelope. In fact, many light- and electron-microscopical observations have provided us with data suggesting that the inner layer as well as the outer layer of oocyte envelope is formed by the oocyte (Yamamoto, 1963a; Anderson, 1967; Flügel, 1967b; Wourms, 1976; Hosokawa, 1985). For example, Yamamoto (1963a) and Anderson (1967) obtained electron microscopical results showing that the inner layer of egg envelope of the medaka, *Oryzias latipes*, the seahorse, *Hippocampus erectus*, and the pipefish, *Syngnathus fuscus*, is thickened by piling up of materials which are transported seemingly by intraoocytic vesicles at the inner surface of the layer. Similarly, in the black scraper, *Novodon modestus*, the inner layer of the growing oocyte envelope becomes thickened inwardly and the thickening is caused as if some materials enclosed in vesicles (dense-cored vesicles) in the oocyte cortex were piled up at the innermost side of the inner layer (Hosokawa, 1985). Similar observations have been made on the formation of the outer layer of oocyte envelope in the annual fishes *Cynolebias melanotaenia* and *Cynolebias ladigesi* (Wourms, 1976). In addition, Tesoriero (1977a,b, 1978) made morphological and autoradiographical examinations of the formation of oocyte envelope in *O. latipes*. By injecting [³H]proline, one of the dominant constituent amino acids of the egg envelope proteins, into the abdominal cavity of the spawning female, he found that the labels which had been once incorporated into the oocytes were gradually transferred to the developing oocyte envelope by way of dense-cored vesicles by 48 or 72 hr. A similar experimental result was obtained for the oocyte envelope formation in the goby, *Pomatoschistus minutus*, by Riehl (1984). In the pipefish, *Syngnathus scovelli*, it was found by Western blotting analysis following SDS-PAGE that two major component proteins of the egg envelope (Z3) were formed in the isolated follicles (Begovac and Wallace, 1989). Moreover, in some vertebrates, such as amphibians and mammals, the egg envelope (vitelline envelope) components are found to be synthesized in the oocyte itself, as revealed by immunochemical and biochemical methods (Yamaguchi et al., 1989; Wassarman, 1988). Thus, most of the results obtained by recent experiments not only on fish but also on some other vertebrates seem to favor the conventional concept that the inner layer of the egg envelope is formed by the oocyte and is a primary egg envelope.

However, it has been very recently reported that some proteins corresponding to the subunits of egg envelope proteins exist outside the ovary, i.e., in liver extracts and blood of cod, *Gadus morhua*, Atlantic salmon, *Salmo salar* (Oppen-Berntsen, 1990), and in blood plasma of the estrogenized rainbow trout, *Oncorhynchus mykiss* (Hyllner et al., 1992), as revealed by Western blotting immunochemical analysis following SDS-PAGE. With *Fundulus heteroclitus*, an experimental result was obtained in which one component protein (61K) of the oocyte envelope was found in the blood and liver of the breeding female and also of the E_2-treated male (Hamazaki et al., 1989b). Thus, our present understanding is that the pattern of egg envelope (vitelline envelope) formation of animals differs in various species.

At present, it is not clear why there is a difference in the pattern of egg envelope formation from species to species among the fishes. One possibility is that it may be caused by differences in the quantity of egg envelope material relative to the egg size and/or the mother's body weight. For example, a mature *O. latipes* female (~600 mg) continues to lay 10 to 20 eggs every day for a few weeks to 2 months during a breeding season (Egami, 1959). As each egg devoid of villi and attaching filaments (~1.2 mg) contains about 20 to 25 μg of egg envelope (inner layer) proteins, a female fish must continue to produce 200 to 500 μg of egg envelope proteins every day during this period of time (Masuda et al., 1991; K. Murata et al., unpublished). Although salmonids and cod do not spawn everyday, they must produce bulky egg envelope proteins during the breeding season, especially in the period of vitellogenesis, which occurs in parallel with oocyte envelope formation. Such bulky oocyte envelope proteins are, like egg yolk proteins and serum proteins, produced possibly in the liver. However, a spawning female of pipefish, whose egg envelope is very thin (2 to 3 μm; Begovac and Wallace, 1989) in comparison with the *O. latipes* oocyte envelope (~20–25 μm), may not need to produce as much egg envelope proteins relative to the egg size and to her body weight, and the material may be of oocyte or follicle cell origin. This explanation may be applicable to some other animal groups, such as amphibians and mammals.

Meanwhile, it seems likely that the discrepancy between the results of some preceding studies and our studies is partly attributable to differences in the analytical methods. Morphological and/or autoradiographical methods were carried out in the preceding experiments and immunoelectrophoretic and biochemical methods were used in our experiments. It is possible that, when labeled amino acids were used as tracers, the injected amino acids were first incorporated into the

SF substances or some other proteins, such as vitellogenin in the liver, and then into the oocyte envelope or the oocyte itself. It would seem difficult to draw conclusions about a dynamic process such as oocyte envelope formation from morphological studies using fixed specimens. It is not evident whether the dense-cored vesicles are actually carriers of the oocyte envelope materials from the inside or, on the contrary, carriers of vitellogenin from the outside, although they *appear* to be the former in section. However, it is evident from the morphological studies that the outer layer of oocyte envelope is formed first and the inner layer is thereafter formed, being thickened inwardly by piling up of materials at the inner surface of the developing oocyte envelope. Although this fact is seemingly incompatible with our view of an extraovarian origin of the envelope materials, we also think that the inward thickening of the envelope is quite reasonable and highly likely to occur even when the materials come from the outside of the oocyte.

Piling up the materials at the inner surface of the envelope, i.e., at the site adjacent to the oocyte surface, seems to be quite reasonable because the construction of the ordered macromolecular architecture of the egg envelope is considered not to occur at random but to be performed under the strict control of cellular (oocytic) activity. The SF substances derived from the liver would reach the small spaces between the inner surface of the thickening envelope and the surface of the oocyte membrane (oolemma) through the pore canals. The SF substances would be associated there to form some insoluble form and piled up, attaching to the inner surface of the envelope. It seems highly probable that the oocyte would participate in this process of forming the highly organized structure of the envelope and that some oocyte-derived substances, including minor constituents of the envelope proteins and/or some enzymes responsible for polymerization, would be involved in this process.

It should be mentioned here that the oocyte envelope increases not only in thickness but also in width, i.e., it grows three-dimensionally. In this connection, the observation by Hosokawa (1985) is of great interest. She found that no increase in the number of microvilli of an oocyte occurred during oocyte growth in the fish *Novodon modestus*. As the oocyte grows, the population of the pore canals in a definite area of the oocyte surface decreases, the total number of pore canals per oocyte remaining unchanged, while the surface area of the egg increases about 11 to 14 times in this fish species (Hosokawa, 1985). Thus, the distance between two pore canals increases during the growth of oocytes. This finding may imply that the constituent glycoproteins of the envelope possibly diffuse from every canal all around and participate in expansion as well as thickening of the envelope. Every pore canal contains a small space around the cytoplasmic processes (microvilli) of the oocyte and follicle cells. Exogenous substances, including vitellogenin and, probably, SF substances, pass through this small space. It is possible that some of the SF substances would diffuse while passing through the canal and participate in extension of the egg envelope, and others would attain to the inner surface of the egg envelope to thicken it inwardly, while all vitellogenins would reach the

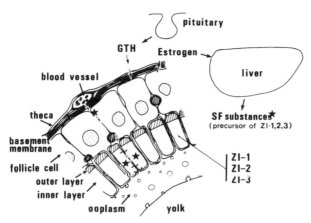

FIG. 8   A hypothetical illustration showing the formation of the inner layer of oocyte (egg) envelope with special reference to the pathway of transportation of SF substances in *Oryzias latipes*.

surface of the oocyte to be incorporated into the oocyte (Hamazaki *et al.*, 1987b). In this context, we could imagine that the SF substances would also be incorporated into the oocyte together with the vitellogenins. In fact, an uptake of various serum proteins other than vitellogenin by growing amphibian oocytes is reported (Wallace, 1985). Considering, however, that no significant amounts of label were found in the oocyte plasm in the ovary of the spawning female that had received ^{125}I-labeled SF (L-SF) substance by injection (Hamazaki *et al.*, 1989a), it seems improbable that L-SF substance is incorporated first into the oocyte and then secreted therefrom to constitute the egg envelope. Analysis of these construction mechanisms of the oocyte envelope in association with vitellogenin incorporation is very important from the viewpoint of cellular regulation of synthesis of an extracellularly organized architecture in biological systems.

IV. Change in the Molecular Architecture of Egg Envelope of *Oryzias latipes* at Fertilization or Activation

As mentioned in the Introduction, the protection of the inside egg or embryo against various harmful influences from the outside is one of the physiologically important roles of the egg envelope. Among these influences, mechanical harm or damage may be best prevented as the egg envelope itself becomes a mechanically hard and tough structure after fertilization, and the egg or embryo is surrounded by a newly formed perivitelline space filled by a "cushion" such as perivitelline fluid and/or spherical bodies in fish (Iwamatsu and Ohta, 1976; Hart and Yu, 1980; Brummett and Dumont, 1981; Kobayashi, 1985). A jelly layer, which is found also in some fish (Flügel, 1967a) as well as in many invertebrates and amphibians, may play a cushioning role outside the egg envelope (vitelline envelope).

A. Previous Studies on Hardening of Egg Envelope

In many animal species, including most fishes, the soft and fragile oocyte (or unfertilized egg) envelope is changed into a strong and tough structure on fertilization or activation. This is called the hardening of egg envelope, although the hardened egg envelope shows toughness rather than hardness in a strict sense. In eggs, a series of physiological as well as morphological changes of their surface and cortex occur following fertilization (fusion with a spermatozoon) or activation. These changes are collectively called the cortical reaction of the egg (Yamamoto, 1961; Monroy, 1965; Schuel, 1978). Among the cortical changes, the most remarkable is the formation of the fertilization envelope, the mechanism of which has been best analyzed in sea urchin eggs (Endo, 1961; Schuel, 1978; Kay and Shapiro, 1985) and in anuran eggs (Wyrick et al., 1974; Wolf et al., 1976; Schmell et al., 1983). In these animal groups, transformation of the vitelline envelope into the fertilization envelope is closely associated with breakdown (or exocytosis) of the cortical granules. In sea urchin eggs, ovoperoxidase is released from the broken cortical granules and the hardening of the vitelline envelope is considered to be at least partly due to the ovoperoxidase-catalyzed formation of cross-links between tyrosyl residues of the neighboring peptide chains in the fibrous protein network of the envelope (Kay and Shapiro, 1985). In *Xenopus* eggs, it has been shown that the fertilization layer (FL) is formed in the inner part of jelly layer (Wyrick et al., 1974); the FL is formed by a precipitation reaction between a lectin liberated from the exocytosed cortical granules and glycoconjugates in the jelly substance of the prefertilization layer (Yoshizaki, 1984).

However, it still remains obscure how the contents of the cortical alveoli, which are considered to be comparable to the cortical granules, contribute to the hardening of the egg envelope in almost fishes. It is well known that the cortical alveoli of fish eggs are derived from the so-called yolk vesicles which first appear in the oocyte during the early phase of oogenesis (Osanai, 1956; K. Yamamoto, 1957; Wallace and Selman, 1981; Masuda et al., 1986; Selman et al., 1988). It seems that the cortical alveolar materials comprise some high-molecular-weight glycoconjugates (Selman et al., 1986; Kitajima et al., 1986; Inoue et al., 1987), whose synthesis occurs throughout vitellogenesis and early maturation (Selman et al., 1986). As a matter of fact, breakdown of cortical alveoli occurs on normal fertilization in many fishes, as has been documented for a long time (Kagan, 1935; Yamamoto, 1939, 1944; Kusa, 1953; Rothschild, 1958), followed by elevation and hardening of the egg envelope. Thus, the breakdown of the cortical alveoli seems to be prerequisite for the hardening of the egg envelope. It is reported, however, that in the eggs of some salmonids, *Oncorhynchus keta*, *Salmo salar*, and *Salmo trutta* (T. S. Yamamoto, 1957; Zotin, 1958), and of cod, *Cadus morhua* (Lönning *et al.*, 1984), no close relationship was observed between the breakdown of cortical alveoli and the egg envelope hardening. On the other hand, Ca^{2+} has been hypothesized by many authors (cf., Yamamoto, 1961) to play some crucial role in the hardening. Besides Ca^{2+}, some reactions or factors such as oxidation of sulfhydryl groups in egg envelope proteins (Ohtsuka, 1957, 1960), "hardening enzyme" (Zotin, 1958), colloidal substances like the contents of cortical alveoli (Nakano, 1956), and an environmental factor such as pH (Iwamatsu, 1969) have been reported to be involved with the hardening. Among them, the idea of "hardening enzyme" was unique and is of great interest, as the data showing that the probable enzymatic formation of cross-links between the constituent peptides of the egg envelope is related to its hardening have also been accumulated for fish (Hagenmaier *et al.*, 1976; Oppen-Berntsen *et al.*, 1990; Masuda *et al.*, 1991; Iuchi *et al.*, 1991). The mechanism of hardening of the egg envelope in fish has not yet been completely determined unlike those in sea urchins and anurans.

One of the problems in the study of egg envelope hardening lies in the difficulty of quantitatively as well as qualitatively estimating hardness (toughness). The simplest and direct method of quantifying hardness (toughness) is to measure the minimum weight which can squash a whole egg under defined conditions (Gray, 1932; Zotin, 1958). In this case, however, the toughness is influenced by turgidity to some extent, and the weight per defined area of the egg envelope varies depending on the turgidness of the egg, although its influence is not so large. Application of weight on a definite area of egg envelope which has been removed from the egg and extended as a flat sheet seems to be a better way of measuring toughness (Zotin, 1958). However, this method is not suitable for a tiny egg. Hardness (toughness) cannot be expressed in terms of a single parameter. More than one parameter should be employed to define hardness (toughness), as the hardening of egg envelope does not seem to be a simple process, but is associated with increasingly complex properties.

B. Changes in Solubility and Subunit Composition of Egg Envelope Proteins as Two Indices of Hardening

1. "Activation Hardening" and "*in Vitro* Ca^{2+}-Hardening" of Egg Envelope

One of the characteristics of the hardened envelopes of fertilized eggs of some fishes such as *O. latipes* and rainbow trout is their insolubility in many solvents, while the unhardened envelopes of unfertilized eggs are more or less soluble in some solvents. Change in solubility of the egg envelope before and after fertilization or activation has been observed, and this change utilized as an index for formation of fertilization envelope (or hardening), in amphibians and sea urchins

(Wolf *et al.*, 1976; Foerder and Shapiro, 1977). When a definite concentration of alkali (e.g., 1 *N* NaOH) is employed as the solvent, proteins of an unhardened and a partially hardened egg envelope of fish can be solubilized, and a time-dependent decrease in solubility of the egg envelope proteins occurs in parallel with the process of its hardening after activation. Although dissolution of proteins in alkali is not a simple phenomenon and the change in alkali-solubility of proteins does not necessarily represent all attributes of the envelope responsible for the hardness (toughness), it can be one of the indices of hardening. An analysis of concurrent change in subunit composition of the egg envelope during hardening would provide more useful information about the degree of hardening.

It is well known that Ca^{2+} ionophore activates unfertilized eggs of animals (Chambers *et al.*, 1974; Steinhardt *et al.*, 1978; Wolf *et al.*, 1979) through an increase in concentration of intracellular Ca^{2+} and an induction of cortical reactions almost comparable to those which occur on normal fertilization. In salmonids, the egg envelope hardened when the eggs were activated by immersion into fresh water. This change, called water hardening, seems to be almost comparable to what occurs after activation by Ca^{2+} ionophore. Following the activation, the egg envelope hardens *in situ* to an extent apparently comparable to that occurring on normal fertilization, and the process is tentatively named "activation hardening or *in situ* hardening" (Masuda *et al.*, 1991). Unhardened envelope which has been removed from the unfertilized egg is also found to harden *in vitro* in the presence of Ca^{2+}, while it remains unhardened if kept under a Ca^{2+}-free condition, e.g., in an ethylenediaminetetraacetic acid (EDTA) or ethylene glycol-bis(β-aminoethyl ether)*N, N'* tetraacetic acid (EGTA) solution. This is termed "Ca^{2+} hardening," though "*in vitro* (Ca^{2+}) hardening" would be preferable (Masuda *et al.*, 1991). A big difference between these two types of hardening process lies in the degree of hardening in terms of decreased alkali-solubility, and in the difference in the SDS-PAGE patterns of the egg envelope proteins (Fig. 9). Activation hardening seems to proceed more intensively than does *in vitro* (Ca^{2+}) hardening, as the alkali-solubility of the egg envelope decreased more markedly in the former than in the latter. During both types of hardening process, the amount of protein that was soluble in SDS also decreased markedly and the number of subunits of the egg envelope as manifested on SDS-PAGE decreased (Fig. 9). This fact strongly suggests increased formation of covalent bonds in the subunit protein network of the envelope, the subunits being polymerized into insoluble forms not only in activation hardening but also in *in vitro* (Ca^{2+}) hardening.

Moreover, it should be mentioned that the molecular weights of some major subunits (especially 150K, 83K, and 78K) of the egg envelope proteins are decreased by activation hardening, while they remain unchanged by *in vitro* (Ca^{2+}) hardening. In the former, the preexisting subunits may undergo a partial proteolysis by protease(s) probably released on cortical change at activation. A release of protease from the egg on fertilization has been reported in sea urchin egg (Vacquier *et al.*, 1973) and amphibian egg (Wolf *et al.*, 1976; Lindsay and

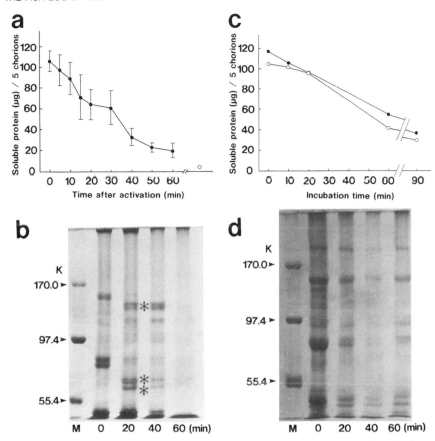

FIG. 9 Changes in alkali-solubility and SDS-PAGE patterns of egg envelope proteins of *O. latipes* during two types of hardening. The activation hardening (a, b) was induced by putting the unfertilized eggs into the medaka saline containing 20 μM Ca^{2+} ionophore A23187. For the *in vitro* (Ca^{2+}) hardening (c, d), the unhardened envelopes that had been removed from the unfertilized eggs in the absence of Ca^{2+} were hardened by the addition of Ca^{2+} at a concentration of 2 (○) or 93 mM (●). Alkali-solubility of the egg envelope proteins decreases more intensively during activation hardening (a) than during *in vitro* (Ca^{2+}) hardening (c). The amounts of SDS-soluble proteins decrease during both types of hardening. Molecular weights of some subunits (150K, 83K, and 78K) decrease only by activation hardening (b). From Masuda *et al.* (1991).

Hedrick, 1989). Therefore, it seems that the polymerization of the partially hydro-lyzed subunits gives rise to the more intensive hardening, probably because more sites responsible for association with other subunits are exposed in the partially hydrolyzed subunits in comparison with the intact subunits. This problem is discussed later. The hardening process of the envelope isolated from the activated egg can be arrested by removal of Ca^{2+} and resumed by readdition of Ca^{2+}. This fact, together with the occurrence of *in vitro* (Ca^{2+}) hardening in the isolated

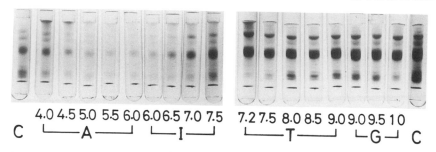

FIG. 10 pH dependence of *in vitro* (Ca^{2+}) hardening of the unfertilized egg envelope of rainbow trout. The optimum pH of the hardening as examined by SDS-PAGE is 5.0–6.0. From Iuchi *et al.* (1991).

unhardened egg envelope, strongly suggests that some hardening system or hardening machinery is incorporated in the egg envelope. This system (or machinery) requires Ca^{2+} for its action, is inactivated by heating, and is dependent on pH (Fig. 10) (Iuchi *et al.*, 1991). At present, no information has been obtained concerning this hardening system (or machinery) in the isolated unfertilized egg envelope. Some circumstantial evidence suggests that it can be an enzyme(s). As mentioned before, the egg envelope is considered, in general, to be an extracellular matrix and to possess almost no physiological activity such as enzymatic activity. However, whether this notion is valid or not remains inconclusive. In this connection, deposition of oviductal pars recta secretions (PR secretions) in the vitelline coat of eggs in the toad (Katagiri *et al.*, 1982; Takamune *et al.*, 1986; Katagiri, 1987) is very interesting. The PR secretions, which may be regarded as part of the so-called tertiary egg envelope, have dual physiological activities toward vitelline coat on the one hand and toward sperm acrosome on the other hand.

2. Formation of Cross-links and Modification of Egg Envelope Proteins during Hardening

There are many kinds of cross-linking in proteins, including disulfide linkage (Folk and Finlayson, 1977), and intermolecular cross-links are known to contribute to the structural strength of protein molecules. In this connection, the hardening of egg envelope in fish was considered to be related to disulfide cross-linking. According to Ohtsuka (1960), the hardening process may be related to oxidation reactions of SH-groups and of polysaccharide α-glycol groups in the component proteins of the egg envelope. As mentioned before, the formation of tyrosine cross-links was first investigated for hardening of the egg envelope in sea urchins (Foerder and Shapiro, 1977; Hall, 1978; Kay and Shapiro, 1985). This type of cross-link was first found in some fluorescent substances of resilin, an elastic ligament protein of some insects (Andersen, 1963), and then identified as di- and trityrosine (Andersen, 1964). In sea urchin eggs, the tyrosine cross-linking is catalyzed by ovoperoxidase released from the cortical granules upon their break-down at fertilization or activation. Besides these major cross-links of di- and trityrosines, another minor cross-link of pulcherosine has been found recently in the fertilization envelope of some sea urchins (Nomura et al., 1990). However, there has not so far been a report that tyrosine cross-links also contribute to the hardening of fish egg envelopes. In O. latipes eggs, di- or trityrosine does not seem to play any role in hardening of the egg envelope, as there is no significant difference in the content of di- and trityrosines between the unhardened and the hardened egg envelope (K. Nomura, personal communication). On the other hand, the possibility of γ-glutamyl-ε-lysine isopeptide cross-linking following fertiliza-tion or activation has become greater in fish egg envelopes. This possibility was probably first suggested by Hagenmaier et al. (1976) in rainbow trout eggs, although the roles of the isopeptide cross-link in biological systems such as the fibrin network formation in blood coagulation were already known (Loewy, 1968; Lorand, 1970). Through an exhaustive proteolytic digestion of the rainbow trout egg envelope, followed by ion-exchange chromatography, Hagenmaier et al. (1976) found that the fertilized egg envelope contained a significant amount of γ-glutamyl-ε-lysine isopeptide, while the unfertilized egg envelope contained neither glutamyl-lysine nor aspartyl-lysine isopeptides. They proposed the idea that the high content of this isopeptide is responsible for the greater mechanical stability of the fertilized egg envelope. Although there have not been any studies since that have investigated this charming idea, some experimental results support-ing this idea were recently obtained in the eggs of cod (Oppen-Berntsen et al., 1990), medaka (Masuda et al., 1991), and of rainbow trout (Iuchi et al., 1991).

In cod eggs, no hardening of the envelope occurs in eggs activated in the presence of dansylcadaverine, which inhibits the transglutaminase activity cata-lyzing the cross-linking reaction between glutamynyl and lysyl residues by re-placing the lysyl residues (Lorand and Campbell, 1971). The 47K subunit of the egg envelope protein was found to be bound to the added dansylcadaverine, as revealed by SDS-PAGE analysis of the envelope proteins (Oppen-Berntsen et al., 1990). Transglutaminase activity is highly dependent on Ca^{2+} (Folk and Chung, 1973). Considering that Ca^{2+} is an indispensable factor for the hardening of egg envelopes in fish, the above-mentioned results strongly suggest that the isopeptide formation catalyzed by transglutaminase is closely related to the hardening of egg envelope following fertilization or activation. Similarly, activation hardening in the eggs of O. latipes and rainbow trout is found to be inhibited by the addition of cadaverine (Masuda et al., 1991; Iuchi et al., 1991). It is of some interest to note that there are cadaverine-sensitive and cadaverine-insensitive hardening processes in the egg envelopes of these fish species. Transglutaminase-related reactions may occur at some definite steps of hardening (Masuda, 1990; K. Masuda et al., unpublished; Iuchi et al., 1991). Although the formation of the dityrosine cross-link is predominant in sea urchin eggs, the participation of transglutaminase has also been reported to occur in the early step of the hardening (Battaglia and Shapiro, 1988). In sea urchin eggs, ovoperoxidase is known to be localized to cortical granules but the localization of transglutaminase is not yet clear. In the case of fish eggs, it is highly probable that the "hardening enzyme" proposed by Zotin (1958) corresponds to transglutaminase. Zotin (1958), however, considered that the cortical alveoli were not involved in the hardening. Thus, the localization of this enzyme, if any, in the unfertilized fish eggs should be examined in the near future.

3. Other Possible Changes Responsible for Hardening

As described above, one of the characteristics of the hardened egg envelope is its insolubility in aqueous solvents. Some features of the unhardened and the hardened egg envelopes, such as their morphological characteristics and their resistance to touch and tear, allow us to compare the unhardened envelope to a sheet of paper and the hardened envelope to a piece of cloth made of some hydrocarbon polymer such as polyvinyl chloride. This impression or "feeling" sometimes leads us to suspect that the hardened egg envelope is a kind of polymer bearing many hydrophobic sites on its surface as well as many cross-links in its molecular architecture. Although the hydrophobic amino acid content of constituent proteins of fish egg envelopes is not very high as compared with other proteins (Young and Smith, 1956; Iuchi and Yamagami, 1976; Hamazaki et al., 1987c; Begovac and Wallace, 1989), their fibrous structure would be favorable, as suggested by Oppen-Berntsen et al. (1990), for external alignment and exposure of the hydrophobic domains, which result in both formation of hydrophobic bonds between the neighboring peptide fibers and insolubility. In the fertilization envelope of amphibians, no increase in hydrophobicity as examined by binding capacity of a hydrophobic dye was observed (Bakos et al., 1990), but this kind of analysis should be carried out for fish egg envelope as well.

Another interesting problem lies in the difference between activation hardening and in vitro (Ca^{2+}) hardening. In both cases, the preexisting subunits are polymerized to form insoluble proteins of higher molecular weights. As described before, the increase in insolubility proceeds more intensively in the former type of hardening than in the latter, and the subunits to be polymerized were partially broken down only in the former. In other words, the partially cleaved subunits seem to be capable of being polymerized more intensively. In connection with the varied information about hardening described above, it is possible to speculate that the preexisting subunits are partially proteolyzed so that the domains containing hydrophobic amino acid residues and/or glutaminyl and lysyl residues are exposed.

In any event, it appears that, on activation, loosely associated subunits of egg envelope proteins are covalently bound to each other due to some hardening machinery which works in the presence of Ca^{2+} to form insoluble polymers. The tough structure thus formed persists through the periods of embryonic development, protecting the embryo inside against mechanical harm, until it is broken down by an enzyme secreted from the enclosed embryo at the time of hatching.

V. Breakdown of Egg Envelope in *Oryzias latipes*

In most animals, after a definite process of development, the embryos change into
the forms which can get nutrients sooner or later from external sources. For this
purpose, they must come into direct contact with their environment by breaking
down or escaping from the egg envelope that has protected them from environmental harm or damage but which is now a barrier between themselves and their
environment. This process of releasing themselves from the egg envelope is called
hatching. In fish, the hatching is performed by the action of the hatching enzyme.
The hatching enzyme solubilizes only the thick and tough inner layer of the egg
envelope; a sheet of thin, fragile outer layer which remains undigested is broken
by movement of the larva. Knowledge of the hatching enzyme and the egg
envelope breakdown (choriolysis) pertaining to fish published up to the mid 1980s
has been reviewed by Ishida (1985) and Yamagami (1988). In this section, information obtained about the hatching enzyme and the mechanism of choriolytic
action in *O. latipes* since 1988 is described.

A. Two Distinct Proteases Responsible for Egg Envelope Breakdown (Choriolysis)

In various animal species, including fish, the hatching enzyme exists in multiple
forms during the course of its purification (Barrett *et al.*, 1971; Yamagami, 1972,
1975; Takeuchi *et al.*, 1979). In spite of these results, however, researchers did not
necessarily suspect that the hatching enzyme consisted of multiple enzyme species
but were apt to conceive that the hatching enzyme interacted with some other
substances, such as hydrolyzed products of the egg envelope, with the result that
the enzyme apparently behaved as if it consisted of different enzymes (Barrett *et
al.*, 1971; Yamagami, 1975). Thus, a prevalent view was that the hatching enzyme
of animals was essentially a single enzyme, although a few investigators still
suspected that different components were contained in the hatching liquid, i.e., the
medium of the hatching larvae, and they possibly constituted the hatching enzyme
(Yamagami, 1972; Takeuchi *et al.*, 1979). As described previously (Yamagami,
1988), a diversity of fish hatching enzymes was documented in *O. latipes* by Ohi
and Ogawa (1968, 1970). They found that the extracts of the isolated hatching
gland cells could be fractionated, on agar-gel electrophoresis, into two groups: one
that caused swelling of the inner layer of egg envelope and the other able to digest
the inner layer of egg envelope. When swollen, the egg envelope increased the
thickness of the inner layer remarkably and it manifested a conspicuous multilamellar structure. The swelling activity of the fraction was not, however, regarded
as the action of some special principle but as a probable limited proteolysis of the
egg envelope by the same hatching protease (Schoots *et al.*, 1983), and the idea
of a "swelling enzyme" did not grow further at that time. As is described below,

the idea of a "swelling enzyme" and the view that the swelling is caused by a limited proteolysis are both found to be correct from our current experimental results.

Although the hatching liquid, a culture medium containing the hatched larvae, of *O. latipes* could be fractionated grossly into two peaks of proteolytic (caseinolytic) activity by Sephadex gel filtration chromatography in our previous experiments (Yamagami, 1972, 1975), five peaks of proteolytic activity can now be obtained from the hatching liquid by an improved fractionation method using Toyopearl gel for molecular sieve chromatography in an alkaline solution (Yasumasu *et al.*, 1988). Some of these peaks also show choriolytic (egg envelope digesting) activity as determined by turbidimetry (Yamagami, 1970). In terms of the relative exclusion volume of each peak and their choriolytic activity relative to the caseinolytic activity, the proteolytic enzymes in the five peaks can be classified ultimately into two different groups: the group that elutes first has a low choriolytic activity and the other, which elutes later, has a high choriolytic activity (Fig. 11). The former and the latter have been named low choriolytic enzyme (LCE) and high choriolytic enzyme (HCE), respectively (Yasumasu *et al.*, 1988). On addition of HCE and LCE together to the isolated egg envelopes, dissolution of the inner layer of the egg envelopes proceeds very rapidly, while HCE alone dissolves the egg envelopes at a moderate rate and LCE alone scarcely dissolves the egg envelopes. The synergism in choriolysis on combined addition of the enzymes indicates cooperative action of HCE and LCE (Fig. 11). This fact, together with the finding of both HCE and LCE in the hatching liquid, strongly suggests that both enzymes are responsible for actual choriolysis as two complementary components of the hatching enzyme.

B. Hatching Enzyme of *Oryzias latipes* as an Enzyme System

1. Some Properties of HCE and LCE

Both HCE and LCE have now been isolated and their enzymological properties as well as their protein–chemical characteristics have been partially clarified (Yasumasu *et al.*, 1989a,b,c). They are both zinc proteases having their optimum pHs slightly on the alkaline side (pH 7.5–9.0). The apparent molecular weights on

---

FIG. 11 Detection of high choriolytic enzyme (HCE) and low choriolytic enzyme (LCE) in the hatching liquid of *O. latipes* and their cooperative choriolytic action. (a–d) Fractionation of the hatching liquid by repeated gel filtration chromatography in an alkaline medium (pH 10) using a Toyopearl HW-50 column. P.A., Proteolytic activity determined by using casein as substrate; C.A., choriolytic activity determined by turbidimetry. Proteases in the hatching liquid are ultimately classified into LCE, which is predominantly present in Pa(1)-1-1 and Pa(234)-1, and HCE, which is predominantly present in Pa(1)-2, Pa(1)-1-2, and Pa(234)-2. (e) Synergistic choriolysis by a combined action of HCE [Pa(234)-1] and LCE [Pa(234)-2]. For details, see Yasumasu *et al.* (1988).

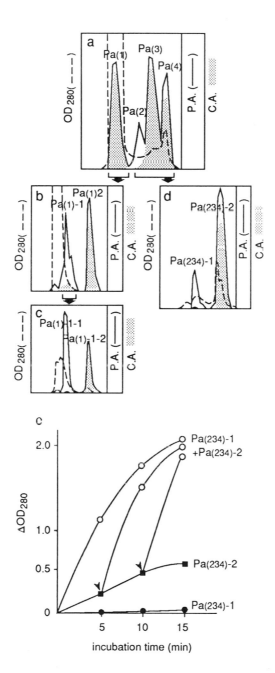

SDS-PAGE are 24K for HCE and 25.5K for LCE. Two isoforms of HCE with microheterogeneity (HCE1 and HCE2) have been obtained from the hatching liquid of an outbred population of *O. latipes* (Yasumasu *et al.*, 1989a). Two kinds of cDNAs for HCE, 940 and 910 bp long, and a cDNA of 936 bp long for LCE have been isolated (Yasumasu *et al.*, 1990, 1991). cDNAs for HCE and LCE were found to contain sequences of a signal peptide of 20 amino acids and a propeptide of 50 and 51 amino acids, respectively. Therefore, both HCE and LCE are considered to be synthesized in the form of a preproenzyme. Recently, proenzymes of HCE and LCE were detected in the secretory granules of the hatching gland cells (Yasumasu *et al.*, 1992). Although they contain amino acid sequences linkable to sugar moieties at the propeptide region (S. Yasumasu *et al.*, unpublished), it is not certain at present whether the mature forms of them are glycoproteins or not. Substrate specificity tests using 11 different MCA-peptides as substrates reveal that both HCE and LCE can hydrolyze Suc-Leu-Leu-Val-Tyr-MCA best of all the synthetic peptides tested, but the relative hydrolytic activity of HCE toward each of these MCA-substrates is different from that of LCE (Yasumasu *et al.*, 1989a,b). HCE and LCE are both basic proteins with p*I*s of 10.2 and 9.8, respectively (Yasumasu *et al.*, 1989a,b). Although the purified and exhaustively dialyzed samples of HCE and LCE are found to contain significant amounts of Ca^{2+} besides Zn^{2+}, the role of Ca^{2+} in their activity remains obscure. It is possible that Ca^{2+} serves as a stabilizer by maintaining the conformation of the enzyme (cf. Barrett, 1986), while a high concentration of Ca^{2+} is found to inhibit the hatching enzyme activity of some fishes (Yamagami, 1973; Hagenmaier, 1974; Schoots and Denucé, 1981; Luberda *et al.*, 1990). Ca^{2+} and Zn^{2+} are also reported to be necessary for the activity of sea urchin hatching enzymes (Barrett and Edwards, 1976; Takeuchi *et al.*, 1979; Nakatsuka, 1979; Roe and Lennarz, 1985). However, the divalent cation chelators as well as high concentrations of these divalent cations are inhibitory to the amphibian hatching enzyme (Katagiri, 1975; Urch and Hedrick, 1981).

Interconversion between HCE and LCE, especially a partial breakdown of LCE into HCE, did not occur when either or both of them were incubated. Each of them is, therefore, considered to be a distinct enzyme (Yasumasu *et al.*, 1989b). Recently, by an immunochemical double staining method employing mouse and rabbit antibodies against HCE and LCE as probes, it was found that both HCE and LCE are colocalized to every secretory granule of the hatching gland cells (see Fig. 12). It has also been reported that plural secretory proteins or peptides are colocated and intermingled in the same secretory granules of some exocrine and neuroendocrine tissues (Laboulenger *et al.*, 1983; Vardell *et al.*, 1984; Laslop *et al.*, 1989). In the case of the hatching gland, however, HCE is located in the central portion (or the inside) of each secretory granule, while LCE is situated separately at its periphery (Fig. 12; S. Yasumasu *et al.*, 1992). These facts strongly support the view that both of them are constituents of the hatching enzyme of this fish. However, copackaging both the enzymes in the same granule but with discrete localization seems puzzling, because both enzyme proteins are found to be synthe-

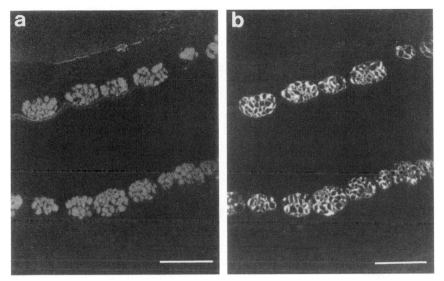

FIG. 12   Colocalization of HCE and LCE to the same secretory granules in the hatching glands of *O. latipes*. A median section of the pharynx of a day 5.5 embryo shows that several gland cells containing many secretory granules are located in the upper and the lower walls of the pharynx. (a) HCE stained with polyclonal anti-HCE antibody and TRTC-conjugated anti-rabbit IgG. (b) LCE stained with monoclonal anti-LCE antibody (B-34), biotin-conjugated anti-mouse IgG, and avidin FITC. HCE and LCE were stained on the same section. Bar, 25 μm.

sized almost concurrently in the gland cells, as revealed by immunoblotting analyses (Yasumasu *et al.*, 1992).

The similarity between HCE and LCE has been substantiated further by experimental data obtained from cloning of their cDNAs. Each of the mature forms of HCE and LCE consists of 200 amino acid residues and the homology between their amino acid sequences is about 54%. The amino acid sequences of and surrounding putative active sites of HCE and LCE are very similar (S. Yasumasu *et al.*, unpublished). These sequences are more similar to those of thermolysin, another zinc protease, than collagenase and/or stromelysin, other zinc proteases, while the sequences of the sea urchin hatching enzyme are reported to be similar to those of the latter (LePage and Gache, 1990). According to Northern blotting analysis, gene expression is found to be initiated in the embryos in the middle of day 2 (just after the stage of lens formation) and sharply increased thereafter (S. Yasumasu *et al.*, unpublished). These results are compatible with some preceding information obtained by morphological, radiation biological, and biochemical studies, which showed that the hatching enzyme activity or the secretory granules of the hatching gland cells appeared in the embryos at stages somewhat more advanced than day 2 (stages from lens formation to retinal pigmentation) (Yamamoto, 1963b; Egami and Hama, 1975; also cf. Yamagami, 1988).

2. Cooperative Choriolytic Action of HCE and LCE

As described above, it is now evident that HCE and LCE are components of the hatching enzyme of *O. latipes* and they are similar to each other in many respects as both enzyme and protein. How do they then work out an efficient choriolysis (egg envelope breakdown) as two components?

In the course of study on their action on the isolated egg envelope (chorion), HCE was found to swell the inner layer of the egg envelope remarkably, while LCE did not (Fig. 13). It was also found that the swelling of the egg envelope was accompanied by the release of low-molecular-weight peptides from it (a partial choriolysis). As shown in Fig. 13, the time course of the swelling of egg envelope fragments proceeds in parallel with that of the partial choriolysis. Light microscopy revealed that the inner layer of the swollen egg envelope was several times as thick as that of the intact egg envelope and that a multilamellar structure of the inner layer became conspicuous as compared with the intact egg envelope. These features strongly suggest that HCE is the "swelling enzyme" reported by Ohi and Ogawa (1968, 1970), and the swelling is, in fact, associated with partial (limited) proteolysis of the inner layer, as reported by Schoots *et al.* (1983) and mentioned above. When HCE exerts the choriolytic swelling action, this enzyme is prone to bind tightly to its natural substrate, the egg envelope. This characteristic seems to be quite important for its choriolytic activity and is discussed later.

On the other hand, LCE seems to be an ordinary protease, showing no such peculiarity. However, it is of great interest to note that LCE can hardly digest the intact hardened egg envelope but is capable of hydrolyzing only the swollen inner layer of the egg envelope very efficiently. In the presence of excess LCE, therefore, the amount of the egg envelope digested is almost proportional to the amount of HCE added. Thus, the egg envelope is considered to be digested through a sequential action by HCE and LCE: it is partly digested and swollen by the action of HCE, and only the swollen portion is solubilized (digested) by LCE. Under a light microscope, however, the egg envelope seems to be digested without preceding swelling in the presence of both HCE and LCE, probably because the inner layer of the egg envelope is digested by LCE as soon as it is swollen by HCE. We can detect, however, a small portion of the swollen inner layer by electron microscopy (Yamamoto *et al.*, 1979; Yamagami, 1988). The amount of HCE secreted from an embryo seems to be two- or threefold more than that of LCE on the basis of their caseinolytic activities and the amounts of HCE and LCE recovered from the hatching liquid. The imbalance between the amounts of HCE and LCE reflects the pattern of their discrete localization in a secretory granule as described above. LCE is supposed to work efficiently, though much less so than HCE, because it does not bind tightly to the egg envelope and it probably acts on only a restricted number of cleavage sites in the inner layer proteins (see later). The sequential action of HCE and LCE and the morphological changes of the egg envelope explain why the synergistic choriolysis described above occurs.

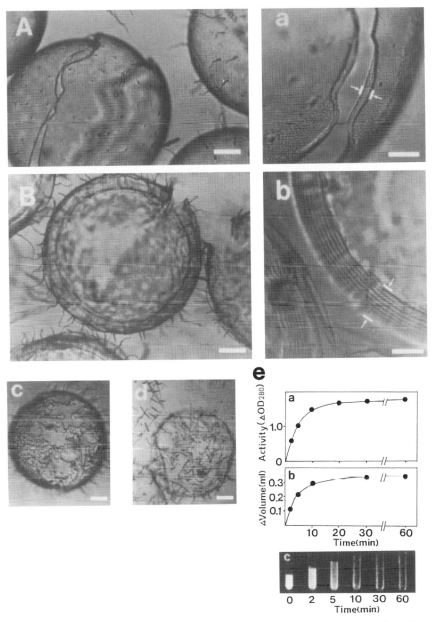

FIG. 13 Choriolytic swelling of the hardened egg envelope (chorion) by HCE. (a–d) (a and b are higher magnifications of A and B, respectively) Morphological changes of the chorion following the enzymatic action. Thickness of the isolated intact egg envelope (A, a) increases remarkably after a 30-min incubation with HCE (3 μg/0.2 ml) (B, b). Combined action of HCE (4.5 μg/0.2 ml) and LCE (0.5 μg/0.2 ml) results in no swelling but a rapid dissolution of the inner layer, leaving the outer layer undigested (c, d). (e) Relationship between the partial choriolysis as expressed in terms of the increase in OD_{280} of the supernatant (e-a) and the swelling of the egg envelope fragments, as expressed in terms of the increase in their volume (e-b,c). Bars, 200 μm for A, B, c, and d; 100 μm for a and b. From Yasumasu *et al.* (1989a, b).

3. Interaction between the Hatching Enzyme and the Egg Envelope

HCE differs markedly from LCE in its extraordinarily high affinity for its natural substrate, the inner layer of egg envelope. This peculiar action of HCE was first realized when dose dependency of the choriolytic swelling activity was examined. As shown in Fig. 14, when a small amount of HCE was added to a relatively large amount of the egg envelope, the process of choriolytic swelling soon leveled off, without proceeding further until all the substrates were swollen. In other words, after HCE swells a definite amount of egg envelope, the catalytic velocity of the HCE is abruptly reduced, with a significant amount of egg envelope remaining unswollen. Thus, the time course of choriolytic swelling by HCE is quite different from that of the catalytic action by ordinary enzymes. The slowed velocity implies a hindered turnover of the HCE. The suppression of turnover could be due to two possible causes: (1) inactivation of the HCE on encounter with the egg envelope, or (2) a tight binding of the HCE to the egg envelope, which results in immobilization of the enzyme. Of these alternatives, the tight binding of HCE has been proved to occur by a "washing experiment," through which we recovered the bound HCE from the swollen egg envelopes by washing with dilute alkali, and by radioimmunoassay, through which we detected the bound HCE (Yasumasu *et al.*, 1989c). As mentioned above, that the hatching enzyme is probably apt to bind

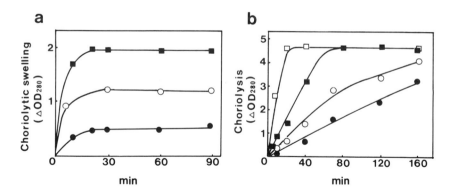

FIG. 14   Time courses of choriolysis by HCE and LCE. (a) Choriolytic swelling induced by incubating 20 mg of coarse fragments of intact egg envelope with 2 (●), 5 (○), or 12 (■) μg of purified HCE. (b) Choriolysis determined by incubating 20 mg of the HCE-swollen egg envelope with 0.05 (●), 0.1 (○), 0.2 (■), or 0.5 (□) μg of purified LCE. In both cases, the activity was expressed in terms of the increase in OD_{280} of the supernatant of the reaction mixture. A peculiar dose-dependency in choriolytic swelling by HCE is considered to be due to its binding to the egg envelope. From Yasumasu *et al.*, (1989a,b).

tightly to the egg envelope or its hydrolyzed products has long been conjectured by some researchers of hatching enzymes in various animals (Barrett et al., 1971; Yamagami, 1972, 1975), and the binding has actually been shown to occur with HCE of O. latipes.

Among some monoclonal anti-HCE antibodies [(Fab')s] we have raised, two antibodies, i.e., A33 and E72, gave us some pieces of valuable information about the binding activity of HCE. The former antibody binds to HCE with no effects on its proteolytic (caseinolytic) and choriolytic (egg envelope digesting) activities or on its binding activity to the egg envelope. The latter antiboy, however, inhibits both the choriolytic activity and the egg envelope-binding activity but not the proteolytic activity Considering that the hydrolysis of both egg envelope proteins and casein is performed through the (same) catalytic site of the HCE molecule, the above-described result of a close correlation between the choriolysis and the binding to egg envelope (chorion) strongly suggests that the choriolytic activity is impaired by suppression of the egg envelope-binding activity by the monoclonal antibody E72. Thus, it seems likely that digestion of the egg envelope by HCE is closely associated with its binding to the egg envelope. The egg envelope-binding property of HCE seems to be necessary for its action on this solid substrate. In contrast, trypsin hardly digests the intact egg envelope of O. latipes; this is possibly due to its inability to bind to the egg envelope. When the egg envelope proteins were changed into soluble forms (or fully hydrated) by alkali, trypsin could digest them very efficiently (O. Tsuchiya et al., unpublished). It seems that HCE has a binding site separate from the catalytic site, and that the binding to the egg envelope through this site is prerequisite for the choriolytic swelling action. A polypeptide fragment with an apparent molecular weight of about 10K on SDS-PAGE has been obtained from a partially digested HCE; this fragment seems to contain a binding site, as it can bind tightly to egg envelopes (Tsuchiya et al., 1990). No information is yet available as to whether the hatching enzymes of some other fish species and/or other animal groups are similarly apt to bind tightly to the egg envelope or not.

Besides the hatching enzyme, there is another group of egg envelope-digesting principles in the animal kingdom, which are named egg envelope (vitelline coat) lysins or sperm lysins from their origin in spermatozoa (Hoshi, 1985). Most of the lysins are considered to be proteases, i.e., trypsin-like proteases such as mammalian acrosin (Stambaugh and Buckley, 1969; Polakoski et al., 1972, 1973; Zaneveld et al., 1972; Parrish and Polakoski, 1979) and an ascidian lysin, spermosin (Sawada et al., 1984), and/or chymotrypsin-like proteases in ascidians (Sawada et al., 1982, 1983) and sea urchins (Harris et al., 1977; Levine and Walsh, 1980). At present, however, we have no information about the detailed mechanism of action of these enzymatic lysins on their substrate, the egg envelope. On the other hand, according to Haino-Fukushima and associates (Haino, 1971; Haino-Fukushima, 1974; Ogawa and Haino-Fukushima, 1984; Haino-Fukushima et al.,

1986; Haino-Fukushima and Usui, 1986; Usui and Haino-Fukushima, 1991) and Lewis *et al.* (1982), the egg envelope lysins of some marine gastropods such as *Tegula, Megathura*, and *Haliotis* are nonenzymatic proteins which solubilize or swell the egg envelope. Like the enzymatic lysins, nonenzymatic lysins are also released from the acrosomal vesicle of a sperm head and they exert lytic or swelling action by binding very tightly to the egg envelope stoichiometrically.

A tight binding of HCE to the substrate seems to favor efficient action on the substrate, especially when the substrate is in a solid form like an egg envelope. Accordingly, it is possible that, in general, such a binding action may be an elaborate mechanism to cause activity on a solid substrate, although the nonenzymatic but stoichiometric action of the gastropod sperm lysins is probably a special case in which only a small distinct hole is necessary for entrance of a spermatozoon (cf. Lewis *et al.*, 1982). In the case of HCE, the binding is weaker than that of the gastropod lysins, i.e., it is not as tight and HCE can be washed away with 50 m*M* bicarbonate buffer (pH 10.2) (Yasumasu *et al.*, 1989c). It is possible that HCE, a basic protein, interacts with anionic domains of egg envelope proteins and/or that hydrophobic interactions may occur between the enzyme and the egg envelope proteins, as they all have hydrophobic domains. At present, it is not clear as to how the the binding occurs and how it leads to the subsequent partial cleavage of the egg envelope proteins, which causes the swelling of the inner layer.

C. The Choriolytic Process: Hydration and Limited Proteolysis

On incubation of the isolated egg envelopes with a sample of the purified *old* hatching enzyme (PII-0.3 enzyme; Yamagami, 1972, 1973) of *O. latipes*, which is now considered to consist of both HCE and LCE, the egg envelopes are digested and changed into a group of water-soluble glycoproteins with apparent molecular weights ranging from 70K to 214K on Ferguson plotting analysis (Iuchi and Yamagami, 1976). Among them, the 70K glycoprotein is almost homogeneous and has been named Fr2 (F2), while the others consist of about six bands of glycoproteins (86K to 214K) on PAGE and they are collectively named Fr1 (F1). Although F1 is thus highly heterogeneous, the components, termed C_1 to C_6 in order from lower to higher molecular weights, are considered to be closely related to each other due to the similarity of their physicochemical properties. All components have similar p*I*s as estimated using Ferguson plotting analysis and are inclined to be loosely associated with each other, eluting as a single peak on gel filtration chromatography and representing a symmetrical Schlieren profile on ultracentrifugation. The average molecular weight difference between any two neighboring components is about 26K. It seems that different components contain different numbers of common repeating units, a polypeptide of about 26K. Thus, they may be placed in the F1 family category (Iuchi and Yamagami, 1976; Yamagami, 1981). A similar group of choriolytic products was found in the hatching liquid, in which the egg envelope was digested by the cooperative action of HCE and LCE. Besides F1 and F2, some lower molecular weight peptides and free amino acids are produced on choriolysis. However, the amount of free amino acids relative to water-soluble proteins and peptides in the choriolytic products is not as large as imagined. In other words, the egg envelope is mostly changed into water-soluble proteins but not into free amino acids (Yamagami, 1988). Most of these choriolytic products are formed by the action of LCE. It is considered, therefore, that LCE performs a limited hydrolysis of egg envelope proteins, yielding the components of F1 and F2.

As described above, prior to undergoing the LCE action, intact egg envelope proteins are first swollen, due to a partial proteolysis by the bound HCE. As a result, relatively low-molecular-weight peptides are released from the inner layer proteins by HCE (Yasumasu, 1989). As shown in Fig. 13, the transparency of the inner layer of egg envelope is found to increase remarkably on swelling. Choriolytic activity of HCE was determined by turbidimetry (Yamagami, 1970), in which an increase in transparency of a turbid suspension of fine fragments of egg envelope due to the action of the added HCE was measured. Thus, HCE exhibits a high "choriolytic" action (clearing action; cf., Yamagami, 1972) in addition to its partial digestion of the inner layer. It is highly probable that the swelling is due to hydration of the inner layer proteins, and that the action of HCE is responsible for hydration of the egg envelope which is, however, normally hydrophobic. Although no experimental evidence has yet been obtained concerning the detailed mechanism of this choriolytic swelling action of HCE, it can be imagined that HCE would remove some hydrophobic regions to expose the hidden hydrophilic domains of the inner layer of the egg envelope and/or that HCE may induce some conformational changes of the inner layer through the partial proteolysis, which makes the hydrophilic regions exposed.

Once the inner layer proteins are hydrated and swollen, they are now highly susceptible to the action of LCE, which is considered to be one of the ordinary proteases and not to have any special device of binding to the inner layer proteins. It seems, however, that a characteristic of the LCE action lies in its limited cleavage of the inner layer proteins. At present, it is uncertain whether the sites of cleavage by LCE are some special bonds including cross-links such as the glutamyl–lysine isopeptide bond or not. The limited proteolysis necessarily leads to an efficient solubilization of the egg envelope in a short period of time. Thus, the hard and tough inner layer of the hardened envelope of O. latipes eggs is transformed into a mass of soluble glycoproteins by the cooperative action of HCE and LCE.

VI. Concluding Remarks

In the present article, some topics of the fish egg envelope from its formation during oogenesis to its breakdown at hatching have been discussed by referring to the experimental results we have obtained recently. As mentioned in the Introduction, this article has not been primarily concerned with the description of the egg envelope itself but has examined the molecular and cellular events associated with or relevant to the changes in the egg envelope. Of our experimental results, some are contradictory to the conventional (or common) concept and others are still incomplete, uncertain, or inconclusive. However, the problems dealt with in each section are closely interrelated with each other. When a problem in one section is solved, the information obtained may be utilized for analysis of a still unsolved problem dealt with in another section. If the subunit composition of the inner layer of egg envelope became clear, based on the study of the precursors, the information obtained would facilitate our understanding of the structure of the egg envelope and the mechanism of egg envelope hardening. This information would, in turn, give us an invaluable clue to the mechanism of choriolytic swelling by HCE. Thus, research aiming to elucidate the mechanisms of molecular and cellular activity in the egg envelope would necessarily result in deepening and widening of our knowledge of the egg envelope itself.

Acknowledgments

The authors wish to thank Dr. H. Kobayashi, Director of the Institute of Zenyaku Kogyo Co., and Professor H. A. Bern of the University of California, Berkeley for their kind advice and encouragement. Thanks are also due to Dr. J. M. Michalec of Sophia University for reading the manuscript. Most of our experiments described in the present article were supported in part by Grants-in-Aid from the Ministry of Education, Science and Culture of Japan to K. Y.

References

Aketa, K., Yoshida, M., Miyazaki, S., and Ohta, T. (1979). *Exp. Cell Res.* **123**, 281–284.
Andersen, S. O. (1963). *Biochim. Biophys. Acta* **69**, 249–262.
Andersen, S. O. (1964). *Biochim. Biophys. Acta* **93**, 213–215.
Anderson, E. (1967). *J. Cell Biol.* **35**, 193–212.
Bakos, M.-A., Kurosky, A., and Hedrick, J. L. (1990). *Biochemistry* **29**, 609–615.
Balfour, F. M. (1880). "A Treatise on Comparative Embryology." Macmillan, London.
Balinsky, B. I. (1965). "An Introduction to Embryology." Saunders, Philadelphia.
Barrett, A. J. (1986). *In* "Proteinase Inhibitors" (A. J. Barrett and G. Salvesen, eds.), pp. 3–22. Elsevier, Amsterdam.
Barrett, D., and Edwards, B. F. (1976). *In* "Methods in Enzymology" (L. Lorand, ed.), Vol. 45B, pp. 354–372. Academic Press, New York.

Barrett, D., Edwards, B. F., Wood, D. B., and Lane, D. J. (1971). *Arch. Biochem. Biophys.* **143**, 261–268.

Barth, L. G. (1953). "Embryology." Holt, Rinehart and Winston, New York.

Battaglia, D. E., and Shapiro, B. M. (1988). *J. Cell Biol.* **107**, 2447–2454.

Begovac, P. C., and Wallace, R. A. (1989). *J. Exp. Zool.* **251**, 56–73.

Bergink, E. W., Wallace, R. A., Van de Berg, J. A., Bos, E. S., Gruber, M., and Ab, G. (1974). *Am. Zool.* **14**, 1177–1193.

Berridge, M. J., and Irvine, R. F. (1989). *Nature (London)* **341**, 197–205.

Bleil, J. D., and Wassarman, P. M. (1980). *Cell (Cambridge, Mass.)* **20**, 873–882.

Bleil, J. D., and Wassarman, P. M. (1983). *Dev. Biol.* **95**, 317–324.

Brummett, A. R., and Dumont, J. N. (1981). *J. Exp. Zool.* **216**, 63–79.

Chambers, E. L., Pressman, B. C., and Rose, B. (1974). *Biochem. Biophys. Res. Commun.* **60**, 126–132.

Chaudhury, H. S. (1956). *Z. Zellforsch. Mikrosk. Anat.* **43**, 478–485.

Dan, J. C. (1956). *Int. Rev. Cytol.* **5**, 365–393.

Dumont, J. N., and Brummett, A. R. (1985). *In* "Developmental Biology: A Comprehensive Synthesis" (R. W. Browder, ed.), Vol. 1, pp. 235-288. Plenum, New York.

Egami, N. (1959). *J. Fac. Sci., Univ. Tokyo, Sec. 4* **8**, 521–538.

Egami, N., and Hama, A. (1975). *Int. J. Radia. Biol.* **28**, 273–278.

Endo, Y. (1961). *Exp. Cell Res.* **25**, 383–397.

Flügel, H. (1967a). *Z. Zellforsch. Mikrosk. Anat.* **77**, 244–256.

Flügel, H. (1967b). *Z. Zellforsch. Mikrosk. Anat.* **83**, 82–116.

Foerder, C. A., and Shapiro, B. M. (1977). *Proc. Natl. Acad. Sci. U.S.A* **74**, 4214–4218.

Folk, J. E., and Chung, S. I. (1973). *Adv. Enzymol.* **38**, 109–191.

Folk, J. E., and Finlayson, J. S. (1977). *Adv. Protein Chem.* **31**, 1-133.

Gray, J. (1932). *J. Exp. Biol.* **9**, 277–299.

Greve, L. C., and Hedrick, J. L. (1978). *Gamete Res.* **1**, 13–18.

Haddon, A. C. (1887). "An Introduction to the Study of Embryology." Charles Griffin, London.

Hagenmaier, H. E. (1974). *Wilhelm Roux' Arch. Entwicklungsmech. Org.* 175, 157–162.

Hagenmaier, H. E., Schmitz, I., and Fohles, J. (1976). *Hoppe-Seyler's Z. Physiol. Chem.* **357**, 1435–1438.

Hagiwara, S., and Jaffe, L. A. (1979). *Annu. Rev. Biophys. Bioeng.* **8**, 385–416.

Haino, K. (1971). *Biochim. Biophys. Acta* **229**, 459–470.

Haino-Fukushima, K. (1974). *Biochim. Biophys. Acta* **352**, 179–191.

Haino-Fukushima, K., and Usui, N. (1986). *Dev. Biol.* **115**, 27–34.

Haino-Fukushima, K., Kasai, H., Isobe, T., Kimura, M., and Okuyama, T. (1986). *Eur. J. Biochem.* **154**, 503–510

Hall, H. G. (1978). *Cell (Cambridge, Mass.)* **15**, 343 355.

Hamazaki, T., Iuchi, I., and Yamagami, K. (1984). *Zool. Sci.* **1**, 148–150.

Hamazaki, T., Iuchi, I., and Yamagami, K. (1985). *J. Exp. Zool.* **235**, 269–279.

Hamazaki, T. S., Iuchi, I., and Yamagami, K.(1987a). *J. Exp. Zool.* **242**, 325–332.

Hamazaki, T. S., Iuchi, I., and Yamagami, K.(1987b). *J. Exp. Zool.* **242**, 333–341.

Hamazaki, T. S., Iuchi, I., and Yamagami, K. (1987c). *J. Exp. Zool.* **242**, 343–349.

Hamazaki, T. S., Nagahama, Y., Iuchi, I., and Yamagami, K. (1989a). *Dev. Biol.* **133**, 101–110.

Hamazaki, T. S., Selman, K., and Wallace, R. A. (1989b). *Dev. Growth Differ.* **31**, 407.

Hansbrough, J. R., and Garbers, D. L. (1981a). *J. Biol. Chem.* **256**, 1447–1452.

Hansbrough, J. R., and Garbers, D. L. (1981b). *J. Biol. Chem.* **256**, 2235–2241.

Harris, E. K., Fox, S. J., Csernansky, J. G., Zimmerman, M., and Troll, W. (1977). *Biol. Bull. (Woods Hole, Mass.)* **153**, 428–429.

Hart, N. H., and Yu, S.-F. (1980). *J. Exp. Zool.* **213**, 137–159.

Hoshi, M. (1985). *In* "Biology of Fertilization" (C. B. Metz and A. Monroy, eds.), Vol. 2, pp. 431–462. Academic Press, Orlando, Florida.

Hosokawa, K. (1985). *Zool. Sci.* **2**, 513–522.
Hyllner, S. J., Oppen-Berntsen, D. O., Helvik, J. V., and Walther, B. T. (1991). *Sarsia*, in press.
Inoue, S., Kitajima, K., Inoue, Y., and Kudo, S. (1987). *Dev. Biol.* **123**, 442–454.
Ishida, J. (1985). *Zool. Sci.* **2**, 1–10.
Iuchi, I., and Yamagami, K. (1976). *Biochim. Biophys. Acta* **453**, 240–249.
Iuchi, I., Masuda, K., and Yamagami, K. (1991). *Dev. Growth Differ.* **33**, 85–92.
Iwamatsu, T. (1969). *Bull. Aichi Univ. Educ.* **18**, 43–56.
Iwamatsu, T., and Ohta, T. (1976). *Wilhelm Roux' Arch. Entwicklungsmech. Org.* **180**, 297–307.
Jaffe, L. A., and Gould, M. (1985). *In* "Biology of Fertilization" (C. B. Metz and A. Monroy, eds.), Vol. 3, pp. 223–250. Academic Press, Orlando, Florida.
Just, E. E. (1919). *Biol. Bull. (Woods Hole, Mass.)* **36**, 1–10.
Kagan, B. M. (1935). *Biol. Bull. (Woods Hole, Mass.)* **69**, 185–201.
Katagiri, C. (1975). *J. Exp. Zool.* **193**, 109–118.
Katagiri, C. (1987). *Zool. Sci.* **4**, 1–14.
Katagiri, C., Iwao, Y., and Yoshizaki, N. (1982). *Dev. Biol.* **94**, 1–10.
Kay, E. S., and Shapiro, B. M. (1985). *In* "Biology of Fertilization" (C. B. Metz and A. Monroy, eds.), Vol. 3, pp. 45–80. Academic Press, Orlando, Florida.
Kishimoto, T., Usui, N., and Kanatani, H. (1984). *Dev. Biol.* **101**, 28–34.
Kitajima, K., Inoue, Y., and Inoue, S. (1986). *J. Biol. Chem.* **261**, 5262–5269.
Kobayashi, W. (1985). *J. Fac. Sci., Hokkaido Univ.* **VI24**, 87–102.
Kusa, M. (1953). *Annot. Zool. Japon.* **26**, 73–77.
Laale, H. W. (1980). *Copeia* **2**, 210–226.
Laboulenger, F., Leroux, P., Tonon, M.-C., Coy, D. H., Vandry, H., and Pelletier, G. (1983). *Neurosci. Lett.* **37**, 221–225.
Laslop, A., Wohlfarter, T., Fischer-Colbrie, R., Steiner, H. J., Humpel, C., Saria, A., Schmid, K. W., Sperk, G., and Winkler, H. (1989). *Reg. Peptide* **26**, 191–202.
LePage, T., and Gache, C. (1990). *EMBO J.* **9**, 3003–3012.
Lereboullet, M. (1862). *Mem. Acad. des Sci. (Sac. Etrang.)* **17**, 449–805.
Levine, A. E., and Walsh, K. A. (1980). *J. Biol. Chem.* **255**, 4814–4820.
Lewis, C., Talbot, C. F., and Vacquier, V. D. (1982). *Dev. Biol.* **92**, 227–239.
Lindsay, L. L., and Hedrick, J. L. (1989). *Dev. Biol.* **135**, 202–211.
Loewy, A. G. (1968). *In* "Fibrinogen" (K. Laki, ed.), pp. 185–223. Dekker, New York.
Lönning, S., Kjorsvik, E., and Davenport, J. (1984). *J. Fish Biol.* **24**, 505–522.
Lorand, L. (1970). *Thromb. Diath. Haemorrh.*, Suppl. **39**, 75–102.
Lorand, L., and Campbell, L. K. (1971). *Anal. Biochem.* **44**, 207–220.
Luberda, Z., Strzezek, J., and Luczynski, M. (1990). *Acta Biochim. Pol.* **37**, 197–200.
Ludwig, H. (1874). *Arb. Zool. Zoot. Inst. (Wurzburg)* **1**, 287–510.
Masuda, K. (1990). *D. Sc. Thesis*, Sophia University, Tokyo, Japan.
Masuda, K., Iuchi, I., Iwamori, M., Nagai, Y., and Yamagami, K. (1986). *J. Exp. Zool.* **238**, 261–265.
Masuda, K., Iuchi, I., and Yamagami, K. (1991). *Dev. Growth Differ.* **33**, 75–83.
Miller, R. L. (1985). *In* "Biology of Fertilization" (C. B. Metz and A. Monroy, eds.), Vol. 2, pp. 275–337. Academic Press, Orlando, Florida.
Monroy, A. (1965). *In* "Biochemistry of Animal Development" (R. Weber, ed.), Vol. 1, pp. 73–135. Academic Press, New York.
Murata, K., Hamazaki, T. S., Iuchi, I., and Yamagami, K. (1991). *Dev. Growth Differ.*, **33**, 553–562.
Nakano, E. (1956). *Embryologia* **3**, 89–103.
Nakatsuka, M. (1979). *Dev. Growth Diff.* **21**, 245–253.
Nomura, K., Suzuki, N., and Matsumoto, S. (1990). *Biochemistry* **29**, 4525–4534.
Ogawa, A., and Haino-Fukushima, K. (1984). *Dev. Growth Differ.* **26**, 345–360.
Ogawa, N., and Ohi, Y. (1968). *Zool. Mag.* **77**, 151–156.
Ohi, Y., and Ogawa, N. (1970). *Zool. Mag.* **79**, 17–18.

Ohtsuka, E. (1957). *Sieboldia Acta Biol. (Fukuoka)* **2**, 19–29.

Ohtsuka, E. (1960). *Biol. Bull. (Woods Hole, Mass.)* **118**, 120–128.

Oppen-Berntsen, D. O. (1990). *PhD Thesis*, Univ. of Bergen.

Oppen-Berntsen, D. O., Helvik, J. V., and Walther, B. T. (1990). *Dev. Biol.* **137**, 258–265.

Osanai, K. (1956). *Sci. Rep. Tohoku Univ. Ser. 4*, **22**, 181–188.

Parrish, R. F., and Polakoski, K. L. (1979). *Int. J. Biochem.* **10**, 391–395.

Polakoski, K. L., Zaneveld, L. J. D., and Williams, W. L. (1972). *Biol. Reprod.* **6**, 23–29.

Polakoski, K. L., McRorie, R. A., and Williams, W. L. (1973). *J. Biol. Chem.* **248**, 8178–8182.

Riehl, R. (1984). *Cytologia* **49**, 127–142.

Roe, J. L., and Lennarz, W. J. (1985). *J. Cell Biol.* **101**, 470A.

Rothschild, L. (1958). *Biol. Rev. Cambridge Philos. Soc.* **33**, 372–392.

Rothschild, L., and Swann, M. M. (1952). *J. Exp. Biol.* **29**, 469–483.

Sawada, H., Yokosawa, H., Hoshi, M., and Ishii, S. (1982). *Gamete Res.* **5**, 291–301.

Sawada, H., Yokosawa, H., Hoshi, M., and Ishii, S. (1983). *Experientia* **39**, 377–378.

Sawada, H., Yokosawa, H., and Ishii, S. (1984). *J. Biol. Chem.* **259**, 2900–2904.

Schmell, E. D., Gulyas, B. J., and Hedrick, J. L. (1983). In "Mechanism and Control of Animal Fertilization" (J. F. Hartmann, ed.), pp. 365–413. Academic Press, New York.

Schoots, A. F. M., and Denucé, J. M. (1981). *Int. J. Biochem.* **13**, 591–602.

Schoots, A. F. M., Sackers, R. J., Overkamp, P. S. G., and Denucé, J. M. (1983). *J. Exp. Zool.* **226**, 93–100.

Schuel, H. (1978). *Gamete Res.* **1**, 299–382.

Selman, K., Wallace, R. A., and Barr, V. (1986). *J. Exp. Zool.* **239**, 277–288.

Selman, K., Wallace, R. A., and Barr, V. (1988). *J. Exp. Zool.* **246**, 42–56.

Somers, C. E., and Shapiro, B. M. (1989). *Dev. Growth Differ.* **31**, 1–7.

Stambaugh, R., and Buckley, J. (1969). *J. Reprod. Fert.* **19**, 423–432.

Steinhardt, R. A., Epel, D., and Carroll, E. J. (1978). *Nature (London)* **252**, 41–43.

Suzuki, N. (1990). *Zool. Sci.* **7**, 355–370.

Suzuki, N., Nomura, K., Ohtake, H., and Isaka, S. (1981). *Biochem. Biophys. Res. Commun.* **99**, 1238–1244.

Suzuki, N., Shimomura, H., Radany, E. W., Ramarao, C. S., Bentley, J. K., and Garbers, D. L. (1984). *J. Biol. Chem.* **259**, 14874–14879.

Takamune, K., Yoshizaki, N., and Katagiri, C. (1986). *Gamete Res.* **14**, 215–224.

Takeuchi, K., Yokosawa, H., and Hoshi, M. (1979). *Eur. J. Biochem.* **100**, 257–265.

Tesoriero, J. V. (1977a). *J. Ultrastruct. Res.* **59**, 282–291.

Tesoriero, J. V. (1977b). *J. Histochem. Cytochem.* **25**, 1376–1380.

Tesoriero, J. V. (1978). *J. Ultrastruct. Res.* **64**, 315–326.

Tsuchiya, O., Yasumasu, S., Iuchi, I., and Yamagami, K. (1990). *Zool. Sci.* **7**, 1127.

Tyler, A. (1948). *Physiol. Rev.* **28**, 180–219.

Tyler, A. (1949). *Am. Nat.* **83**, 195–219.

Urch, U. A., and Hedrick, J. L. (1981). *Arch. Biochem. Biophys.* **206**, 424–431.

Usui, N., and Haino-Fukushima, K. (1991). *Mol. Reprod. Dev.* **28**, 189–198.

Vacquier, V. D., Tegner, M. J., and Epel, D. (1973). *Exp. Cell Res.* **80**, 111–119.

Vardell, I. M., Polak, J. M., Allen, J. M., Terenghi, G., and Bloom, S. T. (1984). *Endocrinology* **114**, 1460–1462.

Waldeyer, W. (1906). In "Handbuch der Vergleichenden und Experimentellen Entwicklungslehre" (O. Hertwig, ed.), pp. 86–476. Fischer, Jena.

Wallace, R. A. (1978). In "The Vertebrate Ovary: Comparative Biology and Evolution" (R. E. Jones, ed.), pp. 469–502. Plenum, New York.

Wallace, R. A. (1985). In "Developmental Biology: A Comprehensive Synthesis" (R. W. Browder, ed.), Vol. 1, pp. 127–177. Plenum, New York.

Wallace, R. A., and Selman, K. (1981). *Am. Zool.* **21**, 325–343.

Wassarman, P. M. (1988). *Annu. Rev. Cell Biol.* **3**, 109–142.

Wassarman, P. M., Florman, H. M., and Greve, J. M. (1985). In "Biology of Fertilization" (C. B. Metz and A. Monroy, ed.), Vol. 2, pp. 341–360. Academic Press, Orlando, Florida.

Wilson, E. B. (1928). "The Cell in Development and Heredity," 3rd Ed. Macmillan, London.

Wolf, D. P., Nishihara, T., West, D. M., Wyrick, R. E., and Hedrick, J. L. (1976). Biochemistry 15, 3671–3678.

Wolf, D. P., Nicosia, S. V., and Hamada, M. (1979). Dev. Biol. 71, 22–32.

Wourms, J. P. (1976). Dev. Biol. 50, 338–354.

Wyrick, R. E., Nishihara, T., and Hedrick, J. L. (1974). Proc. Natl. Acad. Sci., U.S.A. 71, 2067–2071.

Yamagami, K. (1970). Annot. Zool. Jpn. 43, 1–9.

Yamagami, K. (1972). Dev. Biol. 29, 343–348.

Yamagami, K. (1973). Comp. Biochem. Physiol. 46B, 603–616.

Yamagami, K. (1975). J. Exp. Zool. 192, 127–132.

Yamagami, K. (1981). Am. Zool. 21, 459–471.

Yamagami, K. (1988). In "Fish Physiology" (W. S. Hoar and D. J. Randall, eds.), Vol. 11A, pp. 447–499. Academic Press, San Diego, California.

Yamaguchi, S., Hedrick, J. L., and Katagiri, C. (1989). Dev. Growth Diff. 31, 85–94.

Yamamoto, K. (1957). Annot. Zool. Jpn. 29, 91–96.

Yamamoto, M. (1963a). J. Fac. Sci., Univ. Tokyo, Sect. 4 10, 115–121.

Yamamoto, M. (1963b). J. Fac. Sci., Univ. Tokyo, Sec. 4 10, 123–127.

Yamamoto, M., Iuchi, I., and Yamagami, K. (1979). Dev. Biol. 68, 162–174.

Yamamoto, T. (1939). Proc. Imp. Acad. (Tokyo) 15, 269–271.

Yamamoto, T. (1944). Annot. Zool. Jpn. 22, 109–125.

Yamamoto, T. (1961). Int. Rev. Cytol. 12, 361–405.

Yamamoto, T. S. (1957). J. Fac. Sci., Hokkaido Univ., Ser. 6 13, 484–488.

Yasumasu, S. (1989). D. Sc. Thesis, Sophia University, Tokyo, Japan.

Yasumasu, S., Iuchi, I., and Yamagami, K. (1988). Zool. Sci. 5, 191–195.

Yasumasu, S., Iuchi, I., and Yamagami, K. (1989a). J. Biochem. 105, 204–211.

Yasumasu, S., Iuchi, I., and Yamagami, K. (1989b). J. Biochem. 105, 212–218.

Yasumasu, S., Katow, S., Umino, Y., Iuchi, I., and Yamagami, K. (1989c). Biochem. Biophys. Res. Commun. 162, 58–63.

Yasumasu, S., Yamada, K., Akasaka, K., Shimada, H., and Yamagami, K. (1990). Zool. Sci. 7, 1127.

Yasumasu, S., Yamada, K., Akasaka, K., Iuchi, I., Shimada, H., and Yamagami, K. (1991). Zool. Sci. 8, 1080.

Yasumasu, S., Katow, S., Hamazaki, T. S., Iuchi, I., and Yamagami, K. (1992). Dev. Biol. 149, 349–356.

Yoshizaki, N. (1984). Dev. Growth Differ. 26, 191–195.

Young, G., and Smith, D. G. (1956). J. Biol. Chem. 219, 161–164.

Zaneveld, L. J. D., Polakoski, K. L., and Williams, W. L. (1972). Biol. Reprod. 6, 30–39.

Zotin, A. I. (1958). J. Embryol. Exp. Morphol. 6, 546–568.

Ganglia within the Gut, Heart, Urinary Bladder, and Airways: Studies in Tissue Culture

M. J. Saffrey, C. J. S. Hassall, T. G. J. Allen,* and G. Burnstock

Department of Anatomy and Developmental Biology, *Department of Pharmacology, University College London, London WC1E 6BT, United Kingdom

I. Introduction

Autonomic ganglia are present within almost all organs of the body and exhibit considerable differences in their structure and properties. Although such differences probably reflect the diverse roles played by these ganglia, our knowledge of how intrinsic neurons function in the control of their target tissues is limited. The main reasons for this are the particular difficulties associated with the direct study of intrinsic ganglia which are embedded in non-neuronal tissue and, therefore, inaccessible to many experimental techniques. Furthermore, complex denervation procedures may be required to enable the processes of extrinsic neurons to be distinguished from those originating from intrinsic neurons. Hence, the development of techniques for the isolation and maintenance *in vitro* of intrinsic autonomic ganglia is of fundamental importance for investigation of these systems. The use of culture preparations also offers additional advantages such as the ability to examine directly the distribution and properties of membrane-associated molecules on identified cell types under defined conditions, and to analyze the factors, both cellular and molecular, that are responsible for neuronal and glial development and differentiation.

The most well-studied of the intrinsic autonomic ganglia, both in culture and *in situ*, are those of the enteric nervous system (ENS). There are three main reasons for this: enteric ganglia are far more numerous than any other kind of peripheral ganglia; the localization of the ganglia is consistent along almost the entire length of the intestinal tract, and the histological organization of the gut wall facilitates separation of some of its constituent layers. Together, these features have enabled many techniques to be applied to the ENS *in situ*, and have also allowed the development of several different types of culture preparation which have provided valuable information about the ENS. Functionally, enteric neurons are known to regulate motility, secretion, and blood flow within the gut, and to form a complex integrative system that contains a diversity of neuronal phenotypes which is second only to that of the central nervous system (CNS; see Furness and Costa, 1987). Ultrastructurally, the enteric ganglia also resemble central nervous tissue, rather than other autonomic ganglia: they contain no connective tissue or blood supply, but a dense synaptic neuropil and glial cells, which show many similarities to astrocytes, are present (Gabella, 1981). Thus, the ENS has more in common with the CNS than the other autonomic ganglia. Much less is known of the properties of ganglia in organs such as the heart, urinary bladder, and airways, although the development of culture preparations of these ganglia has facilitated studies of various aspects of the intrinsic neurons. However, there is now increasing evidence from studies *in situ* and in culture to indicate that, like the ENS, intracardiac neurons, intramural urinary bladder neurons, and those of the airways constitute complex and specialized local nervous systems.

In this paper, we review the work that has been carried out in culture on the intrinsic neurons of the mammalian gut, heart, urinary bladder, and airways. We do not give detailed information on general tissue culture techniques, as these have been reviewed elsewhere (Jessen, 1982; Hassall *et al.*, 1989). The section on the ENS is subdivided according to the different approaches now available for its study in culture: each subsection provides information about the culture techniques and discusses their application to investigations of enteric neurons and glial cells, including studies of their development, interactions in culture, and neurochemical and electrophysiological properties. The following three sections concern the studies of culture preparations of intracardiac neurons, intramural urinary bladder neurons, and the local neurons in the airways. These sections include information about culture methods and review the findings of studies that have been carried out on these neurons in culture using microscopical, immunocytochemical, autoradiographical, and electrophysiological techniques. Last, the major points arising from the body of work reviewed here are summarized and some of the possible further applications of tissue and cell culture approaches are outlined.

II. Enteric Ganglia

The ganglia of the ENS are grouped into two major plexuses: the myenteric (or Auerbach's), which lies between the outer longitudinal and the circular smooth muscle layers, and the submucous (or Meissner's), which lies in the connective tissue of the submucosa. The myenteric plexus is the greater, containing more neurons and larger ganglia, and extends uninterrupted along the entire length of the gut. The submucous plexus is sometimes subdivided according to the location of the ganglia within the submucosa; it is not present in the esophagus and is very sparse in the stomach. The two plexuses are connected by nerves which pass through the circular muscle layer. The enteric neurons are a highly diverse population functionally, morphologically, and with respect to their neurotransmitter phenotypes (Furness and Costa, 1987).

The ENS was first studied in tissue culture at the beginning of this century, when the technique of explanting tissues and examining their growth *in vitro* was in its infancy. Soon after Harrison's pioneering study of nerve fiber regeneration from the frog neural tube *in vitro* (1907), Lewis and Lewis (1912) performed the first study of enteric nervous tissue in culture by explanting small pieces of embryonic chick intestine and cultivating them in saline solution. However, there were few subsequent studies, and these were concerned mainly with the overall differentiation of the intestine maintained in organ culture (De Jong and De Haan, 1943; Keunig, 1944). Since then, particularly in the past 10–12 years, a variety of different approaches have been developed in order to study the ENS in culture. In the following review, the major types of technique available are outlined, and the particular applications of these different methods are discussed in more detail.

A. Organotypic Cultures

Organotypic cultures are usually described as those in which all the tissue elements of an organ, but not the entire organ itself, are explanted with little disruption to tissue organization. An example of this type of culture is the explantation of small hemisections of the entire gut wall, including the mucosal, submucosal, and muscular layers, together with the intrinsic myenteric and submucous plexus ganglia. This type of culture was first employed to study nerve fiber growth (Lewis and Lewis, 1912). More recently, such cultures, or cultures of the muscularis externa, have been used to investigate the innervation of intestinal smooth muscle by sympathetic neurons in coculture (Cook and Peterson, 1974; Rawdon and Dockray, 1983), to demonstrate the intrinsic nature of serotonergic and peptide-containing neurons in the enteric plexuses (Dreyfus *et al.*, 1977a, b; Schultzberg *et al.*, 1978), to study the appearance of nerve fibers from chick intestine (Cheng and Bjerknes, 1979), and to examine the morphological development of enkephalin (ENK)-containing neurons in the rat gut (Haynes and Zakarian, 1982). Another valuable application of the organotypic culture system is in the study of the projections of enteric neurons. Ascending and descending myenteric neurons projecting to the circular muscle layer have been immunocytochemically identified after application of DiI in organotypic culture (Brookes *et al.*, 1991).

The most significant use of organotypic cultures, however, has been in the study of the embryological development of the ENS. In an elegant series of experiments, Gershon and co-workers have used organotypic cultures to determine the time of arrival, from the neural crest, of undifferentiated neuronal and glial precursors in the embryonic gut (Rothman and Gershon, 1982; Rothman *et al.*, 1984, 1986). These studies have been extended to investigate factors influencing the migration of neural precursors in the gut, and have used the lethal spotted (*Ls/Ls*) mutant mouse, in which the terminal colon, like that in Hirschsprung's disease, is aganglionic. Coculture of the terminal colon from mutant mice or of aneural colon from normal mice (explanted before colonization by undifferentiated precursors) with various sources of migratory neural crest-derived cells showed that the neural precursors were unable to colonize the mutant gut, but readily colonized the normal aneural gut (Jacobs-Cohen *et al.*, 1987). These experiments provide strong evidence that the defect lies in the nonneural intestinal tissue and not in the neural precursors themselves, as had previously been suggested.

Despite their suitability for some types of investigation, organotypic cultures have several inherent disadvantages that make other studies problematic. For example, the size of the explants can cause problems of gas and nutrient exchange and hence cell death and necrosis (Cook and Peterson, 1974). Individual cells are difficult to visualize in living cultures, the dynamic relationships between different constituent cell types cannot be examined, and the growth characteristics and requirements of individual cell types cannot be determined.

B. Mixed Cell Type Dissociates

1. Whole-Gut Dissociates

One way of avoiding the problems associated with organotypic cultures is to use enzymes to dissociate segments of the gut into a cell suspension. This approach has recently been used by Korman *et al.* (1988) in an attempt to select specific subpopulations of enteric neurons from hamster intestine and maintain them in short-term culture. However, without further refinements, this approach has only very limited advantages over the organotypic culture system. The major drawbacks are that both myenteric and submucosal neurons are present and hence cannot be discriminated between, that there are many other different cell types present, and that the nonneural cells of the gut wall greatly outnumber the neural elements. When grown in a dissociated culture system, proliferation of the nonneuronal cells is stimulated so that they dominate the cultures. This can result in a massive reduction in the number of neurons present (Korman *et al.*, 1988). Such cultures are thus unsuitable for studies requiring significant numbers of enteric neurons or glial cells with few other cell types, or for long-term studies of the ENS in culture. Alternative strategies for producing dissociated cell cultures of the ENS have therefore been developed, using tissue enriched in enteric ganglia as a starting substrate for dispersion.

2. Dissociates Enriched for Myenteric Plexus Neurons

Methods of enrichment for myenteric neurons have utilized the fact that the outer muscle layers of the gut wall can easily be separated from the underlying submucosal and mucosal layers. Hence, either longitudinal muscle–myenteric plexus (LM-MP; Maruyama, 1981) or the LM-MP plus circular muscle layer (LM-MP-CM; Nishi and Willard, 1985) has been enzymatically dispersed to produce cultures containing myenteric neurons. Further enrichment can be achieved by a preplating step, although, even in these neuronally enriched cultures, additional procedures such as the use of mitotic inhibitors are necessary to limit the proliferation of the smooth muscle cells, fibroblasts, and glial cells that are also present (Nishi and Willard, 1985). However, the resulting cultures still contain a large and variable number of smooth muscle cells, and may not survive beyond 1–2 weeks. Nevertheless, the neurons present in these enriched mixed cell cultures of the myenteric plexus exhibit many of the characteristics of their counterparts *in situ* (Nishi and Willard, 1985; Willard and Nishi, 1989) and are useful for certain types of investigation, such as electrophysiological studies, in which they have been used to great effect by Willard (see Section II,E).

The ability to perform quantitative studies of cells in dissociated preparations has led to the use of enriched cultures for assessment of environmental influences on the expression of several different neuronal properties by myenteric neurons. The effects of culture medium conditioned by heart cells (heart cell-conditioned medium; HCM) on both electrophysiological and transmitter-related properties of rat myenteric neurons in culture have been examined (Nishi and Willard, 1988). This medium is known to contain a factor which stimulates the expression of cholinergic properties by sympathetic neurons in culture (Furshpan *et al.*, 1986; Potter *et al.*, 1986). In the enriched myenteric neuron culture system, although an increase in the synthesis and storage of acetylcholine (ACh) in response to HCM was demonstrated, the number of cholinergic neurons appeared to be unaffected. Thus, it appears as if either a suitable labile population of neurons or precursors does not exist in the myenteric plexus from the newborn rat small intestine, or that the ability to recruit additional cholinergic neurons from such a population is not retained in this culture system. In contrast, HCM was found to affect the numbers of other populations of myenteric neurons present: increased numbers of serotonin (5-hydroxytryptamine; 5-HT) immunoreactive and decreased numbers of vasoactive intestinal polypeptide (VIP)-immunoreactive neurons were detected in cultures grown in conditioned medium, while there was no change in total neuronal numbers present. These observations may suggest that there is some phenotypic plasticity among myenteric neurons in the rat gut at this stage of development.

3. Dissociates Enriched for Submucosal Ganglion Cells

Barber *et al.* (1989) have recently developed a method for dispersing submucosal ganglion cells from the dog ileum. The ganglion cells were dissociated from submucosa which had been manually separated from the mucosa and outer muscle layers. The dispersed submucosa was further enriched for neurons by elutriator centrifugation, and the resulting cell suspension was grown for up to 72 hr *in vitro*, when 80% of the cells were found to be neurofilament-immunoreactive. These short-term cultures have proved useful for the study of peptide release, particularly that of neurotensin, which is found in some 50% of neurons present. Such cultures may also be useful for other types of study if they can be grown for longer periods *in vitro*, but as most of the neurons present were in large aggregates and characterization of the nonneuronal cells present was not attempted, further modification of these methods may be required.

C. Explant Cultures of Isolated Enteric Ganglia

A major new approach for the culture of the ENS became available in the late 1970s, when methods for the separation of enteric ganglia from the remainder of the gut wall were developed (Jessen *et al.*, 1978, 1983a). The initial techniques, which were employed to isolate the myenteric plexus from beneath the taenia coli of the guinea pig cecum, were subsequently modified to allow isolation of both myenteric and submucous plexus ganglia from different areas of the gut of several species (Jessen *et al.*, 1983a).

The originality of this approach lay with the use of a combination of enzyme treatment and microdissection to separate segments of plexus from the nonneural tissues which form the bulk of the intestinal wall. These segments of plexus, which consist of linked ganglia, could be completely cleaned of all extraneous smooth muscle and almost all connective tissue. Further division of the segments to obtain individual ganglia was also possible. The isolated ganglia could then be used directly for localization of different cell types (Fig. 1) or for biochemical studies (Jessen *et al.*, 1979). Alternatively, they could readily be grown in large numbers as explant cultures, in a similar way to other peripheral ganglia such as sympathetic and sensory ganglia (Chamley *et al.*, 1972), and without many of the disadvantages of the organotypic culture system. This approach thus allowed the first direct observations of the growth and interactions of enteric neurons and glial cells *in vitro*, and comparison of enteric ganglion cells with those of other parts of the peripheral nervous system (PNS) in culture.

1. Identification of Cell Types Present in Guinea Pig Myenteric Plexus Explant Cultures

The cell types present in myenteric plexus cultures have been identified by light and electron microscopy and by immunocytochemical techniques. The two-dimensional nature of the myenteric ganglia, together with the phenomenon of cell migration seen over the first week *in vitro* (see Section II,C,3 below), enabled detailed observations to be made of the living cells by phase-contrast microscopy (Jessen *et al.*, 1978, 1983a). Identification of the live cells was subsequently

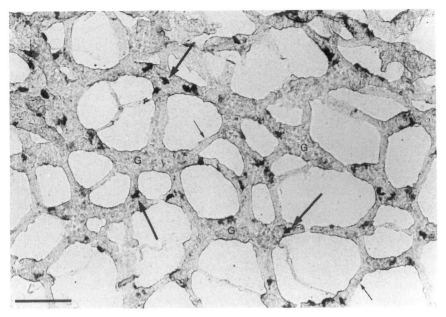

FIG. 1 Bright-field micrograph showing part of a segment of myenteric plexus isolated from newborn guinea pig distal colon. The segment was incubated in tritiated γ-aminobutyric acid ([³H]GABA) and processed for autoradiography to demonstrate the presence of a subpopulation of neurons which express high-affinity uptake sites for GABA (large arrows). Examples of ganglia (G) and interconnecting strands (small arrows) are indicated. Bar, 100 μm. Adapted from Saffrey *et al.* (1983) with permission.

confirmed by electron microscopy (Baluk *et al.*, 1983). Three cell types were identified in the cultures: neurons, glial cells, and a variable number of fibroblasts. *In situ*, the enteric ganglia contain only neurons and glial cells; the fibroblasts seen in culture originated from the thin layer of connective tissue in which the ganglia are embedded and could be virtually eliminated by the use of mitotic inhibitors such as cytosine arabinoside (Bannerman *et al.*, 1987). Type I interstitial cells, described in the mouse gut as lying just outside the myenteric ganglia (Thuneberg, 1982), were not observed, either by light or electron microscopy. However, as the morphology of these cells *in situ* resembles that of fibroblasts, they may not have been recognized in the cultures.

The cultured enteric neurons and glial cells could also be identified by immunocytochemical methods using antibodies to cell-type-specific markers. In common with other neurons, enteric neurons express tetanus toxin receptors (Bannerman *et al.*, 1987) which can be localized by the use of toxin and anti-toxin. An antiserum raised against CNS plasma membranes (anti-CTX) also binds to enteric, as well as to other autonomic neurons (M. J. Saffrey and C. J. S. Hassall, unpublished observations), and has been used as a marker for enteric neurons in explant culture (Figs. 2 and 3; Buckley and Burnstock, 1986). However, since availability of this antiserum is limited, a more convenient marker is protein gene product 9.5 (PGP 9.5), which is also expressed by all enteric neurons in culture (M. J. Saffrey, unpublished observations). In addition, the cell surface glycoprotein Thy-1 and an antigen recognized by the A2B5 antibody are expressed by enteric neurons in culture (Bannerman et al., 1987, 1988a). However, as cultured enteric glial cells also express the A2B5 antigen and fibroblasts express Thy-1, these markers must be examined in combination, or together with other markers for unambiguous cell identification (Bannerman et al., 1988a).

Neurofilament proteins and glial fibrillary acidic protein (GFAP) have so far not proved to be useful markers for neurons and glial cells in guinea pig enteric explant cultures, presumably because there are species variations in these molecules resulting in problems with antibody cross-reactivity (Bannerman et al., 1988a). The calcium-binding protein, S100, however, appears to be conserved in different species and is expressed by all enteric glial cells in culture; hence it is a useful marker for these cells (Bannerman et al., 1988a). The neurotransmitter-associated enzymes acetylcholinesterase and monoamine oxidase are also potential markers for enteric neurons in explant culture. Histochemical studies have shown that both of these enzymes can be demonstrated in all myenteric neurons in culture, provided that prolonged incubation times and appropriate inhibitors of nonneuronal enzymes are incorporated in the technique (Jessen et al., 1983b).

A variety of other molecules have been localized in enteric neurons and glial cells in culture. However, as these molecules are expressed either by both cell types or by subpopulations of cells, they are not useful markers for cell identification and are discussed in Section II,C,5.

2. Morphological Studies of Enteric Ganglion Cells in Explant Culture

Enteric neurons are heterogeneous with respect to size, gross morphology, and ultrastructural features as well as to neurotransmitter content (see Furness and Costa, 1987). This structural diversity is maintained in explant culture. Both phase-contrast microscopy of living cells and intracellular injection of horseradish peroxidase reveal that, as in situ, the cultured neurons vary considerably in size (Jessen et al., 1978, 1983b; Hanani et al., 1982). Intracellular injections of horseradish peroxidase also showed that neurons similar to the type I, II, and III neurons described by Dogiel in in situ preparations were present in culture (Hanani et al., 1982; Jessen et al., 1983b).

The ultrastructural features of the cultured neurons also resembled those seen in situ (Baluk et al., 1983). Nerve profiles containing either small agranular vesicles, large granular vesicles, large opaque vesicles, or mixtures of these types were seen. Synapses were also present in the explant cultures: these may have been retained from the onset of culture, but as some synapses were observed in the outgrowth area (see Section II,C,3), a certain number may have been formed in vitro.

The morphology of the enteric glial cells was found to vary: glial cells at the periphery of the outgrowth area tended to be very flattened and did not express 9 to 11-nm intermediate filaments, those associated with neurons nearer the center of the cultures extended processes and frequently expressed intermediate filaments (Baluk *et al.*, 1983).

3. Cell Interactions in Explant Cultures of the Myenteric Plexus

Cell interactions are fundamental to the formation of the nervous system. How developing neurons and glia interact in the gut, however, is not understood. Explant cultures of isolated myenteric ganglia therefore offered a means of examining interactions of living neurons and glia directly. Detailed observations, including serial photography of selected areas, were made of cultures of the myenteric plexus from the guinea pig cecum over a period of 5 weeks in culture. The growth and interactions of the enteric neurons and glial cells were found to be very different from those of other peripheral ganglia in explant culture. This was consistent for the myenteric plexus from all parts of the gut of all the mammalian species examined (guinea pig, rat, and rabbit; Jessen *et al.*, 1983a).

The development of the mammalian myenteric plexus cultures could be divided into four stages (Jessen *et al.*, 1983a). Typical explants consisted of segments of plexus comprising 20–50 ganglia and their interconnecting strands and remained well-defined during the first day *in vitro* (stage 1). At the end of this stage, glial cells began to migrate from the ganglia into the gaps between and around them, resulting in the loss of the original organization of the explants (stage 2, days 2–3). Enteric neurites regenerated over these glial cells, which continued to migrate and began to proliferate, so that by 4–7 days (stage 3) a broad outgrowth area had formed around the central part of the culture, where the majority of neuronal cell bodies remained as a continuous group, together with some glial cells. The neuronal perikarya were well separated and individual cells were clearly visible at this stage. The enteric glial cells grew as flat sheets of cells rather than as discontinuous areas of bi- or tripolar cells seen in explant cultures of other peripheral ganglia; regenerating neurites were confined to the glial cell surfaces over which they ramified. In the absence of mitotic inhibitors, groups of fibroblasts often formed in the outgrowth area; neurites tended not to regenerate over these cells. In fact, when encroached upon by areas of proliferating fibroblasts, individual neurites and small neurite bundles grouped into thick fiber bundles (Fig. 2). In older cultures (stage IV, days 7–35), the neuronal cell bodies tended to form either flattened groups of clearly defined cells when associated only with glial cells, or compact aggregates connected by thick fiber bundles in areas where there were many fibroblasts. The aggregates tended to form slowly over a period of weeks *in vitro* and thus differed from aggregates seen in other types of neuronal culture (Jessen *et al.*, 1983a). Ultrastructural studies showed that these compact

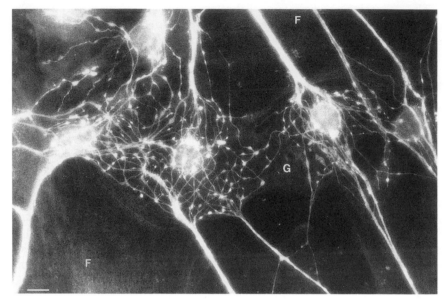

FIG. 2 Fluorescence micrograph showing part of the outgrowth area of a myenteric plexus explant culture from newborn guinea pig cecum grown for 3 weeks *in vitro*. The culture was immunolabeled using an antiserum raised against rat brain synaptosomes (anti-CTX; see text) to facilitate visualization of neuronal processes. Nerve fiber bundles, growing on top of fibroblasts (*F*), split into individual varicose neurites on reaching areas of glial cells (*G*). Bar, 10 μm.

aggregates bore a strong resemblance to intact ganglia *in vivo*. They were found to contain both neuronal and glial cell bodies in a ratio similar to that seen *in situ*, and also a neuropil and synapses (Baluk *et al.*, 1983). It was hence suggested that interactions between enteric neurons, glial cells, and connective tissue elements may be involved in the formation of the network-like arrangement of the ENS seen *in situ* (Jessen *et al.*, 1983a). It should be borne in mind, however, that *in vivo* the enteric ganglia are formed well before birth, at a time when the enteric ganglion cells and other cells of the gut wall are not fully differentiated, so a strict comparison between these events *in vivo* and *in vitro* is not possible. Nevertheless, interactions between these three cell types may be important in the late events in ganglion formation and in the development of nerves supplying smooth muscle or other layers of the gut wall.

4. Growth of Myenteric Plexus Explants in Defined Culture Conditions

Until recently, the vast majority of tissue culture studies have been performed using culture medium supplemented with fetal calf serum (FCS) or other animal sera. However, as well as providing factors essential for cell survival and growth *in vitro*, serum contains many other, largely undefined components, some of which may produce unwanted effects such as cell death and proliferation of nonneuronal cells. Serum may also contain molecules that can hamper studies of the neurochemical properties of neurons in culture. For example, some batches of FCS contain high levels of 5-HT (Saffrey *et al.*, 1984) which, in addition to saturating 5-HT receptors, can be taken up by a low-affinity mechanism into amine-handling, nonserotonergic cells, making the identification of 5-HT-immunoreactive cells in culture problematic. Another disadvantage associated with the use of undefined components in culture medium is that effects of specific agents on cultured cells cannot be unambiguously analyzed, and indeed may even be masked by such components.

To avoid these problems, serum-free, hormone-supplemented culture media have been increasingly used (Barnes and Sato, 1980). Defined media for the growth of primary cultures of neurons were first developed by Bottenstein *et al.* (1980), and have subsequently been widely, although not extensively, used. Myenteric plexus explant cultures may be grown in the absence of FCS in a hormone-supplemented medium based on that described by Bottenstein *et al.* (Saffrey and Burnstock, 1984). The pattern of growth and cell interactions seen in cultures maintained in this medium were found to be different to those seen in cultures grown in serum-supplemented medium. The salient features of these differences were: (1) a reduction in glial cell migration and proliferation, which has subsequently been confirmed using tritiated thymidine and autoradiography (Eccleston *et al.*, 1987); (2) a reduction in the proliferation of fibroblasts, so that few were present in these cultures; (3) the growth of neurites beyond the area of glial cells, on to the culture substratum; and (4) the reduction in the formation of neuronal–glial cell aggregates.

The first observations made on the myenteric plexus explant cultures in defined medium indicated that there was a prerequisite for exposure to serum at the onset of culture (Saffrey and Burnstock, 1984). This had obvious disadvantages for the study of growth factors on enteric neural cells *in vitro*. Later experiments have shown that this serum exposure can be obviated by the use of appropriate culture substrates: plexuses could be explanted directly in serum-free medium if fibronectin or laminin substrates were employed (Saffrey and Burnstock, 1992).

Defined culture media are thus useful in the study of the ENS *in vitro*, and may have specific advantages in this explant culture system. For example, enteric neurite growth may be more readily studied when regeneration occurs in isolation from enteric glial cells.

5. Neurochemical Characterization of Enteric Neurons

a. Neurotransmitter Expression. The ENS contains neurons with a wide range of different neurotransmitter phenotypes (see Furness and Costa, 1987). However, when the first immunohistochemical studies were performed on the gut in the late

1970s, many putative neurotransmitters could only be detected in nerve fibers in the gut wall and their origin was unknown. The explant culture preparations therefore offered a unique means of determining whether these fibers originated in either the myenteric or submucous plexus, or arose from cell bodies which were extrinsic to the gut (see Jessen *et al.*, 1980a,b). Since then, the use of colchicine and microsurgical techniques has not only permitted the routine detection of, for example, peptide-containing neuronal cell bodies in *in situ* preparations, but has also enabled much of the circuitry of the ENS to be elucidated (see Furness and Costa, 1987). The original observations made on the explant cultures of enteric ganglia have thus subsequently been confirmed using *in situ* techniques, with one exception. In early studies, VIP-immunoreactive neurons were not detected in myenteric plexus cultures from the guinea pig cecum, although they were found in such cultures from the colon and ileum and in submucous plexus cultures from the cecum and colon (Jessen *et al.*, 1980a,b). More recently, VIP-immunoreactive neurons have been demonstrated in myenteric plexus cultures from the guinea pig cecum; this difference may have been due to technical reasons, or it may be because this population of enteric neurons has specific properties or requirements *in vitro* (see Saffrey and Burnstock, 1988b). In this respect, it is of interest that VIP-immunoreactive neurons were the only type of peptide-containing neurons that were reduced in number in dissociated cell cultures of the rat gut grown in HCM (Nishi and Willard, 1988). Although no muscle or conditioning factor was present in the explant cultures described above, these observations may suggest that expression of VIP or survival of this particular subpopulation of myenteric neurons is labile in culture. However, it also serves to illustrate that caution is necessary in the interpretation of results obtained in culture.

Subpopulations of cultured myenteric neurons from the newborn guinea pig cecum have also been found to contain the peptides substance P (SP), ENK, and somatostatin (SOM), as has been described in *in situ* preparations from this part of the gut (Saffrey and Burnstock, 1988a,b). Other populations have been found to be immunoreactive for 5-HT, although the nature of these cells has not been unequivocally demonstrated since they were grown in medium supplemented with serum that contained high levels of 5-HT (Saffrey *et al.*, 1984; see Section II,C,4). Even very small subpopulations of neurons, such as those containing catechol-amines in the myenteric plexus of the guinea pig proximal colon, have been found to be present in explant culture (Jessen *et al.*, 1983b).

The techniques for the isolation of the myenteric plexus and its maintenance in culture were instrumental in the first identification of γ-aminobutyric acid (GABA) as an enteric neurotransmitter (Jessen *et al.*, 1979; Jessen, 1981). Isolated myen-teric plexuses from the newborn guinea pig cecum were used to demonstrate unambiguously that the GABA-synthesizing enzyme glutamic acid decarboxylase (GAD) is present within enteric ganglia and that GABA can be synthesized and accumulated by enteric ganglion cells. Myenteric plexus explant cultures were also used, in parallel to *in situ* preparations, to show that a subpopulation of enteric neurons have high-affinity uptake sites for GABA (Jessen et al., 1979). Subsequently, populations of such neurons were also demonstrated in isolated myenteric plexuses and explant cultures from other parts of the guinea pig gut and from the intestines of other species (Fig. 1; Saffrey et al., 1983). Additional evidence for the role of GABA as an enteric neurotransmitter was also obtained by the demonstration of depolarization-induced release of preloaded GABA from parallel culture and in situ preparations (Jessen et al., 1983c).

The ability to clearly visualize myenteric neurons which had taken up [³H]GABA by autoradiographic techniques, at all stages of culture, enabled an estimate of the survival of these cells in vitro to be performed (Jessen et al., 1983b). This could not be considered an accurate analysis, since it was impossible to explant identical segments of plexus to provide sister cultures for different time points (the enteric ganglia are very heterogeneous in the guinea pig cecum). To control for this, the numbers of neurons expected to be present at the onset of culture had to be related to the number of gaps (or holes) in the plexus network of each explant. Nevertheless, this study indicated that there was no loss of this population between days 1 and 5 in vitro, but that the numbers of neurons that took up GABA decreased between days 5 and 15 of culture. Unfortunately, it is not possible to perform similar estimates for other neuronal subpopulations or for the total neuronal population: the antisera available do not consistently permit the resolution of individual cell bodies at the early stages of culture when neurons are closely associated, so the initial numbers of neurons present cannot be determined.

Recent immunohistochemical studies have demonstrated that multiple neurotransmitter molecules are colocalized in single enteric neurons in situ (see Furness and Costa, 1987). While exhaustive studies have not been performed on explant cultures of enteric ganglia, double-labeling experiments have revealed that cultured myenteric neurons may express more than one putative transmitter. For example, a subpopulation of myenteric neurons from the guinea pig cecum has been found to contain the peptides ENK and SP (M. J. Saffrey, unpublished observations). The explant culture system may thus be suitable for analysis of the factors regulating the differential expression of putative transmitters found within single neurons.

Observations to date indicate that myenteric and submucous plexus explant cultures from the guinea pig cecum, ileum, and colon contain subpopulations of enteric neurons present in situ. However, there may be some changes in the proportions of the different neuronal types in culture, either due to selective loss or survival of particular populations, or to changes in neurochemical expression in culture. As an accurate determination of the proportions of neuronal subpopulations in these cultures is not feasible, such possibilities cannot be eliminated.

b. Localization of Muscarinic Receptors.   Enteric neurons in situ express receptors for many different neurotransmitters, frequently with a wide range of receptor subtypes (see Furness and Costa, 1987). While autoradiographic techniques applied to tissue sections can identify ligand-binding sites on enteric ganglia, the resolution of this method is not great enough to localize binding sites to particular cells within the ganglia. The explant cultures of isolated ganglia thus provide a useful system for the study of the expression of receptors by individual ganglion cells. Further, the distribution of receptors on different parts of individual neurons can be examined. The explant culture system has therefore been employed to examine the expression of muscarinic receptors by myenteric neurons from the guinea pig cecum. The irreversible muscarinic ligand propylbenzilylcholine mustard (PrBCM) was used in tritiated form to localize muscarinic receptors on the cultured neurons by autoradiographic techniques. In keeping with electrophysiological studies, approximately 10–20% of the neuronal cell bodies were labeled by this method. Neurites were also labeled, both around neuronal cell bodies in the central part of the cultures and in the outgrowth area (Buckley and Burnstock, 1984). These observations were then extended to examine the distribution and density of muscarinic receptors on single myenteric neurons identified immunocytochemically by the use of an antibody raised against brain synaptic membranes (anti-CTX). This study demonstrated that muscarinic receptors were concentrated over the neuronal somata, where their density was between 30 and 100 receptors per μm^2, and over proximal neurites (Fig. 3). Varicosities and intervaricose regions of distal neurites were discontinuously labeled, whereas receptors were consistently expressed on growth cones (Buckley and Burnstock, 1986).

The application of combined autoradiographic and immunocytochemical techniques to explant culture preparations offers a valuable means of further characterizing the neurochemical properties of neurons which express muscarinic receptors. In a study using antisera to several putative peptide transmitters, it was found that many of the VIP-immunoreactive neurons present in myenteric plexus explant cultures from the newborn guinea pig cecum express muscarinic receptors, while few SP-, ENK-, or SOM-containing cells were autoradiographically labeled. The majority of the cultured myenteric neurons that express muscarinic receptors, however, were not labeled by antisera to these four peptides (Buckley et al., 1988).

6. Characterization of Enteric Glial Cells

The observation that enteric glial cells, unlike other peripheral glia, showed similar features to astrocytes in the CNS was based on ultrastructural studies (Gabella, 1971, 1981). The first evidence that enteric glia, like astrocytes, expressed GFAP, was obtained by immunolabeling explant cultures of the myenteric plexus by Jessen and Mirsky (1980), who suggested that the enteric glia should be classed as a separate class of glial cell. The difference in the morphology and growth patterns of enteric glial cells and other peripheral glial cells in vitro supported this suggestion. The explant cultures of the myenteric plexus were subsequently used to analyze the molecular phenotype of the enteric glia, using antibodies to both

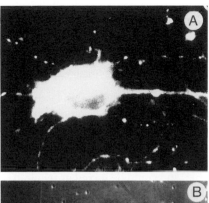

FIG. 3 Fluorescence (A) and corresponding Nomarski (B) micrographs showing the distribution of muscarinic receptors (B) on an immunocytochemically identified (anti-CIX, A) neuronal cell body and its proximal neurites in a myenteric plexus explant culture from newborn guinea pig cecum. Bar, 10 μm. Adapted from Buckley and Burnstock (1986) with permission.

intracellular and cell surface molecules (Jessen and Mirsky, 1983; Bannerman *et al.*, 1987, 1988a). The cultures were particularly suitable for this purpose, since they enabled the expression of intracellular and cell surface molecules to be studied simultaneously and unambiguously in individual cells. Thus, the enteric glia were found to express a number of molecules also expressed by astrocytes, but yet other molecules also expressed by Schwann cells. Furthermore, the antigenic profile of the enteric glial cells changed in culture (Jessen and Mirsky, 1983; Bannerman *et al.*, 1988a). The enteric cultures may thus offer a means of studying the regulation of glial cell phenotypic expression.

The study of enteric glia has since concentrated on factors affecting the proliferation of these cells, and has again employed the explant culture system (Eccleston *et al.*, 1987). Enteric glial cells have been found to divide at a higher rate than Schwann cells in culture, but, like Schwann cells, their proliferation is stimulated by fibroblast growth factor and glial growth factor but not by epidermal growth factor. Both cell types proliferate more rapidly on laminin or fibronectin

and on extracellular matrix produced by corneal endothelial cells than on poly-
L-lysine substrates.

7. Manipulation of Myenteric Plexus Explants *in Vitro* to Produce Enriched Neuronal or Glial Cell Cultures

Unambiguous analysis of the effects of growth factors or other molecules on the
proliferation of enteric glia, or on other properties of either glia or neurons,
requires pure populations of these cells, or as near to pure as can be achieved. In
mixed cell cultures, direct and indirect effects cannot be distinguished. An ideal
situation would therefore be one in which separate, pure populations of all relevant
cell types were available, and could be mixed as required. Isolation of completely
pure populations of enteric neurons or glial cells for primary culture, however, has
not yet been possible. The enriched neuronal populations obtained by dissociation
of LM-MP-CM strips are contaminated by populations of smooth muscle cells and
glia (Nishi and Willard, 1985). Techniques for enriching myenteric plexus explant
cultures for either neurons or glial cells by manipulation of the explants while in
culture have therefore been developed.

The characteristic formation of a central area of neurons and glial cells surrounded by an area of glial cells and neurites in myenteric plexus cultures after
some 7–10 days *in vitro* (see above) has been utilized to produce cultures enriched
for enteric neurons or glial cells (Bannerman *et al.*, 1988b). Purified glial cultures
were obtained by the combined use of three methods: application of mitotic
inhibitors to eliminate fibroblasts, excision of the central neuronal area, and
antibody-mediated cytolysis of remaining neurons and fibroblasts with antibodies
to Thy-1 and complement. The resulting cultures contained >98% glial cells.
Neuronally enriched cultures were obtained by the use of mitotic inhibitors and
antibody-mediated cytolysis alone, utilizing a monoclonal antibody LB1 that
recognizes an antigen present on glial cells. The cultures of purified neurons were
obtained after 4–6 weeks in culture, and consisted of ~87% neurons. The enriched
neuronal cultures were found to include cells that expressed VIP-, SP-, and
SOM-immunoreactivity, and 5-HT-immunoreactive neurons were also present
(Bannerman *et al.*, 1988b).

Cultures enriched for enteric neurons or glial cells have been used to study the
effects of enteric neurons on the proliferation of enteric glial cells and Schwann
cells. The enteric neurons have been found to exert an inhibitory influence on the
proliferation of both these cell types, and thus to differ from sensory neurons,
which stimulate their proliferation (Eccleston *et al.*, 1989).

D. Dissociated Cell Cultures of Isolated Enteric Ganglia

There are a number of disadvantages associated with explant cultures of isolated
enteric ganglia. Although such cultures do not contain smooth muscle, separation

of pure populations of neurons and glial cells at the onset of culture is not possible. Enriched cultures of neurons or glial cells produced by manipulation of explants can only be obtained after several weeks *in vitro*, so that contacts and interactions between the two cell types and between different types of enteric neurons cannot be prevented. Furthermore, accurate quantitation of the number of neurons and glial cells present in the isolated plexuses and in the culture preparations is not possible. An alternative technique for the culture of isolated enteric ganglion cells has therefore recently been developed, in which the isolated plexuses are enzymatically dispersed into a cell suspension (Saffrey *et al.*, 1991). An advantage of this preparation is that it is possible to count the cells present and hence perform accurate quantitative studies on the effects of various agents on, for example, neuronal survival. Separation of enteric neurons and glial cells in suspension may also be possible by preplating and/or antibody-mediated cytolysis.

The dissociated myenteric plexus cultures have been grown in both serum-supplemented and serum-free, hormone-supplemented culture media. Immunocytochemical studies using antiserum to the neuronal marker PGP 9.5 have shown that neurons representative of the different morphological types described *in situ* and in explant cultures are present in the dissociated cell cultures. Neurons expressing VIP, SP, or ENK have also been demonstrated in these cultures (Fig. 4). The ability to dissociate isolated enteric ganglia, count the neurons present in the resulting suspension and then later in culture, and to grow the cells in defined culture conditions promises to be a useful approach for the study of the ENS.

E. Electrophysiological Studies of Enteric Neurons in Culture

Unlike the other ganglia covered in this review, there have been extensive electrophysiological studies of the neurons which comprise the myenteric and submucous plexuses of the mammalian ENS using acutely dissected preparations (by North, 1982; Wood, 1984). Almost all of these studies have been facilitated by the development of very successful intact ganglion preparations and, more recently, acutely dissociated preparations of enteric neurons (Nishi and North, 1973; Hirst *et al.*, 1974; Derkach *et al.*, 1989; Tatsumi *et al.*, 1990). These have provided us with a wealth of data regarding the specific electrophysiological characteristics of the cells within these ganglia and of the pre- and postsynaptic actions of a wide variety of neurotransmitters and neuromodulators.

In addition to these studies, a number of investigations have been carried out using primary cell culture preparations. Both explant and dissociated mixed cell cultures of enteric ganglia have been utilized for electrophysiological recording; studies of the individual neurons in these preparations have been performed to determine their electrophysiological, neurochemical, and synaptic properties in culture.

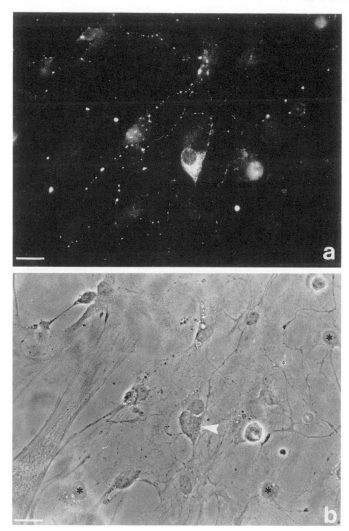

FIG. 4   Immunofluorescence (a) and phase-contrast (b) micrographs showing the same field of a dissociated cell culture of isolated myenteric plexus from the newborn guinea pig cecum. The culture was labeled with antiserum to VIP after 7 days *in vitro*. A single immunopositive neuron and varicose processes are visible in a. The labeled neuron is indicated by an arrowhead in b, where it can be seen to be lying on an area of glial cells (examples of glial cell nuclei are indicated by asterisks:). Bars, 25 µm. Taken from Saffrey *et al.* (1991) with permission.

1. Basic Electrical Properties

The first report of intracellular recordings of cultured enteric neurons was made from myenteric plexus explant cultures from newborn guinea pig cecum (Jessen *et al.*, 1978). In this initial study, two cell types were distinguished, which were similar to those described in intact ganglia. Following this, a brief study of human embryonic myenteric plexus cells in dissociated cell culture was also carried out (Maruyama, 1981). In the latter study, two cell types were again distinguished morphologically, and also on the ionic basis of their action potentials, with one cell type being capable of generating action potentials in the presence of tetrodotoxin (TTX). Interestingly, some of the neurons in these cultures were reported to proliferate, while others were found to be electrically coupled (Maruyama, 1981, 1985). In subsequent, more detailed intracellular studies of neurons in guinea pig myenteric plexus explant cultures (Hanani *et al.*, 1982), 35% of cells were found to display tonic firing characteristics similar to the S/type 1 cells seen in intact adult ganglia (Nishi and North, 1973; Hirst *et al.*, 1974). However, while the remaining cells were quite refractory, like the AH/type 2 cells in ganglia *in situ*, very few exhibited the slow spike afterhyperpolarization (sAHP) characteristic of this cell type, which make up approximately 40% of all neurons in intact ganglia (Hanani *et al.*, 1982). A similar situation has also been reported in dissociated mixed cell cultures containing rat enteric neurons (Willard and Nishi, 1985a). In freshly dissected rat duodenal myenteric plexus, 25% of all neurons display a sAHP of up to 10 sec duration (Brookes *et al.*, 1988), whereas in the cell culture preparation, sAHPs were never seen after a single action potential, although they could be evoked in approximately 50% of cells following a train of action potentials (Willard and Nishi, 1985a). The reason for the lower incidence of cells exhibiting a sAHP in these cultures is not known. In explanted guinea pig myenteric neurons it seems unlikely that it was the result of a low survival rate for AH/type 2 cells, as very little neuronal cell death was observed during prolonged periods in culture (Jessen *et al.*, 1983b). One possibility may be the culture medium in which the cells were maintained. It has been observed that altering the culture conditions has a marked effect on the frequency with which sAHPs are observed in rat sympathetic neurons (O'Lague *et al.*, 1978). Alternatively, it may be that a particular local extracellular environment, which may be disrupted by the changes that occur in culture, is required for a normal sAHP.

2. Neurochemical Properties and Synaptic Interactions

In the guinea pig, investigation of the synaptic interactions occurring between cultured enteric neurons has shown that, as in intact ganglia, almost 50% of the neurons display cholinergic fast excitatory postsynaptic potentials (epsps). Nicotinic fast epsps and, occasionally, slow inhibitory postsynaptic potentials (ipsps) could also be evoked by focal stimulation of nerve tracts in culture, but no slow

epsps could be elicited (Hanani and Burnstock, 1985a). Slow epsps have, however, been elicited from cultured rat enteric neurons (see below). While no slow
epsps could be evoked in cultured guinea pig myenteric plexus neurons, direct
application of SP, a prime neurotransmitter candidate for the slow epsp in these
cells, evoked a slow depolarization similar to that seen in intact ganglia (Fig.
5A,B; Hanani and Burnstock, 1984, 1985a,b). This suggests that the underlying
deficit may have a pre-, rather than a postsynaptic origin and may possibly result
from inadequate recruitment of the relevant synaptic inputs or from down-regula-
tion of transmitter synthesis in the presynaptic neuron.

In addition to evoking a slow depolarization in cultured guinea pig myenteric
neurons, SP also elicited a short-latency (30–50 msec) rapid depolarization in
approximately 30% of the cells studied (Fig. 5C,D; Hanani and Burnstock, 1984).
Similar novel fast peptidergic depolarizations have not been reported in intact
ganglia, and it has been suggested that such responses may have been revealed as
a result of the more rapid, unimpeded access that exogenously applied drugs have
to the cells in culture (Hanani and Burnstock, 1984).

The most extensive studies of the neurochemical and synaptic properties of
cultured enteric neurons have been made in rat preparations. Exogenous application of a variety of neurotransmitters, including ACh, 5-HT, GABA, SP, VIP, and
ENK (Willard and Nishi, 1985b), evoked essentially the same responses as those
described in freshly dissected ganglia (Brookes et al., 1988), indicating that the
cells retain their receptor properties in culture. Furthermore, the similarity between
the electrophysiological and neurochemical properties of the rat neurons and those
of both cultured and intact guinea pig myenteric neurons indicates considerable
homology between the two species (Katayama et al., 1979; Cherubini and North,
1984; Hanani and Burnstock, 1984, 1985a,b; Mihara et al., 1985; Zafirov et al.,
1985).

In the rat, the improved visualization afforded by dissociated culture preparations has also been used to enable pairs of clearly identified rat enteric neurons to
be impaled and their synaptic interactions investigated (Willard and Nishi, 1985b).
These studies have revealed several patterns of synaptic excitation. In 40% of
cells, nicotinic fast epsps can be elicited in response to stimulation of one of the
pair of cells, thereby demonstrating the cholinergic nature of these presynaptic
neurons. The cholinergic neurotransmission in these cells can be modulated by a
variety of transmitter candidates, including SP, VIP, GABA, ENK, and 5-HT
(Willard and Nishi, 1985b). When stimulated repetitively at frequencies between
5 and 20 Hz, a small proportion of these cholinergic neurones (5%) also evoked
slow noncholinergic synaptic potentials in the target cell. In a further 10–20% of
the cells in which no fast epsps could be evoked, high-frequency stimulation
evoked just a slow noncholinergic depolarization (Willard and Nishi, 1985b). In
an elegant series of experiments, Willard has attempted to correlate the peptide
content of the presynaptic neuron (driver cell) with the different types of synaptic
potential elicited in the postsynaptic cell. Three of the potential candidates for

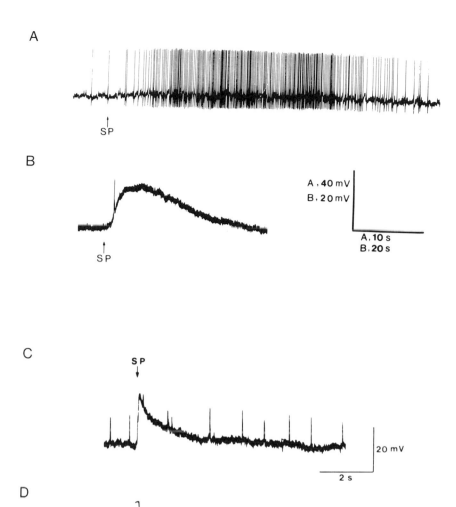

A

B

A , 40 mV
B , 20 mV

A , 10 s
B , 20 s

C

S P

20 mV

2 s

D

10 mV

0.2 s

FIG. 5   Slow (A and B) and fast (C and D) responses of cultured myenteric neurons to SP pressure-injected from micropipettes. (A) SP ($3 \times 10^{-5} M$) evoked a small depolarization which was accompanied by a large increase of spike frequency. Duration of SP injection was 50 msec. (B) In another cell SP ($5 \times 10^{-5} M$) evoked a larger depolarization but only a single spike. Duration of injection was 100 msec. The concentrations given are the original values in the micropipettes. (C) A 50-msec pulse of SP ($5 \times 10^{-5} M$) evoked a fast response. This cell showed spontaneous activity of subthreshold potentials. (D) Recording from another cell at a faster time scale. The top trace indicates the duration and timing of injection of SP ($5 \times 10^{-5} M$). Adapted from Hanani and Burnstock (1984) with permission.

mediating the slow noncholinergic epsp (SP, 5-HT, and VIP) were investigated. In these studies, driver cells, whose stimulation elicited cholinergic fast epsps followed by slow noncholinergic epsps in the same target cell, were termed dual-function neurons (Willard and Nishi, 1987). Immunocytochemical staining revealed that all the presynaptic dual-function neurons displayed VIP-immunoreactivity. Furthermore, the slow epsp in the target cells was mimicked by VIP and reversibly inhibited by VIP but not SP antisera. This indicates that, in cells innervated by dual-function neurons, the slow epsp is mediated by VIP and that VIP is a cotransmitter with ACh in dual-function neurons (Willard, 1990a).

In a similar study of driver cells that evoke only a slow synaptic potential in the target neuron, it was found that 60% displayed SP-immunoreactivity, while none were immunoreactive for VIP or 5-HT (Willard, 1990b). As in other cells, the slow epsp could be mimicked by exogenous VIP and/or SP. However, in contrast to the cells innervated by dual-function neurons, the slow epsp in these cells was selectively inhibited by SP but not VIP antisera. Furthermore, in all cases where the slow epsp could be inhibited by exogenous SP antisera, the driver cells were immunoreactive for SP, indicating that a proportion of noncholinergic driver cells utilize SP to mediate slow synaptic excitation (Willard, 1990b).

3. Whole-Cell Voltage-Clamp and Single Channel Studies

Recent examination of the calcium current in dissociated rat myenteric plexus neurons using the patch-clamp technique has revealed that these cells display two distinct types of calcium channel. One has a slope conductance of 27 pS, is sensitive to dihydropyridines, and shows little inactivation at −40 mV. The second channel type has a unitary conductance of 14 pS, is insensitive to dihydropyridines, but is very sensitive to ω-conotoxin. Unlike the high-conductance channel, this channel displays almost complete inactivation at potentials of less than −40 mV (Hirning et al., 1990). These findings, together with the whole-cell current data, are consistent with there being L- and N-, but not T-type calcium channels in these cells (see review by Bean, 1989) and contrast with what has been described in acutely dissociated submucous plexus neurons, where only a single small-conductance, N-type channel of 12 pS has been described (Surprenant et al., 1990).

Examination of the actions of exogenously applied dynorphin, noradrenaline (NA) and neuropeptide Y (NPY) on the calcium current in myenteric neurons has revealed that they all strongly inhibit the transient component of the current, with NA and NPY acting almost exclusively on the N-type current (Sturek et al., 1987; Hirning et al., 1990). More detailed analyses of the action of NPY, using a combination of patch-clamp and the Fura II microfluorometric technique for measuring changes in intracellular calcium, indicate that inhibition of the calcium current by NPY is mediated via a pertussis toxin-sensitive G-protein by a mechanism that does not appear to involve the generation of a freely diffusible second messenger (Hirning *et al.*, 1990).

III. Intracardiac Neurons in Dissociated Cell Culture

It is surprising that so little is known of the properties and roles of the intrinsic neurons of the mammalian heart since this organ has been the subject of intensive scientific study. However, many investigators have either ignored the presence of intracardiac neurons or have assumed that they function exclusively as simple nicotinic relays of the extrinsic vagal input to the heart. There is now increasing evidence to support the view that the intracardiac neurons play an important role in regulating both the extrinsic parasympathetic and sympathetic nerves in the heart; that they have regional specializations that correlate with their functional innervation of the nodal and conducting tissues; that they innervate the ventricles and coronary vasculature; and that they may even mediate local reflexes (Smith, 1970; Nozdrachev and Pogorelov, 1982; Ardell and Randall, 1986; Blomquist *et al.*, 1987). Unfortunately, many studies of the intrinsic ganglia of the mammalian heart are limited because the neurons are inaccessible to many types of direct investigation. Furthermore, they are quite widely distributed in relatively small subepicardial ganglia throughout the atria and in the interatrial septum, although most occur around the nodal and conducting tissues, the points of venous entry and the coronary sinus (King and Coakley, 1958).

A. Methods

Since the intracardiac ganglia are not easily dissected free from the surrounding cardiac tissue, the range of approaches for their study in culture is limited. While explant cultures of isolated ganglia would be difficult to prepare, organotypic cultures may be more feasible for the routine study of intracardiac neurons. To date, however, dispersal techniques, consisting of enzymatic and mechanical treatments, have been used to prepare cultures for the direct investigation of intracardiac neurons. Dissociated, mixed-cell culture preparations from fetal human and mouse, newborn guinea pig and rat hearts contain intracardiac neurons along with the other non-neuronal cell types that are present in the atria (Lane *et al.*, 1976; Hassall and Burnstock, 1986; Hassall *et al.*, 1988, 1990a; Fieber and Adams, 1991a). Examination of guinea pig preparations showed that vigorous dissociation procedures led to an increased proportion of single intracardiac neurons, but to a lower neuronal yield overall (Hassall and Burnstock, 1986). The method was, therefore, tailored to optimize the yield and survival of intracardiac neurons in culture. This preparation has been the subject of several different types of direct investigation which will now be reviewed.

B. Identification of Intracardiac Neurons in Culture

Intracardiac neurons can be distinguished from the other atrial cells in living cultures by phase-contrast microscopy; the accuracy of the criteria used for the visual identification of these cells has been confirmed at the ultrastructural level and by the use of cell type-specific immunocytochemical markers (Hassall and Burnstock, 1986; Kobayashi et al., 1986a,b; Hassall et al., 1990a,b).

Both mononucleate and binucleate intracardiac neurons are present in cultures from newborn guinea pig and rat hearts (Hassall and Burnstock, 1986; Kobayashi et al., 1986a; Fieber and Adams, 1991a). The occurrence of binucleate neurons is not a result of the conditions of culture, nor does it indicate that the cells are immature since they have also been described in the mature heart in situ and are quite common in other autonomic ganglia (Tay et al., 1984; Baluk and Gabella, 1989a). Studies to date do not suggest that these morphologically distinct neurons necessarily express different neurochemical properties; although there does appear to be a degree of association between the basic electrophysiological characteristics of some guinea pig intracardiac neurons and their size and shape in culture (Allen and Burnstock, 1987).

An ultrastructural study of newborn guinea pig intracardiac neurons in culture has shown that a population of these cells appears to undergo a type of degeneration that may well be due to the loss of their extrinsic nervous input (Kobayashi et al., 1986a). Similar changes have been reported in the intracardiac ganglia of the monkey Macaca fascicularis after vagotomy in vivo (Tay et al., 1984; Wong et al., 1987). This suggests that similar processes may be occurring in the neurons after their extrinsic denervation in vivo and upon isolation in culture. These observations may be of importance for surgical transplantation, where extrinsically denervated hearts are used, since the intracardiac neurons in the donor heart may also be affected in some way by this procedure. With respect to this possibility, there is evidence from the human heart that, although the intracardiac neurons survive after transplantation, they may lose some of their capability for neuropeptide synthesis (Wharton et al., 1990).

All of the human and guinea pig intracardiac neurons observed in culture are imunoreactive for the specific neuronal marker-protein gene product 9.5 (Fig. 6; Hassall et al., 1990a; C.J.S. Hassall, unpublished observations), which facilitates their identification. However, many of these neurons in culture are present in large, rounded groups in which it is impossible to count the number of individual cells accurately. Therefore, in most cases, only estimations of the proportion of intracardiac neurons expressing a particular phenotype in culture can be made. Furthermore, it can often be difficult to assess the survival of specific subpopula-

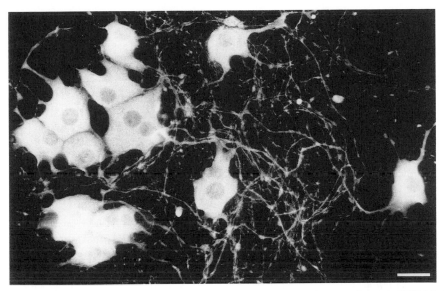

FIG. 6   Fluorescence micrograph of a group of intracardiac neurons that were intensely PGP 9.5-immunoreactive in a 7-day-old culture from newborn guinea pig heart. Bar, 25 μm.

tions of neurons in culture because quantitative studies have not been carried out *in situ* and subtle changes in the properties of the cultured neurons are possible.

The groups of intracardiac neurons in culture also contain non-neuronal cells, such as glia, and probably consist of only partially dissociated ganglia since aggregation of these cells has not been observed (Hassall and Burnstock, 1986; Kobayashi *et al.*, 1986a,b; Hassall *et al.*, 1990a). All of the neurons in cultures prepared from the newborn guinea pig heart appear to be ensheathed by glial cells that are also present in the intracardiac ganglia *in situ*, and therefore may remain under the influence of these supporting cells (Hassall and Burnstock, 1986; Kobayashi *et al.*, 1986a,b). However, human fetal intracardiac neurons in culture are not in such close association with glial cells (Hassall *et al.*, 1990a,b). This probably reflects a difference in the efficacy of the dissociation procedure on the relatively immature ganglia of the fetal human heart compared with the ganglia of the newborn guinea pig heart, rather than a species variation in itself.

C. Neurochemical Characterization

As *in situ*, intracardiac neurons from the newborn guinea pig and rat contain the enzyme acetylcholinesterase (AChE) in culture (see Hassall and Burnstock, 1986; Seabrook *et al.*, 1990). A population of small intensely fluorescent (SIF) or granule-containing cells is present in culture, as *in situ* (Hassall and Burnstock, 1986, 1987a; Kobayashi *et al.*, 1986a). In contrast, catecholamines have not been detected in intracardiac neurons in culture or *in situ* using established techniques, despite the expression of dopamine β-hydroxylase (DBH)-immunoreactivity and catecholamine-handling properties by some intracardiac neurons *in situ* (Jacobowitz, 1967; Dalsgaard *et al.*, 1986; Hassall and Burnstock, 1984, 1986, 1987b; Baluk and Gabella, 1990). Interestingly, however, DBH-immunoreactivity has been shown to be lost from newborn guinea pig intracardiac neurons under the conditions of culture and has not been detected in intracardiac neurons cultured from the fetal human heart (Hassall and Burnstock, 1987b; Hassall *et al.*, 1990a). The reason for this change is not clear, but it may be related to the production of an adrenergic suppressing/cholinergic inducing factor by the non-neuronal cardiac cells that are also present in culture (Potter *et al.*, 1986; Furshpan *et al.*, 1986; Matsumoto *et al.*, 1987).

The presence of a large population of intracardiac neurons that were immunoreactive for the indoleamine 5-HT in culture prompted an investigation of their amine-handling properties, since 5-HT-immunoreactive neurons have not been demonstrated in the heart *in situ* and the serum used to supplement the growth medium contains high levels of 5-HT (Saffrey *et al.*, 1984). By employing a serum-free, defined culture medium, it was found that the intracardiac neurons were 5-HT-immunoreactive as a result of the uptake of exogenous 5-HT from the growth medium, although they were also capable of synthesizing 5-HT from its immediate precursor, 5-hydroxytryptophan (Hassall and Burnstock, 1987a). It is not clear whether this feature of some newborn guinea pig intracardiac neurons is confined to cells in culture because a recent study of these neurons *in situ* was not able to demonstrate the uptake and synthesis of 5-HT (Baluk and Gabella, 1990). Unfortunately, these data cannot easily be compared with the results from the culture experiments, where it was found that exogenous 5-HT levels of at least 10^{-6} M were required for the visualization of 5-HT-immunoreactivity in intracardiac neurons (Hassall and Burnstock, 1987a). Nevertheless, since some intracardiac neurons are amine-handling, but do not appear to store endogenous amines in culture or *in situ*, care should be taken in the interpretation of studies at the ultrastructural level where 5-hydroxydopamine is used to label cells because the presence of such a population of intracardiac neurons does not necessarily indicate that they are adrenergic.

There are interesting species differences in the expression of neuropeptides by intracardiac neurons. As *in situ*, many newborn guinea pig intracardiac neurons in culture contain NPY and its sister peptide, the C-terminal peptide of NPY (C-PON), and no SOM-immunoreactive neurons have been demonstrated (Hassall and Burnstock, 1984, 1986, 1987b; Dalsgaard *et al.*, 1986; C. J. S. Hassall, unpublished observations). Conversely, in the human heart in culture and *in situ*, SOM-immunoreactive neurons are numerous, while NPY/C-PON-immunoreactive neurons are rarely seen in culture and have not been reported *in situ* (Fig. 7;

Franco-Cereceda *et al.*, 1986; Hassall *et al.*, 1990a). These two neuropeptides may be used by intracardiac neurons of the different species to perform similar roles.

Another difference in the neurons from guinea pig and human fetal heart in culture is that approximately 40% of guinea pig intracardiac neurons contain both 5-HT- and NPY-immunoreactivities (Hassall and Burnstock, 1987b), while 5-HT and NPY coexist in less than 1% of human fetal intracardiac neurons (Fig. 7; Hassall *et al.*, 1990a). In both cases, it appears that many, if not all, of the intracardiac neurons that contain NPY are also 5-HT-immunoreactive, so that the difference in the proportions of this type of neuron may simply reflect the comparatively low incidence of NPY-immunoreactive neurons in the human fetal cultures (Hassall *et al.*, 1990a). In the guinea pig cultures, 5-HT is taken up by intracardiac neurons from the growth medium (Hassall and Burnstock, 1987a); similarly, the human fetal intracardiac neurons in culture may also have an uptake mechanism for 5-HT.

The apparent absence of SP- and calcitonin gene-related peptide (CGRP)-immunoreactive neurons in culture is consistent with the rare occurrence of neurons containing these peptides in the guinea pig heart *in situ* (Gerstheimer and Metz, 1986; Hassall and Burnstock, 1986; Baluk and Gabella, 1989c; C. J. S. Hassall, unpublished observations). Similarly, a very small proportion of newborn guinea pig intracardiac neurons are immunoreactive for VIP in culture, and this reflects the low incidence of neurons that contain this peptide *in situ* (Forssmann *et al.*, 1982; Hassall and Burnstock, 1986). Although ENK- and neurotensin-immunoreactive nerve fibers have been demonstrated in the guinea pig heart, they are likely to be of extrinsic origin since intracardiac neurons that contain these peptides have not been observed *in situ* or in culture (Weihe *et al.*, 1983, 1984; Hassall and Burnstock, 1986). These findings illustrate the use of culture preparations to help to determine the origin of particular nerve fibers within the heart: since only neurons intrinsic to the heart are present in culture, the properties of intracardiac neurons can be distinguished from those of extrinsic neurons that project to the heart.

D. Autoradiographic Localization of Receptors

Use of the irreversible muscarinic receptor antagonist PrBCM has revealed that all intracardiac neurons in dissociated cell cultures of the newborn guinea pig heart express muscarinic receptors (Fig. 8; Hassall *et al.*, 1987). As discussed in the next section, these receptor sites have been shown to be heterogeneous and to mediate several different responses to muscarinic agonists (Allen and Burnstock, 1990a). Since the drugs currently available to distinguish muscarinic receptor subtypes pharmacologically are not selective enough to identify the five cloned muscarinic receptors unequivocally (see Hulme *et al.*, 1990), studies to determine whether these five genes are expressed by intracardiac neurons in culture using *in situ*

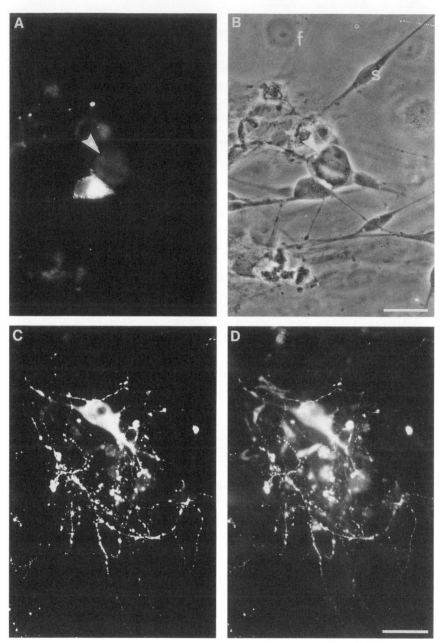

FIG. 7 (A and B) Fluorescence (A) and corresponding phase-contrast (B) micrographs of a large nonimmunoreactive intracardiac neuron (arrowhead) and a relatively small, brightly SOM-immuno-reactive intracardiac neuron from 20-week human fetal atria in culture for 6 days. Fibroblasts (*f*) and Schwann cells (*s*) were not immunoreactive. (C and D) Fluorescence micrographs of NPY- (C) and 5-HT- (D) immunoreactivities present in the same intracardiac neuron from a 20-week human fetal heart in culture for 9 days. Bars, 25 μm. Adapted from Hassall *et al.* (1990a) with permission.

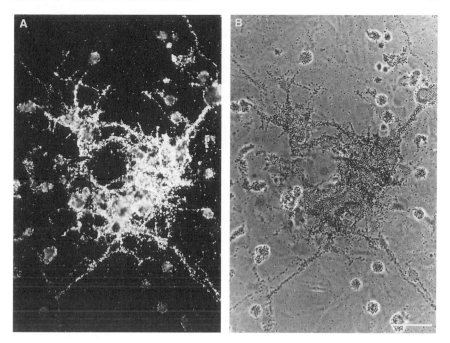

FIG. 8 Dark-field (A) and corresponding phase-contrast (B) micrographs to show the autoradiographic grains that demonstrate the presence of muscarinic receptors on a group of newborn guinea pig intracardiac neurons in culture for 14 days before incubation with [³H]PrBCM. Bar, 25 μm.

hybridization are currently underway. Preliminary results indicate that most, if not all of the intracardiac neurons in culture express four of the five muscarinic receptor genes (C. J. S. Hassall, unpublished observations). Furthermore, these four genes are also expressed by rat intracardiac neurons *in situ* (N.J. Buckley, personal communication), where muscarinic receptors have been also been demonstrated by autoradiography (Hancock *et al.*, 1987). These findings, together with electrophysiological data (Allen and Burnstock, 1990a), indicate that muscarinic mechanisms are extremely important in the control of the activities of the intracardiac neurons.

The presence of non-neuronal cardiac cells in the mixed-cell culture preparations facilitates concurrent studies of their receptors as well as those of the intracardiac neurons. Hence, muscarinic receptors are also present on atrial myocytes in culture (Hassall *et al.*, 1987). Conversely, subpopulations of glial cells, fibroblasts, and endothelioid cells, but not myocytes or intracardiac neurons, express binding sites for atrial natriuretic peptide in culture (James *et al.*, 1990a).

There has been a brief report that SP acts on guinea pig intracardiac neurons (Konishi *et al.*, 1985) and, correspondingly, SP-binding sites have been visualized in association with guinea pig intracardiac ganglia *in situ* (Hoover and Hancock, 1988). However, the proportion of intracardiac neurons that express these receptors *in situ* has not yet been determined. Recent work indicates that a only subpopulation of newborn guinea pig intracardiac neurons express SP-binding sites in culture (Hassall *et al.*, in preparation). Both labeled and unlabeled intracardiac neurons are seen in the same group of cells in culture and, since these groups probably consist of only partially dissociated ganglia, this finding would suggest that intracardiac neurons that express SP receptors are present in the same ganglia *in situ* as neurons that do not. This corresponds to the observation that, within individual ganglia of the guinea pig heart *in situ*, some intracardiac neurons are not innervated by SP-immunoreactive fibers while others are associated with such fibers (Baluk and Gabella, 1989c).

E. Electrophysiological Studies of Intracardiac Neurons and Their Receptors

Until comparatively recently, almost nothing was known of the electrophysiological characteristics or receptor subtypes expressed by intracardiac neurons of any species. The first electrophysiological studies of these cells *in situ* were made in the amphibian heart, where ganglia can be clearly visualized within the transparent interatrial septum (see Topchieva, 1965; Dennis *et al.*, 1971). In mammals, visualization of the intracardiac ganglia is far more difficult and there are few direct electrophysiological studies of these neurons either *in vivo* or *in situ*. Extracellular recordings have been made from cat and dog intracardiac ganglia (Nozdrachev and Pogorelov, 1982; Gagliardi *et al.*, 1988), and limited intracellular studies have been carried out in the guinea pig, rat, and dog (Konishi *et al.*, 1984, 1985; Seabrook *et al.*, 1990; Xi *et al.*, 1991). The use of culture preparations affords a unique opportunity to overcome many of the difficulties associated with the study of intracardiac ganglia *in situ* by allowing direct visualization and recording from single neurons free from overlying tissue and without the problems of recording from ganglia lying on a contractile bed of cardiac muscle.

In dissociated cell cultures from the guinea pig, three types of intracardiac neuron have been identified (Allen and Burnstock, 1987). One type, comprising 65–75% of the neurons studied, is highly refractory and generally discharges only a single action potential in response to prolonged intrasomal current stimulation, these have been termed AH_s cells. Action potentials in these cells are sodium- and calcium-dependent and have a characteristic calcium-dependent plateau on their repolarization phase. In the presence of tetraethylammonium (1–3 mM) and tetrodotoxin (TTX; 300 nM), they are capable of firing regenerative calcium action potentials. The entry of calcium during the action potential results in activation of a calcium-dependent potassium conductance ($g_{K,Ca}$), which results in the generation of a pronounced (8–22 mV) and prolonged (0.2–3 sec) afterhyperpolarization. A further 10–15% of guinea pig intracardiac neurons, termed AH_m cells, have similar calcium-dependent action potentials and spike afterhyperpolarizations, but are capable of brief (100–400 msec) bursts of firing at the onset of current stimulation. A third type of neuron, termed M cells, which constitutes 10–15% of the cells, is highly excitable and discharges nonaccommodating trains of TTX-sensitive, non-calcium-dependent action potentials. The action potentials in these cells have no plateau on their repolarization phase and are not followed by prolonged calcium-dependent afterhyperpolarizations.

The basic spiking characteristics of rat intracardiac neurons in culture have yet to be fully investigated. However, limited recordings from the ganglia *in situ* have detected rat intracardiac neurons with delayed spike repolarization and calcium-dependent afterhyperpolarizations, which may indicate that they have some similarity with the predominant cell types described in the guinea pig (T. G. J. Allen, unpublished observations).

Preliminary studies of the underlying membrane currents involved in regulating and determining the firing characteristics of cultured guinea pig intracardiac neurons have been carried out (T. G. J. Allen, unpublished observations). These reveal that, in addition to a calcium-dependent potassium current ($I_{K,Ca}$) responsible for the sAHP, the majority of AH-type cells also exhibit an M-current (Fig. 9; I_M; Brown and Adams, 1980) which acts to oppose prolonged depolarization. Together, the actions of I_M and $I_{K,Ca}$ are likely to be responsible for much of the observed refractory firing behavior of these cells. In addition, a small population of AH-type cells also display a transient outward current, similar to the A-current (Connor and Stevens, 1971), which would act to inhibit further the rapid discharge of multiple action potentials in these cells. In a number of AH-type intracardiac neurons, membrane hyperpolarization reveals a time- and voltage-dependent inwardly rectifying current, similar to the Q-current described in other neurons (Halliwell and Adams, 1982; Allen and Burnstock, 1990b). This current acts to oppose prolonged membrane hyperpolarization and, together with the M-current, may, at least in part, be responsible for setting the resting membrane potential in these cells.

Further evidence of the complex nature of mammalian intracardiac ganglia has been revealed from studies made in culture of the actions of exogenously applied neurotransmitters.

1. Cholinergic Receptors

Focal application of cholinergic agonists to the soma of cultured guinea pig and rat intracardiac neurons evokes both nicotinic and multiple muscarinic receptor-mediated responses (Allen and Burnstock, 1990a; Fieber and Adams, 1991a).

a. Nicotinic Receptors.   The nicotinic depolarization produced in intracardiac neurons by ACh is similar to that reported in other autonomic ganglion cells, both in culture and *in situ* (Dennis *et al.*, 1971; Nishi and North, 1973; Hirst and McKirdy, 1975). Detailed analysis of the ACh-evoked nicotinic currents in cultured rat intracardiac neurons (Fieber and Adams, 1991a) showed that the whole-cell current has a brief latency (approximately 10 msec) that is consistent with integral receptor/ion channel coupling, displays marked inward rectification, and has a reversal potential of −3 mV. Single-channel studies made from cell-attached and excised patches in symmetrical CsCl-containing solutions revealed slope conductances for the underlying channels of 32 and 38 pS, respectively (Fieber and Adams, 1991a). These values are similar to those reported for other neurons (Mathie *et al.*, 1987). Correlation of the amplitude of the whole-cell currents with that of the unitary current in rat cultured neurons indicates a nicotinic receptor density of approximately 0.5 channels/μm^2, which corresponds to >10^3 receptor channels per cell (Fieber and Adams, 1991a).

An interesting feature of the individual nicotinic receptor channels of these intracardiac neurons is their permeability to calcium ions. In common with nicotinic channels at both the neuromuscular junction and also in peripheral and central neurons, the channels are highly selective for cations, with a relative permeability for chloride ions with respect to sodium ions (P_{Cl}/P_{Na}) of 0.05. However, unlike the receptor at the neuromuscular junction, calcium permeability is high, with a P_{Ca}/P_{Na} value of 0.93 (Fieber and Adams, 1991a), compared with a value of 0.2 at the frog end plate (Adams *et al.*, 1980). This may indicate a possible role for calcium influx during nicotinic receptor activation in the regulation of second messenger pathways and other neuronal functions in these ganglia.

b. Muscarinic Receptors.

b. Muscarinic Receptors.   Receptor localization studies have demonstrated the presence of muscarinic receptors on all guinea pig intracardiac neurons in culture (Hassall *et al.*, 1987). In agreement with this, electrophysiological studies of these neurons have shown that, out of a population of more than 200 AH-type cells studied, which is the predominant cell type in these cultures (>80%), all displayed functional muscarinic receptors (Allen and Burnstock, 1990a). The overall effect of muscarinic receptor activation was to greatly enhance prolonged action potential discharge in these otherwise highly refractory cells: cells could frequently be induced to fire for up to 30 sec, with sustained rates of firing of 2–12 Hz in response to low-intensity current stimulation (50–200 pA).

The actions of muscarinic agonists on individual guinea pig neurons in culture are mediated by different receptor subtypes and result in activation and/or inhibition of several distinct membrane ion conductances. Under current-clamp conditions, focal application of muscarine to the cell soma typically elicits an initial brief membrane hyperpolarization, which is associated with an increase in membrane conductance; this is followed by a slower, more prolonged depolarization and a fall in membrane conductance resulting from inhibition of an M-current that is partially activated at rest (Fig. 9; Allen and Burnstock, 1990a). The initial inhibitory response has a similar ionic basis to that of the slow ipsp described in

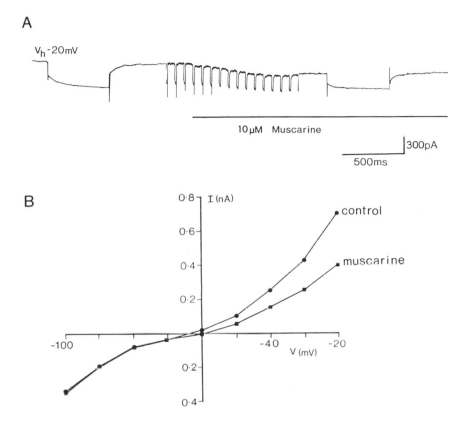

A

V_h -20mV

10 µM Muscarine

300pA

500ms

B

0·8 ┐ I (nA)

• control

0·6 ┤

0·4 ┤   ■ muscarine

0·2 ┤

-100   -40   -20
                              V (mV)

0·2 ┤

0·4 ┘

FIG. 9   Inhibition of the M-current by muscarine in an AH-type guinea pig intracardiac neuron that had been maintained in culture for 7 days. (A) The cell was voltage-clamped at –20 mV and subjected to 0.5-sec duration voltage jumps to –40 mV at 5-sec intervals. The expanded traces show evoked current responses under control conditions and in the presence of 10 µM muscarine. Under control conditions, the hyperpolarizing voltage steps evoked a near instantaneous flow of current through a combination of leak and open M-channels, followed by a characteristic slow inward current relaxation as M-channels gradually closed in response to membrane hyperpolarization. Application of muscarine caused closure of a proportion of the M-channels open at –20 mV, resulting in a decrease in the steady outward holding current, a fall in membrane conductance, and a consequent decrease in the size of the instantaneous current flow at the start of the voltage step. In addition, the amplitude of the slow inward current relaxation during the voltage step was smaller, as many of the M-channels had already been closed as a result of muscarinic receptor activation. (B) The steady-state current–voltage relationship for the cell in A under control conditions and in the presence of muscarine (10 µM). Muscarine clearly reduced the outward rectification which resulted from the voltage-dependent opening of M-channels at depolarized membrane potentials.

amphibian cardiac ganglia (Hartzell *et al.*, 1977) and may indicate the presence of similar slow ipsps in intact mammalian intracardiac ganglia. The muscarine-induced membrane hyperpolarization in cultured cells is inhibited by AF-DX 116 and 4-DAMP, but is only weakly antagonized by pirenzepine, indicating that it is mediated through an M_2 cardiac receptor subtype (M_2). In contrast, the subsequent M-current inhibition and resulting slow depolarization are selectively antagonized by the M_1 receptor antagonist pirenzepine. Although commonly found in sympathetic and central neurons, the presence of a muscarine-sensitive M-current has not previously been reported in any mammalian intramural ganglion cell. In sympathetic ganglia, M-current inhibition has been linked to the generation of the slow epsp (Brown *et al.*, 1986), and the presence of this muscarine-sensitive current in guinea pig intracardiac neurons may indicate that there is a similar slow synaptic excitatory pathway involved in vagal innervation of the mammalian heart.

A further, as-yet-uncharacterized muscarinic receptor-mediated depolarization, associated with a decrease in input resistance was observed in 5% of cultured guinea pig intracardiac neurons (Allen and Burnstock, 1990a). This response may be similar to those reported in bullfrog and rabbit sympathetic ganglia, which were found to result from increases in a mixed cation or chloride conductance, respectively (Akasu *et al.*, 1984; Mochida and Kobayashi, 1986).

In addition to its effects on the resting potential/conductance of cells, muscarinic receptor activation has an inhibitory effect on the calcium-dependent increase in the potassium conductance which underlies the prolonged sAHP (Allen and Burnstock, 1990a). It is likely that this effect is secondary to a reduction in calcium entry during the action potential because inhibition of the sAHP is invariably associated with a concomitant reduction in the amplitude of the calcium action potential. Furthermore, focal application of muscarinic agonists immediately following a train of action potentials fails to inhibit the sAHP, whereas application at any period during the train results in marked inhibition. The muscarinic receptor responsible for mediating this response appears to be of the M_2 glandular (M_3) subtype as it displays a low sensitivity to pirenzepine and AF-DX 116, but is potently antagonized by 4-DAMP (Allen and Burnstock, 1990a).

2. Receptors for ATP and Related Purines

The direct effects of purine compounds on intracardiac neurons have been investigated in guinea pig and rat cultures (Allen and Burnstock, 1990d; Fieber and Adams, 1991b). Adenosine 5′-triphosphate (ATP) and adenosine have long been known to have powerful actions on the mammalian heart: their effects include negative chronotropic and dromotropic actions at the nodes and potent dilation of coronary blood vessels (see Burnstock, 1980). Intracardiac ganglia may play a role in mediating some of these actions since they are associated with the nodes and coronary vasculature; also, large numbers of nerve fibers and intramural neurons in guinea pig and rabbit atria show a positive reaction to quinacrine, indicating a high ATP content (Crowe and Burnstock, 1982).

Approximately 75% of all guinea pig and rat intracardiac neurons in culture are responsive to exogenous application of ATP. In 40% of the guinea pig and all of the ATP-responsive rat intracardiac neurons, focal application of ATP results in a rapid excitatory depolarization similar to the nicotinic fast epsp (Allen and Burnstock, 1990d; Fieber and Adams, 1991b). Under voltage-clamp, ATP evokes a brief latency, rapidly activating inward current, suggestive of close receptor/ion channel coupling, with a reversal potential of -11 mV in the guinea pig and $+10$ mV in the rat. In the guinea pig, the amplitude of the evoked current is linearly related to membrane potential between -90 and -25 mV, however, in the rat, although the ATP-evoked current I/V-relationship is fairly linear in this region, the current displays marked inward rectification at depolarized membrane potentials (up to $+40$ mV) and outward rectification at strongly hyperpolarized potentials. The single ATP-activated channels underlying this response in the rat have a linear I/V-relationship with a slope conductance of 60 pS, with no evidence of any substates to the main channel conductance (Fieber and Adams, 1991b). As in sensory neurons (Krishtal et al., 1988), the ATP-activated channels are selective for cations. In the guinea pig, reducing extracellular sodium or calcium ion concentrations markedly attenuates the ATP-evoked current, whereas reducing extracellular chloride concentration is without effect (Allen and Burnstock, 1990d). In the rat, reduced extracellular sodium concentrations also attenuate the size of the ATP-induced current; however, despite the high relative permeability of the channel to calcium (with a P_{Ca}/P_{Na} value of 1.48 compared with a value of 0.3 for bullfrog sympathetic neurons; Bean et al., 1990), calcium does not readily carry current through the open channel. This is similar to the condition seen in rat sensory neurons, where calcium has also been observed to be a poor carrier of the ATP-evoked current (Krishtal et al., 1988). At present, the relative permeabilities of the ATP-activated channels for calcium and sodium have not been determined for guinea pig intracardiac neurons, but the sensitivity of the ATP-evoked current to removal of calcium in these cells may indicate a greater similarity to the channels found in some smooth muscle cells, where about 10% of the current may be carried by calcium ions (Benham, 1989).

Pharmacologically, the ATP-evoked current in rat and guinea pig intracardiac neurons in culture is very similar (Allen and Burnstock, 1990d; Fieber and Adams, 1991b). The observed rank order of agonist potency of ATP > ADP, with AMP and adenosine being ineffective, together with weak desensitization by α,β-methylene ATP, indicates a receptor belonging to the P_{2Y} subtype (Burnstock and Kennedy, 1985).

In addition to a rapidly activating cation-selective current, a population of AH-type guinea pig intracardiac neurons displayed slower, multiphase responses to ATP in culture (Allen and Burnstock, 1990d). Similar multicomponent responses have not been reported in the rat. However, this may be due to problems associated with the whole-cell patch-clamp recording technique which, by dia-lyzing intracellular contents, is likely to disrupt second messenger-mediated path-ways. The multicomponent responses evoked from guinea pig intracardiac neur-ons generally consisted of an initial transient depolarization followed by a brief hyperpolarization and a slow depolarization. Analysis of the first two components suggests that they result from increases in chloride and potassium conductance, respectively. The mechanism underlying the slow depolarization has yet to be fully elucidated.

Further to these multicomponent responses to ATP, adenosine produced a dose-dependent inhibition of the spike afterhyperpolarization and the underlying calcium-activated potassium current in 40% of guinea pig AH-type neurons. This was associated with a reduction in spike width, in particular the calcium-depen-dent shoulder on the repolarizing phase of the action potential, indicating a possible direct effect on calcium currents through activation of P_1 purinoceptors (Allen and Burnstock, 1990d).

3. Receptors for Noradrenaline

A brief report of the action of noradrenaline (NA) on cultured rat intracardiac neurons has described a shortening of action potential duration as a result of a voltage-dependent decrease in the calcium current (Adams et al., 1987). In these cells, the calcium current appears to be of the predominantly high-threshold type and is potently inhibited by ω-conotoxin (up to 90% inhibition at 10 nM; Seabrook and Adams, 1989).

A similar effect on action potential duration has been observed in guinea pig intracardiac neurons in culture: in particular, NA abolished the calcium-dependent hump in the repolarization phase of the spike and decreased the amplitude and duration of the calcium-activated potassium current underlying the sAHP. Wheth-er NA also has a direct effect on the calcium current has not yet been tested. NA also reduces the amplitude of the M-current and reduces the voltage-independent leak conductance of guinea pig intracardiac neurons (Allen and Burnstock, in preparation).

The finding that there are functional noradrenergic receptors as well as cho-linergic receptors on many individual intracardiac neurons is of considerable importance because it indicates that there may be integration of vagal and sympa-thetic inputs to the heart at the level of the intrinsic ganglia.

4. Receptors for Peptides

To date, studies of the actions of peptides on mammalian intracardiac ganglia have been very limited. In a brief abstract, SOM has been reported to inhibit the calcium current in cultured rat intracardiac neurons (Adams et al., 1987). Guinea pig intracardiac neurons *in situ* respond to SOM and SP, indicating further similarities in the receptors expressed by intracardiac neurons from different mammalian species (Konishi *et al.*, 1984, 1985).

IV. Intramural Ganglia of the Urinary Bladder in Dissociated Cell Culture

Intramural ganglia are not present in the urinary bladder of all mammals; for example, while intrinsic bladder neurons occur in the guinea pig, cat, and man, they are absent from rat and mouse bladders and are very rare in the ferret and rabbit (see Gabella, 1990). However, neurons are present on the surface of the urinary bladder in vesical pelvic ganglia (e.g., de Groat and Kawatani, 1989; Kawatani *et al.*, 1989).

A. Method

As with intracardiac neurons, dissociated cell cultures are required for the direct study of intramural neurons from the newborn guinea pig urinary bladder, vesical neurons are excluded by preparing the cultures from the region of detrusor muscle where the surface ganglia are absent (Pittam *et al.*, 1987). The culture preparation consists of both single neurons and groups, along with nonneuronal cells such as glia, smooth muscle cells, fibroblasts, and epithelial cells (Crowe *et al.*, 1986; Pittam *et al.*, 1987; James and Burnstock, 1988). Several different types of direct study have been carried out on these culture preparations in order to learn more of the properties of the intramural neurons of the urinary bladder.

B. Ultrastructure

A preliminary study (Fig. 10; J. Cavanagh and G. Burnstock, unpublished observations) of the ultrastructure of cells in the dissociated bladder preparations has shown that, as *in situ* (Gabella, 1990), glial cells remain in association with the neurons, but that the glial cell covering is sometimes discontinuous in culture. Mononucleate and binucleate neurons can clearly be seen under phase-contrast optics in culture (James and Burnstock, 1988, 1989, 1990), and their presence has been confirmed at the ultrastructural level. To our knowledge, no binucleate intramural neurons have been observed in the bladder *in situ*, but this could be due to problems in the identification of such cells in sections of ganglia. The cytoplasm of the intramural bladder neurons in culture contains many mitochondria, neurofilaments, rough endoplasmic reticulum, and free ribosomes. Some Golgi

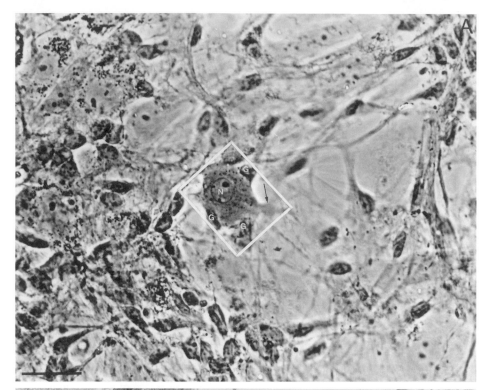

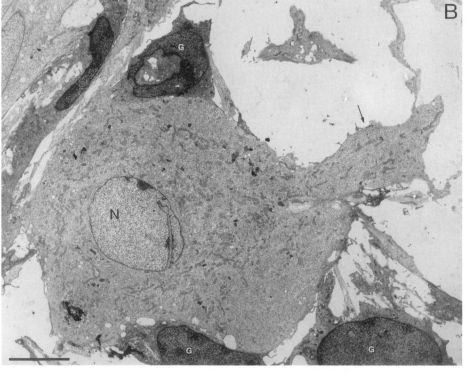

elements are also present. The processes of the cultured neurons contain many microtubules and vesicles, which are sometimes concentrated in terminal regions of the neurons that are closely apposed to other neurons in culture; these areas may represent the synapses that have been detected electrophysiologically (Pittam *et al.*, 1987).

C. Neurochemical Characterization

The neurochemical properties of the intramural bladder neurons in culture have been compared with the properties of these neurons *in situ*, and, as *in situ*, the cultured neurons contain AChE, as well as DBH-, NPY-, VIP-, SOM-, and SP-immunoreactivities (Crowe *et al.*, 1986; James and Burnstock, 1988). Interestingly, many of the intramural bladder neurons contain both NPY- and SOM-immunoreactivities, some contain NPY alone, or NPY together with DBH, while others contain SOM, but not NPY or DBH in culture. Furthermore, there appears to be a degree of association between the morphology of the neurons and their neuropeptide content: neurons that contain NPY (with or without DBH) tend to be binucleate and, conversely, those that contain SOM alone are mononucleate (James and Burnstock, 1990).

The small number of DBH-immunoreactive neurons that are present in the guinea pig bladder *in situ* do not appear to synthesize or store catecholamines, since no catecholamine-containing neuronal cell bodies have been detected *in situ* (Crowe *et al.*, 1986; James and Burnstock, 1988), although catecholamine-containing intramural bladder neurons have been demonstrated in other species (e.g., Hamberger and Norberg, 1965). However, in dissociated bladder cultures, both catecholamine-containing and DBH-immunoreactive neurons are present (Crowe *et al.*, 1986; James and Burnstock, 1988, 1990). Therefore, it is possible that the intramural neurons from the guinea pig bladder only synthesize and store catecholamines when in culture. These findings require further investigation, but they may be related to the loss of extrinsic nervous input to the neurons that occurs with culture, as there is evidence to indicate that local catecholamine-containing bladder neurons in the cat are sensitive to the removal of their extrinsic parasympathetic input *in situ* (Sundin and Dahlström, 1973; see also de Groat and Kawatani, 1989).

Other properties of the intramural bladder neurons in culture also require further investigation: 5-HT-immunoreactive intramural bladder neurons have only been observed in culture and so may be related to the culture conditions and indicate

---

FIG. 10 Phase-contrast (A) and electron (B) micrographs of the same neuron in a dissociated bladder culture. The neuron (*N*) and three closely associated glial cells (*G*) were identified by phase-contrast microscopy (A) and subsequently examined by electron microscopy (B). A neuronal process is indicated by an arrow. Bars: A, 5 μm; B, 25 μm. Micrographs courtesy of J. F. R. Cavanagh.

the presence of a population of amine-handling neurons in the bladder, as in other organs (see Hassall and Burnstock, 1987a; Baluk and Gabella, 1989b, 1990). In addition, the technique of quinacrine fluorescent staining, to visualize cells containing high levels of ATP, labeled neurons in whole mounts of the bladder, but the reaction was nonspecific in the culture preparations (Crowe et al., 1986). Thus, an alternative method will be required to label neurons that may be purinergic in the dissociated cell cultures of the urinary bladder. However, it is already clear from the relatively few studies of the intramural neurons of the urinary bladder in culture carried out so far that they comprise a heterogeneous population containing a variety of neurochemicals with several different patterns of coexistence. If these patterns prove to be exclusive to particular populations of neurons, they may be used to identify and characterize the projections and roles of the different intramural bladder neurons in situ.

D. Autoradiographic Localization of Receptors

The expression of muscarinic ACh receptors by a large population (>75%) of intramural bladder neurons in culture has been demonstrated (James and Burnstock, 1989). The neurons that express these receptors have been further characterized by their SOM content in that 40–60% of the neurons that were SOM-immunoreactive in culture were also labeled with PrBCM. Since these SOM-immunoreactive neurons can be further subdivided into those that are also NPY-immunoreactive and those that do not contain NPY (James and Burnstock, 1990), this raises the possibility that there are at least four different classes of intramural bladder neuron in culture, based on their SOM and NPY content and their expression of muscarinic receptors alone. This further illustrates the heterogeneity of these neurons in culture, which may well reflect their properties and role in situ.

E. Electrophysiology

To date, electrophysiological studies of intramural bladder ganglia in culture have been confined to the guinea pig (Pittam et al., 1987). Intracellular recordings made from the neurons in these cultures indicate that most cells have tonic firing characteristics and are capable of sustaining high rates of discharge (up to 60 Hz) for prolonged periods in response to low-intensity current stimulation. A number of these neurons also displayed spontaneous electrical activity which appeared to be intrinsic rather than synaptic in origin. This activity included slow wave changes in membrane potential and input resistance, as well as spontaneous action potential discharge (Pittam et al., 1987). Many of the neurons also formed nicotinic cholinergic synapses with other neurons in culture; however, it is not known whether similar intrinsic synapses occur between the cells *in situ*. While it has not been determined whether these cells display slow epsps, slow muscarinic depolarizations can be evoked by exogenous application of ACh (B. S. Pittam and G. Burnstock, unpublished observation). In addition, preliminary studies of the actions of exogenously applied neuropeptides indicate that there are populations of intramural bladder neurons that display slow excitatory or inhibitory responses to VIP, SP, SOM, and ENK (Burnstock and Pittam, 1984; Burnstock *et al.*, 1987b).

There is evidence to indicate that purines may have a role in bladder ganglionic transmission. The bladder contains large numbers of quinacrine-positive neuronal cell bodies and nerve fibers in association with the intrinsic ganglia (Crowe *et al.*, 1986), and it has been suggested that ATP is involved in the nonadrenergic, noncholinergic innervation of the urinary bladder smooth muscle (Burnstock *et al.*, 1978). Direct application of ATP on to the cultured postganglionic intramural bladder neurons evokes a brief latency, transient depolarization, similar to the nicotinic epsp, in a significant proportion of the cells (Burnstock *et al.*, 1987b).

V. Airway Ganglia in Culture

In some species, paratracheal ganglia lie in connective tissue on the dorsal surface of the trachealis muscle; similarly, ganglia are present in the adventitial layer of the bronchi (Chiang and Gabella, 1986; Baluk and Gabella, 1989a; Dey *et al.*, 1991). Therefore, these ganglia should not strictly be classified as intramural. However, they are very closely associated with the main body of the airways as a whole, and since intramural neurons have been observed in the ferret trachea (Baker *et al.*, 1986), we are including the ganglia of the airways in the present review. In the adult ferret, paratracheal ganglia can be located in living, unstained tissue at the light microscope level because they are surrounded by fat (Cameron and Coburn, 1984), but often visualization of the ganglia requires the use of techniques such as staining with vital dyes or observation under modulation contrast optics (e.g., Skoogh *et al.*, 1983; Allen and Burnstock, 1990b). Furthermore, as with all of the other ganglia discussed here, there are problems in distinguishing the properties of the nerves of intrinsic and extrinsic origin in the airways *in situ* without the use of surgical or chemical denervation procedures. Therefore, culture preparations provide a useful tool to supplement and extend studies of local airway neurons *in situ*.

A. Methods

Since paratracheal ganglia are not deeply embedded in the trachealis muscle and they lie in a network that can be followed fairly easily, it should be possible to prepare explant cultures of single, isolated paratracheal ganglia. In practice, the relatively small size of most of the ganglia precludes this approach, but a dissociated cell culture preparation has been used successfully (Fig. 11; Burnstock *et al.*, 1987a; James *et al.*, 1990b). The dissociation procedure is slightly different to those developed for the atria and the bladder because less tissue is taken and the number of neurons in the trachea are approximately an order of magnitude lower

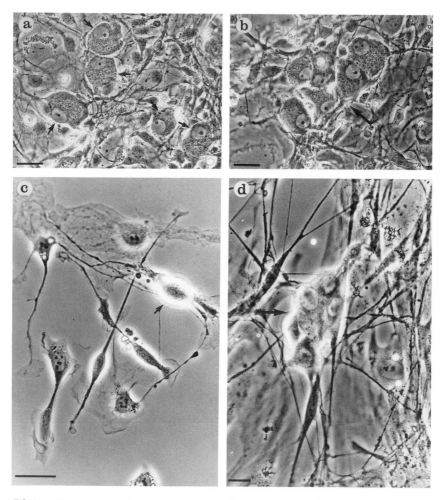

FIG. 11   Phase-contrast micrographs of paratracheal neurons in culture. (a) Ferret paratracheal neurons (small arrows) in culture for 6 days. (b) A group of six ferret paratracheal neurons (large arrow) in culture for 10 days. (c) A single mouse paratracheal neuron (small arrow) in culture for 24 hr. (d) A group of six rat paratracheal neurons (large arrow) in culture for 8 days. Bars: a, b, d, 20 μm; c, 12 μm. Taken from Burnstock *et al.* (1987a) with permission.

than in, for example, the urinary bladder (compare Baluk and Gabella, 1989a with Gabella, 1990). However, as with the procedures for the atria and bladder, mixed-cell cultures consisting of paratracheal neurons, glia, smooth muscle cells, fibroblasts, and ciliated and nonciliated epithelial cells result from the dispersal of the trachealis muscle and overlying ganglia from 12 to 13-day-old rats. These different cell types can be distinguished in living cultures under phase-contrast microscopy; in addition, their identity can be confirmed with a number of cell type-specific markers (James *et al.*, 1990b; James and Burnstock, 1991).

Recently, another culture technique for the study of airway ganglia has been developed whereby pieces of cat bronchial wall, consisting of airway epithelium, smooth muscle, glands, blood vessels, and adherent ganglia, are explanted. This organotypic preparation maintains its structural organization for at least 7 days in culture, and thus the different tissue elements can be identified and studied (Dey *et al.*, 1991).

B. Immunocytochemical Studies

VIP-, peptide histidine-isoleucine-, SOM-, galanin (GAL)-, SP-, and some CGRP-immunoreactive neuronal cell bodies have been localized in the airways *in situ* (Christofides *et al.*, 1984; Cheung *et al.*, 1985; Dayer *et al.*, 1985; Cadieux *et al.*, 1986; Dey *et al.*, 1988; Luts and Sundler, 1989). Accordingly, populations of newborn rat paratracheal neurons in dissociated cell culture are also VIP- (Fig. 12), SOM-, and GAL-immunoreactive; furthermore, some NPY-immunoreactive neurons have been detected in culture see Burnstock *et al.*, 1987a; C. J. S. Hassall, unpublished observations). SP- and VIP-immunoreactive neurons have also been observed in organotypic preparations of the cat bronchi; the projections of these neurons remained in association with the smooth muscle of the airways and blood vessels, as well as glands in culture (Dey *et al.*, 1991). However, there appeared to be a change in the pattern of coexistence of SP- and VIP-immunoreactivity in the bronchial neurons in culture when compared to *in situ*. VIP-immunoreactivity was lost from many, but not all, SP-containing neurons in culture. This alteration in the expression of VIP by neurons that retained their SP-immunoreactivity in culture is interesting and may be related to the changes in the phenotype of local neurons in the airways that occur as a consequence of heart–lung transplantation in man (Springall *et al.*, 1990), as there is extrinsic denervation of the airway neurons in both cases. Nevertheless, the loss of VIP from some neurons in culture also illustrates the need for caution when employing culture preparations as models: for example, the absence of SP-immunoreactive fibers in the airway epithelium in culture could indicate that these nerves were of extrinsic origin in the bronchi (Dey *et al.*, 1991), but, alternatively, it could represent the selective loss in culture of SP in specific, intrinsic neurons of bronchial ganglia that innervate this tissue.

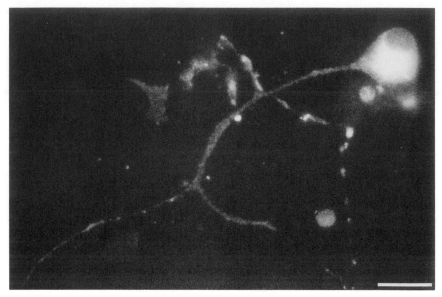

FIG. 12   Fluorescence micrograph to show a single VIP-immunoreactive rat paratracheal neuron in culture for 7 days. Bar, 20 µm.

C. Autoradiographic Localization of Receptors

Muscarinic ACh receptors are expressed by a number of different cell types in dissociated cell cultures of the rat trachealis: most, if not, all paratracheal neurons, smooth muscle cells, and nonciliated epithelial cells are labeled with autoradiographic grains after incubation with PrBCM (James *et al.*, 1990b). This correlates well with studies of the bovine and human trachea *in situ* (van Koppen *et al.*, 1987; Mak and Barnes, 1990). However, paratracheal neurons in culture are not labeled by iodinated atrial or brain natriuretic peptides, although subpopulations of smooth muscle cells, fibroblasts, and glial cells expressed binding sites that may represent clearance receptors for the natriuretic peptides (James and Burnstock, 1991).

D. Electrophysiology

Electrophysiological investigation of paratracheal ganglia has been almost exclusively confined to *in situ* preparations where individual neurons can be visualized directly (see, for example, Cameron and Coburn, 1984; Mitchell *et al.*, 1987; Allen and Burnstock, 1990b,c).

Very preliminary studies of rat paratracheal neurons in dissociated cell culture have revealed the presence of phasic and tonic firing cell types (Burnstock *et al.*, 1987a). In contrast to these cells when they were studied in intact ganglia (Allen and Burnstock, 1990b), no burst firing cells were observed in culture. The reason for this difference is not known, but it may be related to the observation that the cultured cells display smaller sAHPs than those *in situ*. The sAHP might be expected to punctuate periods of intense action potential discharge and thus produce burst firing behavior in otherwise tonic firing neurons. A similar, as yet unexplained, reduction in the amplitude of the sAHP in cultured cells has been observed in enteric neurons in culture (see Section II,E). However, action potentials recorded from *in situ* paratracheal neurons show strong TTX resistance, whereas action potentials in the cultured cells could be totally abolished by TTX; this may indicate that the reduced amplitude of the sAHP seen in culture arises as a result of a decrease in calcium entry during the action potential.

As in intact ganglia, all paratracheal neurons in culture display nicotinic responses to ACh (Burnstock *et al.*, 1987a) and, in some cells, muscarine has been found to reduce the afterhyperpolarization following a single action potential (T. G. J. Allen, unpublished observation). Studies to characterize the actions of muscarine and other neurotransmitters more thoroughly have yet to be carried out.

VI. Summary and Future Directions

The increasing use of tissue culture techniques for the study of autonomic ganglia that are closely associated with their target organs has provided valuable information about the properties of these neurons, much of which would have been impossible to achieve in *in situ* preparations. Since ultrastructural, immunocytochemical, autoradiographic, and electrophysiological studies have demonstrated that such neurons retain many of their differentiated properties *in vitro*, culture preparations are useful models for the study of these ganglia. However, subtle changes may occur in culture and the importance of caution in the interpretation of results obtained *in vitro* cannot be overstressed. Nevertheless, detailed analysis of the factors responsible for changes that occur *in vitro* may, in turn, provide additional information about the factors that maintain a stable system *in vivo*.

Increasingly sophisticated techniques have become available for the study of the nervous system, particularly at the molecular level. A number of these techniques can be applied to tissue culture preparations, allowing further analysis of the properties of intramural neurons. For example, *in situ* hybridization enables individual cells expressing particular genes to be visualized, and cultured neurons are particularly amenable to this approach. The use of this type of technique should resolve the question of whether the apparent lability of certain populations of peptide-containing neurons in culture is due to a true change in neuronal phenotype or to posttranslational changes in processing, storage, or release of these molecules.

Preparations of isolated intramural ganglion cells may also have functional applications. The use of grafts for the treatment of neurological disorders has recently received considerable attention (e.g., Dunnett and Richards, 1990). Isolated preparations of enteric ganglion cells may provide a valuable source of donor tissue for transplantation. The use of autologous enteric ganglia would overcome the major problem of incompatability and would be ethically acceptable.

Although there is now a vast body of information about the ENS, there are still many unanswered questions about enteric neurons and glial cells. A major point concerns the development of the ENS and how the differentiated characteristics of enteric ganglion cells are regulated throughout life. In other parts of the nervous system, growth factors such as nerve growth factor have been found to play an important role in neuronal survival during development. Nothing is known, however, of the roles of this or other growth factors on enteric neurons. It is possible that specific growth factors exist for enteric neurons and such factors would probably originate in the developing gut itself. Tissue culture techniques, ideally applied to the embryonic gut, should play a major role in this type of investigation, since it is impossible to separate the developing enteric neurons from this potential source of trophic factors *in vivo*. Further aspects of ENS development which might be studied in culture include the processes of neurite growth and the formation of the appropriate connections with target cells. The latter poses a problem of considerable complexity: cells within individual ganglia are of diverse neurochemical phenotypes, containing multiple neuroactive molecules and possessing a variety of projections to different target cells both within and outside the gut. The recent development of culture systems of purified neuronal and glial cell populations and of dissociated enteric ganglia should prove valuable for such studies.

The particular problems inherent in the study of the ganglia described in this review, such as the difficulty in distinguishing the processes of intrinsic neurons from those of extrinsic origin, also hold true for other autonomic ganglia that are closely associated with their target organs. Tissue culture techniques are therefore also likely to prove fruitful in the study of such ganglia. For example, the diverse population of neurons present in the pancreas merits further study, particularly since evidence suggests that they are involved in the control of exocrine pancreatic secretion and the integration of nervous input to the pancreas and that they are affected by alloxan-induced diabetes (Luiten et al., 1986; King et al., 1989; Mulvihill et al., 1990).

The unique advantages of culture preparations, which include the increased accessibility of the cultured cells, and the ability to observe cells over extended periods of time and to control their physical and chemical environment, are pertinent in the study of the regulation of phenotypic differentiation. For example, this type of study would be particularly relevant for the study of pelvic ganglia, which are also closely associated with their target organs and which exhibit sex differences in anatomy and function. Moreover, complex changes in pelvic neuronal phenotype and patterns of innervation occur during pregnancy (Alm and Lundberg, 1988; Mione *et al.*, 1990). Culture preparations of pelvic ganglion cells would greatly facilitate direct studies of the roles of hormones and other factors on the differentiation of these cells, and help to determine the mechanisms effecting sex differences and pregnancy-induced changes in pelvic neurons.

In conclusion, although it has long been recognized that the intramural nervous system of the gut is, to a large extent, independent of CNS control (Langley, 1921), it has generally been assumed that other intramural neurons only serve as simple, nicotinic relays of the parasympathetic input to their target tissue. While it is clearly inaccurate to make generalizations about the properties of different intramural ganglia, studies of these neurons in tissue culture have shown that, contrary to earlier assumptions, the intramural neurons of the autonomic nervous system are a very diverse population that are likely to be specialized with respect to their particular targets. They exhibit complex electrophysiological properties that enable information from several sources to be integrated and that involve a number of neurotransmitters, neuromodulators, and their receptors. It now seems likely that, like enteric ganglia, other intramural ganglia play a role in the local control of their target tissue and may even mediate reflexes independent of the CNS. It is of obvious importance to extend studies of intramural ganglia in culture and *in situ*; the results from culture preparations performed to date will be instrumental to the design and interpretation of these future experiments.

Acknowledgments

The authors would like to acknowledge the support of the Medical Research Council, The Wellcome Trust, The British Heart Foundation, and The Asthma Research Council.

References

Adams, D. J., Dwyer, T. M., and Hille, B. (1980). *J. Gen. Physiol.* **75**, 493–510.
Adams, D. J., Fieber, L. A., and Konishi, S. (1987). *J. Physiol. (London)* **394**, 154P.
Akasu, T., Gallagher, J. P., Koketsu, K., and Shinnick-Gallagher, P. (1984). *J. Physiol. (London)* **351**, 583–593.
Allen, T. G. J., and Burnstock, G. (1987). *J. Physiol. (London)* **388**, 349–366.
Allen, T. G. J., and Burnstock, G. (1990a). *J. Physiol. (London)* **422**, 463–480.
Allen, T. G. J., and Burnstock, G. (1990b). *J. Physiol. (London)* **423**, 593–614.
Allen, T. G. J., and Burnstock, G. (1990c). *Br. J. Pharmacol.* **100**, 261–268.
Allen, T. G. J., and Burnstock, G. (1990d). *Br. J. Pharmacol.* **100**, 269–276.
Allen, T. G. J., and Burnstock, G. In preparation.
Alm, P., and Lundberg, L.-M. (1988). *Cell Tissue Res.* **254**, 517–530.
Ardell, J. L., and Randall, W. C. (1986). *Am. J. Physiol.* **251**, H764–H773.

Baker, D. G., McDonald, D. M., Basbaum, C. B., and Mitchell, R. A. (1986). *J. Comp. Neurol.* **246**, 513–526.

Baluk, P., and Gabella, G. (1989a). *J. Comp. Neurol.* **285**, 117–132.

Baluk, P., and Gabella, G. (1989b). *Neurosci. Lett.* **102**, 191–196.

Baluk, P., and Gabella, G. (1989c). *Neurosci. Lett.* **104**, 269–273.

Baluk, P., and Gabella, G. (1990). *Cell Tissue Res.* **261**, 275–285.

Baluk, P., Jessen, K. R., Saffrey, M. J., and Burnstock, G. (1983). *Brain Res.* **262**, 37–47.

Bannerman, P. G., Mirsky, R., and Jessen, K. R. (1987). *Dev. Neurosci.* **9**, 201–227.

Bannerman, P. G., Mirsky, R., and Jessen, K. R. (1988a). *Brain Res.* **440**, 87–98.

Bannerman, P. G., Mirsky, R., and Jessen, K. R. (1988b). *Brain Res.* **440**, 99–108.

Barber, D. L., Buchan, A. M. J., Leeman, S. E., and Soll, A. H. (1989). *Neuroscience* **32**, 145–153.

Barnes, D., and Sato, G. (1980). *Anal. Biochem.* **102**, 255–270.

Bean, B. (1989). *Annu. Rev. Physiol.* **51**, 367–384.

Bean, B., Williams, C. A., and Ceelen, P. W. (1990). *J. Neurosci.* **10**, 11–19.

Benham, C. D. (1989). *J. Physiol. (London)* **419**, 689–701.

Blomquist, T. M., Priola, D. V., and Romero, A. M. (1987). *Am. J. Physiol.* **252**, H638–H644.

Bottenstein, J. E., Skaper, S. D., Varon, S. S., and Sato, G. M. (1980). *Exp. Cell Res.* **125**, 183–190.

Brookes, S. J. H., Ewart, W. R., and Wingate, D. L. (1988). *Neuroscience* **24**, 297–307.

Brookes, S. J. H., Steele, P. A., and Costa, M. (1991). *Neuroscience* **42**, 863–878.

Brown, D. A., and Adams, P. R. (1980). *Nature* **283**, 673–608.

Brown, D. A., Gähwiler, B. H., Marsh, S. J., and Selyanko, A. A. (1986). *Trends Pharmacol. Sci., Suppl.* pp. 66–71.

Buckley, N. J., and Burnstock, G. (1984). *Brain Res.* **310**, 133–137.

Buckley, N. J., and Burnstock, G. (1986). *J. Neurosci.* **6**, 531–540.

Buckley, N. J., Saffrey, M. J., Hassall, C. J. S., and Burnstock, G. (1988). *Brain Res.* **445**, 152–156.

Burnstock, G. (1980). *Circ. Res.* **46** Suppl. I, I175–I182.

Burnstock, G., and Kennedy, C. (1985). *Gen. Pharmacol.* **16**, 433–440.

Burnstock, G., and Pittam, B. S. (1984). *Proc. Int. Congr. Pharmacol., 9th, 1984* p. 151.

Burnstock, G., Cocks, T., Crowe, R., and Kasakov, L. (1978). *Br. J. Pharmacol.* **631**, 125–138.

Burnstock, G., Allen, T. G. J., and Hassall, C. J. S. (1987a). *Am. Rev. Respir. Dis.* **136**, S23–S26.

Burnstock, G., Allen, T. G. J., Hassall, C. J. S., and Pittam, B. S. (1987b). *Exp. Brain Res. Ser.* **16**, pp. 323–328.

Cadieux, A., Springall, D. R., Mulderry, P. K., Rodrigo, J., Ghatei, M. A., Terenghi, G., Bloom, S. R., and Polak, J. M. (1986). *Neuroscience* **19**, 605–627.

Cameron, A. R., and Coburn, R. F. (1984). *Am. J. Physiol.* **246**, C450–C458.

Chamley, J. H., Mark, G. E., Campbell, G., and Burnstock, G. (1972). *Z. Zellforsch. Mikrosk. Anat.* **135**, 387–314.

Cheng, H., and Bjerknes, M. (1979). *J. Cell Biol.* **83**, 109a.

Cherubini, E., and North, R. A. (1984). *Br. J. Pharmacol.* **82**, 93–100.

Cheung, A., Polak, J. M., Bauer, F. E., Cadieux, A., Christofides, N. D., Springall, D. R., and Bloom, S. R. (1985). *Thorax* **40**, 889–896.

Chiang, C.-H., and Gabella, G. (1986). *Cell Tissue Res.* **246**, 243–252.

Christofides, N. D., Yiangou, Y., Piper, P. J., Ghatei, M. A., Sheppard, M. N., Tatemoto, K., Polak, J. M., and Bloom, S. R. (1984). *Endocrinology (Baltimore)* **115**, 1958–1963.

Connor, J. A., and Stevens, C. F. (1971). *J. Physiol. (London)* **213**, 21–30.

Cook, R. D., and Peterson, E. R. (1974). *J. Neurol. Sci.* **22**, 25–38.

Crowe, R., and Burnstock, G. (1982). *Cardiovasc. Res.* **16**, 384–390.

Crowe, R., Haven, A. J., and Burnstock, G. (1986). *J. Auton. Nerv. Syst.* **15**, 319–339.

Dalsgaard, C.-J., Franco-Cereceda, A., Saria, A., Lundberg, J. M., Theodorsson-Norheim, E., and Hökfelt, T. (1986). *Cell Tissue Res.* **243**, 477–485.

Dayer, A. M., De Mey, J., and Will, J. A. (1985). *Cell Tissue Res.* **239**, 621–625.

de Groat, W. C., and Kawatani, M. (1989). *J. Physiol. (London)* **409**, 431–449.
De Jong, B. J., and De Haan, J. (1943). *Acta Neerl. Morphol. Norm. Pathol.* **5**, 26–51.
Dennis, M. J., Harris, A. J., and Kuffler, S. W. (1971). *Proc. R. Soc. London, Ser. B* **177**, 509–539.
Derkach, V., Surprenant, A., and North, R. A. (1989). *Nature (London)* **339**, 706–709.
Dey, R. D., Hoffpauir, J., and Said, S. I. (1988). *Neuroscience* **24**, 275–281.
Dey, R. D., Altemus, J. B., and Michalkiewicz, M. (1991). *J. Comp. Neurol.* **304**, 330–340.
Dreyfus, C. F., Sherman, D., and Gershon, M. D. (1977a). *Brain Res.* **128**, 109–123.
Dreyfus, C. F., Bornstein, M. B., and Gershon, M. D. (1977b). *Brain Res.* **128**, 125–139.
Dunnett, S. B., and Richards, S.-J. eds. (1990). *Prog. Brain Res.* **82**.
Eccleston, P. A., Jessen, K. R., and Mirsky, R. (1987). *Dev. Biol.* **124**, 409–417.
Eccleston, P. A., Bannerman, P. G. C., Pleasure, D. E., Winter, J., Mirsky, R., and Jessen, K. R. J. (1989). *Development (Cambridge, UK)* **107**, 107–111.
Fieber, L. A., and Adams, D. J. (1991a). *J. Physiol. (London)* **434**, 215–237.
Fieber, L. A., and Adams, D. J. (1991b). *J. Physiol. (London)* **434**, 239–256.
Forssmann, W. G., Reinecke, M., and Weihe, E. (1982). *In* "Systemic Role of Regulatory Peptides" (S. R. Bloom, J. M. Polak, and E. Lindenlaub, eds.), pp. 329–349. Schattauer Verlag, Stuttgart.
Franco-Cereceda, A., Lundberg, J. M., and Hökfelt, T. (1986). *Eur. J. Pharmacol.* **132**, 101–102.
Furness, J. B., and Costa, M. (1987). "The Enteric Nervous System." Churchill-Livingstone, Edinburgh.
Furshpan, E. J., Landis, S. C., Matsumoto, S. G., and Potter, D. D. (1986). *J. Neurosci.* **6**, 1061–1079.
Gabella, G. (1971). *Z. Naturforsch. B: Anag. Chem., Org. Chem., Biochem., Biophys., Biol.* **266**, 244–245.
Gabella, G. (1981). *Neuroscience* **6**, 425–436.
Gabella, G. (1990). *Cell Tissue Res.* **261**, 231–237.
Gagliardi, M., Randall, W. C., Bieger, D., Wurster, R. D., Hopkins, D. A., and Armour, J. A. (1988). *Am. J. Physiol.* **255**, H789–H800.
Gerstheimer, F. P., and Metz, J. (1986). *Anat. Embryol.* **175**, 255–260.
Halliwell, J. V., and Adams, P. R. (1982). *Brain Res.* **250**, 71–92.
Hamberger, B., and Norberg, K.-A. (1965). *Int. J. Neuropharmacol.* **4**, 41–45.
Hanani, M. and Burnstock, G. (1984). *Neurosci. Lett.* **48**, 19–23.
Hanani, M., and Burnstock, G. (1985a). *J. Auton. Nerv. Syst.* **14**, 49–60.
Hanani, M., and Burnstock, G. (1985b). *Brain Res.* **358**, 276–281.
Hanani, M., Baluk, P., and Burnstock, G. (1982). *J. Auton. Nerv. Syst.* **5**, 155–164.
Hancock, J. C., Hoover, D. B., and Houghland, M. W. (1987). *J. Auton. Nerv. Syst.* **19**, 59–66.
Harrison, R. G. (1907). *Proc. Soc. Exp. Biol. Med.* **4**, 140–143.
Hartzell, H. C., Kuffler, S. W., Stickgold, R., and Yoshikami, D. (1977). *J. Physiol. (London)* **271**, 817–846.
Hassall, C. J. S., and Burnstock, G. (1984). *Neurosci. Lett.* **52**, 111–115.
Hassall, C. J. S., and Burnstock, G. (1986). *Brain Res.* **364**, 102–113.
Hassall, C. J. S., and Burnstock, G. (1987a). *Neuroscience* **22**, 413–423.
Hassall, C. J. S., and Burnstock, G. (1987b). *Brain Res.* **422**, 74–82.
Hassall, C. J. S., Buckley, N. J., and Burnstock, G. (1987). *Neurosci. Lett.* **74**, 145–150.
Hassall, C. J. S., Wharton, J., Gulbenkian, S., Anderson, J. V., Frater, J., Bailey, D. J., Merighi, A., Bloom, S. R., Polak, J. M., and Burnstock, G. (1988). *Cell Tissue Res.* **251**, 161–169.
Hassall, C. J. S., Allen, T. G. J., Pittam, B. S., and Burnstock, G. (1989). *In* "Regulatory Peptides" (J. M. Polak, ed.), pp. 113–136. Birkhäuser Verlag, Basel.
Hassall, C. J. S., Penketh, R., Rodeck, C., and Burnstock, G. (1990a). *Anat. Embryol.* **182**, 329–337.
Hassall, C. J. S., Penketh, R., Rodeck, C., and Burnstock, G. (1990b). *Anat. Embryol.* **182**, 339–346.
Hassall, C. J. S. *et al.* In preparation.
Haynes, L. W., and Zakarian, S. (1982). *Regul. Pept.* **3**, 73.
Hirning, L. A., Fox, A. P., and Miller, R. J. (1990). *Brain Res.* **532**, 120–130.

Hirst, G. D. S., and McKirdy, H. C. (1975). *J. Physiol. (London)* **249**, 369–385.

Hirst, G. D. S., Holman, M. E., and Spence, I. (1974). *J. Physiol. (London)* **236**, 303–326.

Hoover, D. B., and Hancock, J. C. (1988). *J. Auton. Nerv. Syst.* **23**, 189–197.

Hulme, E. C., Birdsall, N. J. M., and Buckley, N. J. (1990). *Annu. Rev. Pharmacol. Toxicol.* **30**, 633–673.

Jacobowitz, D. (1967). *J. Pharmacol. Exp. Ther.* **158**, 227–240.

Jacobs-Cohen, R. J., Payette, R. F., Gershon, M. D., and Rothman, T. P. (1987). *J. Comp. Neurol.* **255**, 425–438.

James, S., and Burnstock, G. (1988). *Regul. Pept.* **23**, 237–245.

James, S., and Burnstock, G. (1989). *Neurosci. Lett.* **106**, 13–18.

James, S., and Burnstock, G. (1990). *Regul. Pept.* **28**, 177–188.

James, S., and Burnstock, G. (1991). *Cell Tissue Res.* **265**, 555–565.

James, S., Hassall, C. J. S., Polak, J. M., and Burnstock, G. (1990a). *Cell Tissue Res.* **261**, 301–312.

James, S., Bailey, D. J., and Burnstock, G. (1990b). *Brain Res.* **513**, 74–80.

Jessen, K. R. (1981). *Mol. Cell. Biochem.* **38**, 69–76.

Jessen, K. R. (1982). *In* "Systemic Role of Regulatory Peptides" (S. R. Bloom, J. M. Polak, and E. Lindenlaub, eds.), pp. 35–49. SchattenerVerlag, Stuttgart.

Jessen, K. R., and Mirsky, R. (1980). *Nature (London)* **286**, 736–737.

Jessen, K. R., and Mirsky, R. (1983). *J. Neurosci.* **3**, 2206–2218.

Jessen, K. R., McConnell, J. D., Purves, R. D., Burnstock, G., and Chamley-Campbell, J. (1978). *Brain Res.* **152**, 573–579.

Jessen, K. R., Mirsky, R., Dennison, M. E., and Burnstock, G. (1979). *Nature (London)* **281**, 71–74.

Jessen, K. R., Polak, J. M., Van Noorden, S., Bloom, S. R., and Burnstock, G. (1980a). *Nature (London)* **283**, 391–393.

Jessen, K. R., Saffrey, M. J., Van Noorden, S., Bloom, S.R., Polak, J. M., and Burnstock, G. (1980b). *Neuroscience* **5**, 1717–1735.

Jessen, K. R., Saffrey, M. J., and Burnstock, G. (1983a). *Brain Res.* **262**, 17–35.

Jessen, K. R., Saffrey, M. J., Baluk, P., Hanani, M., and Burnstock, G. (1983b). *Brain Res.* **262**, 49–62.

Jessen, K. R., Hills, J. M., Dennison, M. E., and Mirsky R. (1983c). *Neuroscience* **10**, 1427–1442.

Katayama, Y., North, R. A., and Williams, J. T. (1979). *Proc. R. Soc. London, Ser. B* **206**, 191–208.

Kawatani, M., Shioda, S., Nakai, Y., Takeshige, C., and de Groat, W. C. (1989). *J. Comp. Neurol.* **288**, 81–91.

Keunig, F. J. (1944). *Acta Neerl. Morphol. Norm. Pathol.* **5**, 237–247.

King, B. F., Love, J. A., and Szurszewski, J. H. (1989). *J. Physiol. (London)* **419**, 379–403.

King, T. S., and Coakley, J. B. (1958). *J. Anat.* **92**, 353–376.

Kobayashi, Y., Hassall, C. J. S., and Burnstock, G. (1986a). *Cell Tissue Res.* **244**, 595–604.

Kobayashi, Y., Hassall, C. J. S., and Burnstock, G. (1986b). *Cell Tissue Res.* **244**, 605–612.

Konishi, S., Okamoto, T., and Otsuka, M. (1984). *IUPHAR Proc.* p. 1604.

Konishi, S., Okamoto, T., and Otsuka, M. (1985). *In* "Substance P: Metabolism and Biological Actions" (C. C. Jordan and P. Oehme, eds.), pp. 121–136. Taylor & Francis, London.

Korman, L. Y., Nylen, E. S., Finan, T. M., Linnoila, R. I., and Becker, K. L. (1988). *Gastroenterology* **95**, 1003–1010.

Krishtal, O. A., Marchenko, S. M., and Obukhov, A. G. (1988). *Neuroscience* **27**, 995–1000.

Lane, M.-A., Sastre, A., and Salpeter, M. M. (1976). *Proc. Natl. Acad. Sci. U.S.A.* **73**, 4506–4510.

Langley, J. N. (1921). "The Autonomic Nervous System." Heffner, Cambridge, England.

Lewis, W. H., and Lewis M. R. (1912). *Anat. Rec.* **6**, 7–31.

Luiten, P. G. M., ter Horst, G. J., Buijs, R. M., and Steffens, A. B. (1986). *J. Auton. Nerv. Syst.* **15**, 33–44.

Luts, A., and Sundler, F. (1989). *Cell Tissue Res.* **258**, 259–267.

Mak, J. C. W., and Barnes, P. J. (1990). *Am. Rev. Respir. Dis.* **141**, 1559–1568.

Maruyama, T. (1981). *Neurosci. Lett.* **25**, 143–148.

Maruyama, T. (1985). *Brain Res.* **336**, 368–371.

Mathie, A., Cull-Candy, S. G., and Colquhoun, D. (1987). *Proc. R. Soc. London, Ser. B* **232**, 239–248.

Matsumoto, S. G., Sah, D., Potter, D. D., and Furshpan, E. J. (1987). *J. Neurosci.* **7**, 380–390.

Mihara, S., Katayama, Y., and Nishi, S. (1985). *Neuroscience* **16**, 1057–1068.

Mione, M. C., Cavanagh, J. F. R., Lincoln, J., Milner, P., and Burnstock, G. (1990). *Cell Tissue Res.* **259**, 503–509.

Mitchell, R. A., Herbert, D. A., Baker, D. G., and Basbaum, C. B. (1987). *Brain Res.* **437**, 157–160.

Mochida, S., and Kobayashi, H. (1986). *Brain Res.* **383**, 299–304.

Mulvihill, S. J., Bunnett, N. W., Goto, Y., and Debas, H. T. (1990). *Metab. Clin. Exp.* **39**, 143–148.

Nishi, R., and Willard, A. L. (1985). *Neuroscience* **16**, 187–199.

Nishi, R., and Willard, A. L. (1988). *Neuroscience* **25**, 759–769.

Nishi, S., and North, R. A. (1973). *J. Physiol. (London)* **231**, 471–491.

North R. A. (1982). *Neuroscience* **7**, 315–325.

Nozdrachev, A. D., and Pogorelov, A. G. (1982). *J. Auton. Nerv. Syst.* **6**, 73–81.

O'Lague, P. H., Potter, D. D., and Furshpan, E. J. (1978). *Dev. Biol.* **67**, 384–403.

Pittam, B. S., Burnstock, G., and Purves, R. D. (1987). *Brain Res.* **403**, 267–278.

Potter, D. D., Landis, S. C., Matsumoto, S. G., and Furshpan, E. J. (1986). *J. Neurosci.* **6**, 1080–1098.

Rawdon, B. B., and Dockray, G. J. (1983). *Dev. Brain Res.* **7**, 53–59.

Rothman, T. P., and Gershon, M. D. (1982). *J. Neurosci.* **2**, 381–393.

Rothman, T. P., Nilaver, G., and Gershon, M. D. (1984). *J. Comp. Neurol.* **225**, 13–23.

Rothman, T. P., Tennyson, V. M., and Gershon, M. D. (1986). *J. Comp. Neurol.* **252**, 493–506.

Saffrey, M. J., and Burnstock, G. (1984). *Int. J. Dev. Neurosci.* **2**, 591–602.

Saffrey, M. J., and Burnstock, G. (1988a). *Cell Tissue Res.* **253**, 105–114.

Saffrey, M. J., and Burnstock, G. (1988b). *Cell Tissue Res.* **254**, 167–176.

Saffrey, M. J., and Burnstock, G. (1992). In preparation.

Saffrey, M. J., Marcus, N., Jessen, K. R., and Burnstock, G. (1983). *Cell Tissue Res.* **234**, 231–235.

Saffrey, M. J., Legay, C., and Burnstock, G. (1984). *Brain Res.* **304**, 105–116.

Saffrey, M. J., Bailey, D. J., and Burnstock, G. (1991). *Cell Tissue Res.* **265**, 527–534.

Schultzberg, M., Dreyfus, C. F., Gershon, M. D., Hökfelt, T., Elde, R. P., Nilsson, G., Said, S., and Goldstein, M. (1978). *Brain Res.* **155**, 239–248.

Seabrook, G. R., and Adams, D. J. (1989). *Br. J. Pharmacol.* **97**, 1125–1136.

Seabrook, G. R., Fieber, L. A., and Adams, D. J. (1990). *Am. J. Physiol.* **259**, H997–H1005.

Skoogh, B.-E., Grillo, M. A., and Nadel, J. A. (1983). *J. Neurosci. Methods* **8**, 33–39.

Smith, R. B. (1970). *Br. Heart J.* **32**, 108–113.

Springall, D. R., Polak, J. M., Howard, L., Power, R. F., Krausz, T., Manickam, S., Baner, N. R., Khagani, A., Rose, M., and Yacoub, M. (1990). *Am. Rev. Respir. Dis.* **141**, 1538–1546.

Sturek, M., Hirning, L. A., and Miller, R. J. (1987). *Soc. Neurosci. Abstr.* **13**, 32.5.

Sundin, T., and Dahlström, A. (1973). *Scand. J. Urol. Nephrol.* **7**, 131–149.

Surprenant, A., Shen, K.-Z., North, R. A., and Tatsumi, H. (1990). *J. Physiol. (London)* **431**, 585–608.

Tatsumi, H., Costa, M., Schimerlik, M., and North, R. A. (1990). *J. Neurosci.* **10**, 1675–1682.

Tay, S. S. W., Wong, W. C., and Ling, E. A. (1984). *J. Anat.* **138**, 67–80.

Thuneberg, L. (1982). *Adv. Anat. Embryol. Cell Biol.* **71**, 1–130.

Topchieva, E. P. (1965). *Fiziol. Zh. im. I.M. Sechenova SSSR* **51**, 1231; *Fed. Proc., Transl. Suppl.* **25**, T739–T742 (1966).

van Koppen, C. J., Blankesteijn, W. M., Klaassen, A. B. M., Rodrigues de Miranda, J. F., Beld, A. J., and van Ginneken, C. A. M. (1987). *Neurosci. Lett.* **83**, 237–240.

Weihe, E., McKnight, A. T., Corbett, A. D., Hartschuh, W., Reinecke, M., and Kosterlitz, H. W. (1983). *Life Sci.* **33**, Suppl. 1, 711–714.

Weihe, E., Reinecke, M., and Forssmann, W. G. (1984). *Cell Tissue Res.* **236**, 527–540.

Wharton, J., Polak, J. M., Gordon, L., Banner, N. R., Springall, D. R., Rose, M., Khagani, A., Wallwork, J., and Yacoub, M. H. (1990). *Circ. Res.* **66**, 900–912.

Willard, A. L. (1990a). *Neuroscience* **10**, 1025–1034.

Willard, A. L. (1990b). *J. Physiol. (London)* **426**, 453–471.

Willard, A. L., and Nishi, R. (1985a). *Neuroscience* **16**, 201–211.

Willard, A. L., and Nishi, R. (1985b). *Neuroscience* **16**, 213–221.

Willard, A. L., and Nishi, R. (1987). *Brain Res.* **422**, 163–167.

Willard, A. L., and Nishi, R. (1989). *In* "Handbook of Physiology" (S. G. Shultz, J. D. Wood, and B. B. Rauner, eds.), Sect. 6, Vol. 1, Part 1, pp. 331–347. Oxford Univ. Press, New York.

Wong, W. C., Ling, E. A., Yick, T. Y., and Tay, S. S. W. (1987). *J. Anat.* **150**, 75–88.

Wood, J. D. (1984). *Am. J. Physiol.* **247**, G585–G598.

Xi, X., Thomas, J. X., Randall, W. C., and Wurster, R. D. (1991). *J. Auton. Nerv. Syst.* **32**, 177–182.

Zafirov, D. H., Palmer, J. M., Nemeth, P. R., and Wood, J. D. (1985). *Eur. J. Pharmacol.* **115**, 103–107.

The Nuclear Envelope of the Yeast *Saccharomyces cerevisiae*

Eduard C. Hurt, Ann Mutvei, and Maria Carmo-Fonseca

European Molecular Biology Laboratory, D-6900 Heidelberg, Germany

I. Introduction

A genuine cell nucleus surrounded by a nuclear envelope is the typical achievement of eukaryotic life. Prokaryotes lack a nucleus and their DNA is still exposed within the cytoplasm. Like other compartmentalization steps which took place inside the eukaryotic cytoplasm, the acquisition of a nuclear organelle allowed the cell more accurately and specifically to segregate nuclear processes from cytoplasmic reactions. The nuclear envelope enabled the eukaryotic cell to control nucleocytoplasmic transport and, by doing so, it could regulate gene expression in a more sophisticated way during the cell cycle and development. On the other hand, the formation of an enclosed nuclear compartment created many new problems for the evolving eukaryotic cell which were not evident in prokaryotic life. Nuclear import and export mechanisms had to be developed, nuclear pores to be constructed, and ribosome biogenesis had to be redesigned. Furthermore, biochemical reactions that were not compartmentalized in prokaryotes suddenly became spatially and temporally separated from each other. Whereas in bacteria ribosomes already bind to nascent transcripts and begin to translate the as-yet-unfinished mRNA, in eukaryotes the mRNA has to leave the nucleus in order to get access to the translational machinery. One tends to think that the prokaryote tries to maximize gene expression to grow at optimal speed; a nuclear membrane barrier would only hamper such a strategy. In contrast, the existence of a nuclear boundary forces the eukaryotic organism to use up more time for expressing genes and, accordingly, it grows more slowly, but gains new tools for controlling gene expression. It is therefore tempting to speculate that progression to a more complex life only became possible after the development of the nuclear envelope created a regulatory site for nucleocytoplasmic movement and gene expression.

Many fundamental functions are fulfilled by the nuclear envelope, such as transport of molecules and macromolecules between the cytoplasm and the nucleus, nuclear division, chromatin organization, nuclear architecture, regulation of

145

gene expression, and RNA processing. Most of these reactions and functions are conserved between lower and higher eukaryotes, suggesting that the key components of these reactions are also evolutionarily conserved. In order to study these processes on a molecular level, it is important to identify the molecules involved in these nuclear envelope functions. The lower eukaryote yeast *Saccharomyces cerevisiae* offers a suitable experimental system to study these complex reactions because of the powerful genetics with which essential genes can be manipulated (Botstein and Fink, 1988). Accordingly, the yeast nuclear envelope has become increasingly attractive to scientists and different aspects of nuclear envelope functions have already been studied successfully in this organism.

It is the aim of this article to review the progress which has been achieved over the past few years in identifying the structures, functions, and molecules of the yeast nuclear envelope. Hopefully, such a survey may stimulate further research on the nuclear envelope of the yeast *S. cerevisiae,* which could become a good model system to study elementary functions of the eukaryotic nuclear envelope in general.

II. Structural Organization of the Nuclear Envelope

The nuclear envelope of higher eukaryotes consists of an inner and outer membrane, nuclear pore complexes, and the nuclear lamina (for recent reviews, see Newport and Forbes, 1987; Gerace and Burke, 1988; Nigg, 1988, 1989). A double nuclear membrane and nuclear pore complexes are readily visible in electron micrographs of yeast cells, but the existence of a nuclear lamina remains uncertain. In addition, and unlike that of most eukaryotes, the yeast nuclear envelope does not break down during mitosis. Spindle microtubules emanate from the spindle pole bodies (SPB), which are microtubule-organizing centers embedded in the nuclear envelope (Byers, 1981).

A. Purification Methods for Yeast Nuclei and Subnuclear Fractions

Yeast nuclei can be efficiently isolated by a variety of methods based on the enzymatic digestion of the cell wall followed by lysis in a buffer complemented with 18% Ficoll as an osmotic stabilizer (Mann and Mecke, 1980, 1982; Hurt *et al.,* 1988; Aris and Blobel, 1988, 1990; Amati and Gasser, 1988; Allen and Douglas, 1989; Kalinich and Douglas, 1989; Cardenas *et al.,* 1990). Each laboratory has its preferred purification protocol for yeast nuclei and therefore methodological advice is difficult to give. Purified yeast nuclei have been used successfully to measure *in vitro* nuclear binding (Stochaj *et al.,* 1991) or nuclear transport of proteins (Kalinich and Douglas, 1989; Garcia-Bustos *et al.,* 1991b), *in vitro* transcription, and mRNA elongation (Cardenas *et al.,* 1990) and have served as a source from which to isolate nuclear envelopes (Aris and Blobel, 1989, 1990) and nuclear membrane proteins (Hurt *et al.,* 1988; Schimmang and Hurt, 1989), nuclear scaffolds (Amati and Gasser, 1990; Cardenas *et al.,* 1990), nuclear pore complex/lamina preparations (Allen and Douglas, 1989), and SPBs (Rout and Kilmartin, 1990). Electron microscopy of isolated nuclei allows a better evaluation of the nuclear envelope and its substructures such as nuclear pores, SPBs, and the mitotic spindle, which in whole cells are partly masked by the densely packed cytoplasmic and nuclear matrix (Figs. 1–3). A partial swelling of the perinuclear space, which occurs during nuclear isolation, causes the outer and inner nuclear membrane to separate better and, accordingly, nuclear pores can be more easily identified (Figs. 2 and 3).

B. Architecture of the Yeast Nuclear Envelope

1. Nuclear Pore Complexes

The nuclear pore complexes are elaborate, supramolecular assemblies that span the nuclear envelope at sites where the outer and inner membrane are joined. Using freeze-fracture EM analysis, nuclear pores can be easily seen in *S. cerevisiae* (Moor and Mühlethaler, 1963) (Fig. 4). The number of nuclear pores in the envelope can vary depending on the cell type (Franke, 1974, Maul, 1977). Yeast nuclei have been estimated to contain 10–15 pores per μm^2 of surface, occupying 6–8% of the envelope area (Moor and Mühlethaler, 1963; Jordan *et al.,* 1977; Maul and Deaven, 1977). Since the interphase nucleus has a roughly spherical shape of about 2.3 μm diameter, the total number of pores per nucleus is approximately 200. The distribution of pores on the nuclear surface does not appear to be regular and, particularly in old cells (derived from starved cultures), the nuclear pores may be concentrated in some areas of the nuclear envelope, while other areas are devoid of pores. The frequency of pores can vary during the yeast cell cycle (Jordan *et al.,* 1977). The number increases in the early G0 phase and a second time around nuclear division. In parallel with these two peaks in pore biogenesis, the nuclear surface enlarges rapidly and, accordingly, the pore frequency per unit nuclear membrane remains quite constant throughout the cell cycle (Jordan *et al.,* 1977). Interestingly, in yeast there is no relationship between pore number and DNA synthesis. This makes it less likely that nuclear pores have an additional function related to DNA synthesis or chromatin organization, as suggested from work on higher eukaryotes. Growth-arrested yeast cells (e.g., arrested by starvation) have nuclear pores with a wider pore diameter, and a correlation between the nuclear pore and nuclear envelope size exists, larger pores tending to be present in smaller nuclear envelopes (Willison and Johnston, 1978).

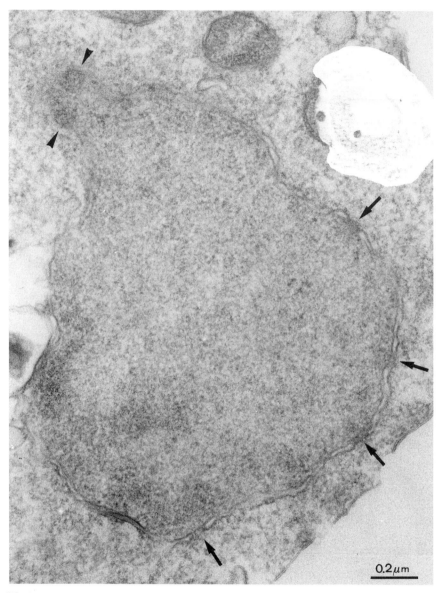

FIG. 1   Electron micrograph of a thin section through the nucleus of a yeast cell *(Saccharomyces cerevisiae)*. The nucleus is surrounded by a double membrane interrupted by pores. The nuclear pore complexes are observed both in cross-section (arrows) and tangential view (arrowheads).

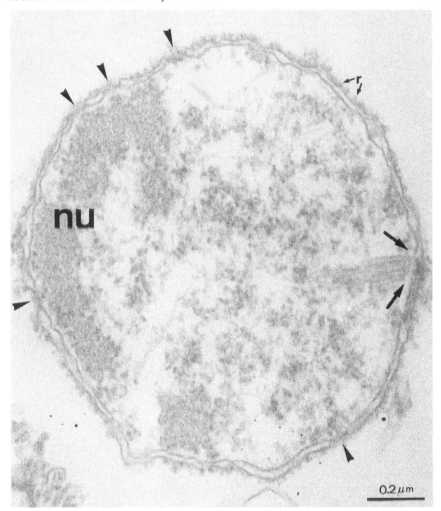

FIG. 2   Thin-section electron micrograph of an isolated nucleus from *Saccharomyces cerevisiae* showing nuclear pores and the nucleolus. Structural features of the nuclear periphery are easily identified in thin-section electron micrographs of isolated yeast nuclei. Ribosomes (r) associate with the outer nuclear membrane and the nucleolus (nu) is apposed to the inner nuclear membrane. Nuclear pore complexes are indicated by arrowheads. Note the presence of pores in close vicinity to the nucleolus. Microtubules emanate from the spindle pole body (arrows).

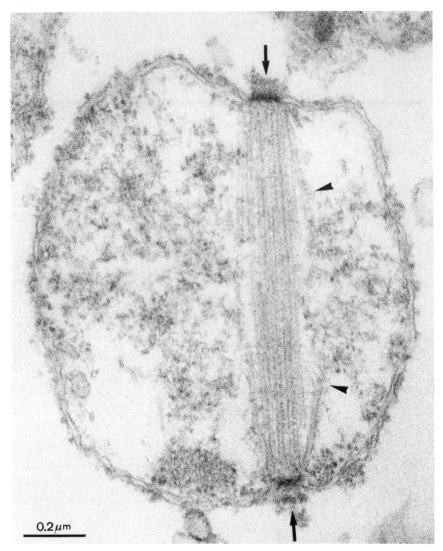

0.2 μm

FIG. 3   Electron micrograph of an isolated nucleus from *Saccharomyces cerevisiae* showing spindle pole bodies and the mitotic spindle. Spindle microtubules span the nucleoplasma, interconnecting the two spindle pole bodies (arrows). In the spindle pole bodies two distinct plaques are identified: a lightly staining cytoplasmic plaque and a dense plaque at the plane of the nuclear envelope. The spindle microtubules originate from a third, inner plaque which is better observed in negatively stained isolated spindles or in meiotic cells. Microtubules are either continuous, extending from pole to pole, or discontinuous (chromosomal microtubules; arrowheads).

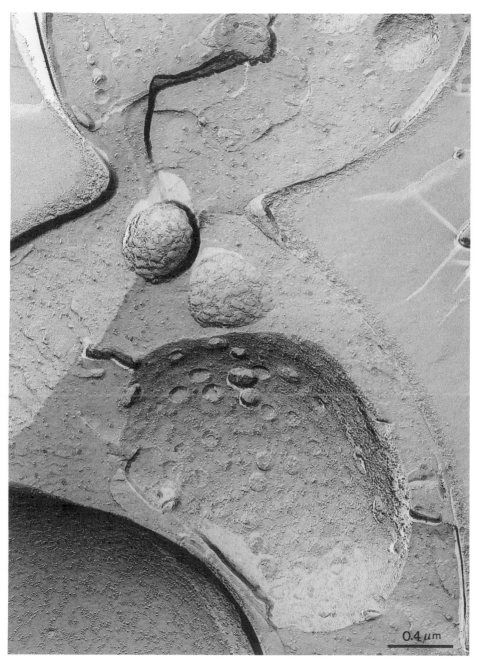

FIG. 4 Freeze-fracture electron micrograph of a budding yeast cell. Clusters of nuclear pores and pore-free areas of the nuclear envelope are observed within the nuclear membrane of *Saccharomyces cerevisiae*.

Several architectural models have been proposed for the nuclear pore complex (NPC), based on EM analysis (Franke *et al.*, 1981; Unwin and Milligan, 1982; Akey, 1989; Reichelt *et al.*, 1990). In general, all models point to an eightfold symmetrical arrangement of nuclear pore components: nuclear and cytoplasmic "rings" or "annuli," central "spokes," and a "plug" or "central granule." In addition, fibrils have been seen to extend from the annuli toward the nucleoplasm and cytoplasm (Franke, 1974; Scheer *et al.*, 1988; Richardson *et al.*, 1988). In the cytoplasm some of these fibrils may interact with intermediate filaments (Carmo-Fonseca *et al.*, 1987; Carmo-Fonseca and David-Ferreira, 1990).

The overall structure and diameter (80–120 nm) of the NPC are highly conserved throughout evolution (Franke, 1974; Maul, 1977). In a detailed ultrastructural study, Allen and Douglas analyzed the organization of the NPC in *S. cerevisiae* and presented convincing evidence that yeast NPCs exhibit a conserved structure (Allen and Douglas, 1989). Yeast NPCs were either analyzed in purified nuclei which retained an intact nuclear envelope or in a so-called NPC/lamina preparation which was obtained by extraction of purified nuclei with detergent, nucleases, and high-salt buffers. Electron microscopy of this insoluble nuclear fraction revealed a network of filamentous structures and remnants of NPCs, suggesting a similarity with the NPC/lamina fraction from rat liver (Allen and Douglas, 1989).

In yeast, isolated NPCs are associated with different types of filaments and the EM image resembles a "wagon wheel." A 5- to 9-nm filament radiates from the pore complex center ("hub") to the "rim" of the wheel which is built up by an 8- to 11-nm-thick filament. In molecular terms, these filaments are undefined but could be part of a nuclear lamina. Nuclear fractions enriched in NPCs have been solubilized so that NPCs can be isolated free of most filamentous attachments (Allen and Douglas, 1989). Isolated pore complexes still reveal eightfold symmetry, but the central plus and other structural details are lost. SDS-PAGE analysis of the NPC preparation shows approximately 15 abundant proteins, ranging from 27 to 300 kDa, but the nature of these polypeptides is not known. Lamins, which are prominent protein bands in corresponding higher eukaryotic fractions, could not be seen in the yeast preparation.

2. Nuclear Lamina

The nuclear lamina is an intermediate filament network that is attached to the nucleoplasmic surface of the nuclear envelope in higher eukaryotic cells (Gerace, 1986; McKeon, 1987). Several important functions have been assigned to the nuclear lamina: the lamina may be a structural support for the nuclear envelope, an attachment site for other intermediate filaments, involved in higher order chromatin organization, and, finally, may mediate nuclear envelope disassembly and assembly during mitosis (Aebi *et al.*, 1986; Benavente and Krohne, 1986; Georgatos and Blobel, 1987; Worman *et al.*, 1988; Glass and Gerace, 1990; Peter *et al.,* 1990). Some of these functions are also expected to be important in yeast and, accordingly, yeast may contain a nuclear lamina. On the other hand, yeast undergoes a closed mitosis and the nuclear envelope is not disassembled/assembled during mitotic division (Byers, 1981). If a nuclear lamina is only involved in this process, it may not exist in yeast.

In the process of purifying nuclear pores from yeast, Allen and Douglas (1989) obtained a NPC fraction associated with a filamentous structure resembling a nuclear lamina. The diameter and the properties of these yeast 10-nm lamina filaments attached to the nuclear pore filaments come close to those of other intermediate filament-forming proteins. These studies therefore suggest that yeast has the morphological equivalent of the higher eukaryotic lamina, but this can still only be inferred until sequence analysis is completed (see also Section IV,B). Homologs of either the cytoplasmic or nuclear family of intermediate filament proteins have not yet been identified in yeast.

Cytoplasmic filaments with a diameter characteristic of intermediate filaments (10 nm) but of unknown biochemical nature can be observed at the mother-bud neck of dividing cells (Byers and Goetsch, 1976; Haarer and Pringle, 1987). A filamentous interaction between the nuclear envelope and the plasma membrane is seen in EM photographs of nuclear preparations from yeast (Allen and Douglas, 1989; Cardenas *et al.,* 1990). This has been taken as evidence that a network of cytoskeletal filaments extends from the nuclear periphery to the plasma membrane.

3. The Nuclear Envelope During Cell Division

Saccharomyces cerevisiae divides by budding and, as a bud forms and enlarges on the parental cell, the nucleus becomes elongated and extends a process into the daughter cell (Matile *et al.,* 1969; Byers, 1981; Pringle and Hartwell, 1981). When the nucleus extends maximally between mother and daughter cells it has a dumbbell shape, with a constriction at the bud neck. The use of antibodies that recognize nuclear pore proteins as immunofluorescent markers has allowed a direct assessment of the nuclear envelope in unsectioned, whole cells (Davis and Fink, 1990). The observations confirm the earlier conclusion, reached by EM, that the yeast nuclear envelope does not break down during mitosis. The nuclear envelope first protrudes toward the bud neck and then invades the bud as a narrow, elongated structure. Extension of the nuclear envelope into the bud is closely followed by migration of chromatin, as visualized by fluorescent staining of DNA. Unlike that of most eukaryotic cells, yeast chromatin is not seen to condense into discrete chromosomes, at least not in the mitotic cell cycle.

The mitotic spindle is composed of microtubules that emanate from the spindle plaques at the nuclear envelope (Byers, 1981; King and Hyams, 1982; Kilmartin and Adams, 1987). The spindle is first oriented obliquely to the budding axis, but later it reorients along the axis of the nucleus. Finally, the spindle breaks down and the two daughter nuclei become separated by a fusion of the nuclear envelope at the bud neck. Following nuclear division cytokinesis quickly ensues.

4. Spindle Pole Body

When thin sections of dividing yeast cells were first examined in the EM, spindle microtubules were seen spanning the nucleoplasm (Robinow and Marak, 1966). The microtubules emanate from two dense plaques, the spindle pole bodies (SPBs), which are embedded in the nuclear envelope (Byers, 1981; Pringle and Hartwell, 1981; King and Hyams, 1982). The SPB duplicates at the beginning of the cell cycle and the two SPBs remain juxtaposed in the nuclear envelope connected by a bridge structure. Then, spindle microtubules start to assemble from the SPBs. The two short half-spindles interdigitate to form a bundle of continuous microtubules that separate the two SPBs (Fig. 3). In addition, there are discontinuous microtubules which arise from the intranuclear surface of the SBP and terminate in the nucleoplasm, and astral or cytoplasmic microtubules which emanate from the outer surface of the SPB and extend into the cytoplasm. The SPB thus acts as a microtubule-organizing center analogous to the centrosome of animal cells.

Electron microscopic evaluation of temperature-sensitive cell division cycle (cdc) mutants has shown that emergence of the bud is coincident with duplication of the SPB. Mutants unable to replicate the SPB do not form a bud and mutants blocked in SPB separation can form multiple buds (Byers and Goetsch, 1974). Thus, duplication and separation of SPBs appear to play a key role in nuclear division. More recent data suggest that, in addition to SPBs, microtubules may be directly involved in migration and proper orientation of the yeast dividing nucleus (Jacobs *et al.,* 1988).

Electron microscopic analysis of both thin sections of glutaraldehyde–osmium-fixed yeast cells and negatively stained isolated spindles revealed that the SPB is a multilaminar structure embedded in the nuclear envelope (Peterson and Gray, 1972; Byers and Goetsch, 1975; King and Hyams, 1982). At least three distinct plaques or layers can be identified. The lightly staining outer cytoplasmic plaque anchors the cytoplasmic microtubules. The inner nucleoplasmic plaque appears to be the origin of the spindle microtubules. In addition, there is a densely staining central plaque in the plane of the nuclear envelope (see also Fig. 3). Several groups have tried to identify components of the SPB using biochemical, immunological, and genetic approaches (see Section IV,C).

5. Association of the Nucleolus with the Nuclear Envelope

When viewed in the electron microscope and by indirect immunofluorescence, the nucleolus of *S. cerevisiae* appears as a "dense crescent" region which lacks the distinct subcompartmentalization (Sillevis Smitt *et al.*, 1972, 1973; Hurt *et al.*, 1988; Aris and Blobel, 1988; Yang *et al.*, 1989; Potashkin *et al.*, 1990) observed in higher eukaryotic nucleoli (see, for example, Scheer and Benavente, 1990). However, as in all other eukaryotes, the yeast nucleolus appears to be functionally engaged in ribosome biogenesis (Sillevis Smitt *et al.*, 1972, 1973; Trapman *et al.*, 1975). Ultrastructural and immunofluorescence analysis of yeast nuclei reveals that the crescent-shaped nucleolus, which occupies approximately one-third of the nucleus, is tightly associated with the nuclear envelope (Molenaar *et al.*, 1970; Sillevis Smitt *et al.*, 1972) (Fig. 2). A spatial relationship between the nucleolus and the nuclear envelope has also been observed in higher eukaryotes (Bourgeois and Hubert, 1988). Antibodies directed against NSP1 (see Section IV,A) as a marker for the nuclear pores and NOP1 (Aris and Blobel, 1988; Schimmang *et al.*, 1989; Henriquez *et al.*, 1990) as a marker for the nucleolus have allowed a direct visualization by immunofluorescence of the spatial organization and interaction of the nucleolus and nuclear envelope; clusters of NPCs as visualized by anti-NSP1 antibodies are often observed in regions of the nuclear envelope directly apposed to the nucleolus (Fig. 5). Because nuclear pores are involved in transport between the nucleus and the cytoplasm, the presence of pores close to the nucleolus may reflect a kinetically efficient short-circuit pathway for transport of ribosomal precursor particles into the cytoplasm without passage through the nuclear matrix compartment.

Using antibodies against nucleolar and SPB antigens, Yang *et al.* (1989) demonstrated that the yeast nucleolus preferentially localizes at the opposite site of the SPB in the interphase nucleus. The functional significance of this phenomenon is not known, but it is tempting to speculate that a structural communication between the SPB and nucleolus is essential for the organization of interphase nuclear structure, nuclear division, and/or nucleolar segregation. Taken together, these observations point to the presence of higher-order spatial organization of the yeast nucleus. How this is achieved on a molecular level is a challenging topic for future research, and yeast genetics should help in addressing this structural problem.

6. Nucleoskeleton and Nuclear Scaffold

The higher order organization and compaction of interphase chromatin into looped domains are maintained by nonhistone proteins which are part of an internal nuclear structure called the nuclear scaffold or nuclear matrix (Gasser *et al.*, 1989). Nuclear scaffolds can be isolated from *S. cerevisiae* and they retain the capability to specifically bind scaffold-attached regions (SAR) which, in yeast, are known to be centromeric elements and origins of replication (ARS elements). The morphological organization of the isolated yeast nuclear scaffold (obtained by extracting the histones from nuclei with lithium diiodosalicylate) is filamentous, similar to preparations from higher eukaryotes (Jackson and Cook, 1988). The yeast preparation appears to lack the peripheral nuclear lamina meshwork (Cardenas *et al.,* 1990). In contrast, the corresponding higher eukaryotic scaffold fraction contains, in addition, the characteristic peripheral ring of nuclear lamins. Furthermore, the yeast scaffold preparation does not have the highly conserved lamin proteins, because many tested antilamin antibodies which are cross-reactive among many species did not cross-react in yeast.

Therefore, the questions of whether and how a yeast nuclear scaffold is associated with the nuclear periphery, and whether a nuclear lamina forms a bridging meshwork of filaments between a nuclear matrix and the nuclear envelope are still controversial and need further clarification (see also Section II,B,2).

7. Karmellae

In most cells, the outer nuclear membrane is studded with ribosomes and presents continuities with both the smooth and rough endoplasmic reticulum (ER) (Gerace and Burke, 1988) (see also Fig. 2). In consequence, the perinuclear space is directly connected to the lumen of the ER. Lipids and integral membrane proteins could in principle diffuse from the ER to the outer and inner nuclear membranes through membrane continuity at the pore level. Exchange of proteins between the outer and inner nuclear membranes has in fact been demonstrated in virus-infected mammalian cells (Torrisi and Bonnatti, 1985). The outer nuclear membrane is a major site of protein synthesis and it can be considered as a "minimal" ER for any cell (Franke, 1974). Although functionally very similar, the distribution of specific enzymes varies between the outer nuclear membrane and the more peripheral ER (Fahl *et al.,* 1978). Also, the translation of mRNAs encoding the inducible smooth ER enzyme 3-hydroxy-3-methylglutaryl-CoA (HMG-CoA) reductase occurs primarily at the outer nuclear membrane (Pathak *et al.,* 1986).

In yeast, the overproduction of the HMG-CoA reductase (an enzyme which catalyzes the rate-limiting step in sterol biosynthesis) leads to a drastic increase in intracellular membranes, which are associated with the nuclear envelope as lamellar stacks of paired membranes (Wright *et al.,* 1988). These nuclear-associated membranes, called karmellae, contain high amounts of HMG-CoA reductase, as shown by immuno-EM. The karmellae do not completely surround the nuclear envelope. The nucleus always keeps an akarmellar region consisting only of the authentic nuclear membrane, which at this site is enriched in nuclear pore complexes. One can expect that most of the nucleocytoplasmic exchange reactions occur in these akarmellar regions of the nuclear envelope. Surprisingly, karmellae stay associated with the mother nucleus in mitosis, whereas the daughter nucleus inherits a nuclear envelope free of karmellae stacks. Furthermore, the akarmellar region of the nucleus is always oriented toward the site where the bud emerges. How the asymmetry of nuclear envelope/nuclear pore orientation and membrane distribution is achieved and maintained in these karmellae-containing yeast cells remains to be shown.

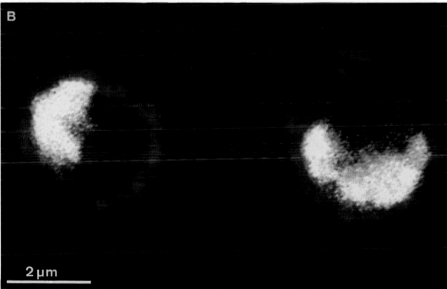

FIG. 5 Immunolabeling of the yeast nuclear envelope as analyzed by confocal fluorescence microscopy. (A) Indirect immunofluorescence was performed on whole yeast cells (*Saccharomyces cerevisiae*) using an anti-NSP1 affinity-purified polyclonal antibody. A punctate staining of the nuclear envelope is observed. (B) Double immunofluorescence was performed on a tetraploid yeast strain using anti-NSP1 antibody (red staining) as a marker for the nuclear pores and anti-NOP1 monoclonal antibody (green-yellow staining) as a marker for the nucleolus. The nucleolus is tightly associated with a region of the nuclear envelope rich in pore complexes.

III. Functions of the Nuclear Envelope

A. Nucleocytoplasmic Transport

Nuclear pores are the major sites of entry and exit for nucleocytoplasmic transport. The mechanism of nuclear transport and components involved in this process have been extensively reviewed (Dingwall, 1985; Dingwall and Laskey, 1986; Silver and Goodson, 1989; Burke, 1990; Nigg, 1990; Garcia-Bustos *et al.*, 1991a; Silver, 1991). Most of our knowledge of how proteins are imported into the nucleus stems from *in vivo* systems involving microinjection of nuclear reporter molecules into intact cells (Bonner, 1975a,b; Goldfarb *et al.*, 1986; Yoneda *et al.*, 1987; Dingwall *et al.*, 1988), from *in vitro* import systems using nuclei assembled from *Xenopus* egg interphase extracts and isolated DNA (Newmeyer *et al.*, 1986a,b; Finlay *et al.*, 1987), and recently also from permeabilized mammalian cells able to import exogenously added fluorescently labeled nuclear proteins (Adam *et al.*, 1990).

1. *In Vivo* Assays

In vivo approaches in the *Xenopus* oocyte system were the first to measure protein migration into the nucleus (Bonner, 1975b). With the advent of modern yeast molecular biology, artificial nuclear proteins (generally fusion proteins consisting of a nuclear-targeting signal attached to a nonnuclear reporter protein) can be expressed in yeast and their subsequent nuclear accumulation analyzed by indirect immunofluorescence (Hall *et al.*, 1984, 1990; Silver *et al.*, 1984; Moreland *et al.*, 1985, 1987; Pinkham *et al.*, 1987; Underwood and Fried, 1990). This gene fusion approach in yeast to measure *in vivo* nuclear transport led to the identification of nuclear localization sequences (NLS) in various yeast nuclear proteins such as transcription factors (Hall *et al.*, 1984; Silver *et al.*, 1984; Pinkham *et al.*, 1987), ribosomal proteins (Moreland *et al.*, 1985; Underwood and Fried, 1990), and histones (Moreland *et al.*, 1987). From this analysis, it subsequently became clear that yeast and higher eukaryotic NLS are structurally similar, containing a high portion of positively charged amino acids and resembling the prototype of a NLS, the nuclear-targeting sequence of SV40 large T antigen (Kalderon *et al.*, 1984). Furthermore, the NLS from SV40 large T antigen is also functional in the yeast system, directing corresponding fusion proteins into the nucleus (Nelson and Silver, 1989). This is further evidence for an evolutionary conservation of the nuclear import machinery between species.

Some peculiar features concerning the mechanism of nuclear import were detected in yeast. Whereas most of the fusion proteins between Matα2 (a transcriptional repressor) and *Escherichia coli* β-galactosidase were localized inside the nucleus, some particular constructs caused cell death in yeast; this was attributed to a partial mislocation of the hybrid protein (e.g., at the nuclear pore complex) and thereby a jamming up of the import machinery (Hall *et al.*, 1984). Furthermore, it was shown that Matα2 contains two distinct nuclear import signals, one at the amino-terminal end and the second far away within the protein sequence, each of them being able to direct *lacZ* into the nucleus (Hall *et al.*, 1990). The amino-terminal signal resembles other NLSs; the internal one is part of the Matα2 homeodomain. Deletion of the internal NLS reduces nuclear transport and the mutant protein accumulates at discrete sites on the nuclear envelope, most likely the nuclear pores. It was suggested that the two different signals are not functionally equivalent but acting in series: the amino-terminal NLS binds first to the import receptor on the nuclear envelope and the internal NLS mediates translocation through the NPC (Hall *et al.*, 1990). Although such a two-step translocation mechanism (binding to the nuclear pore, which is ATP-independent, followed by an ATP- and temperature-dependent translocation) has been shown to exist in the higher eukaryotes (Newmeyer and Forbes, 1988; Richardson *et al.*, 1988; Breeuwer and Goldfarb, 1990), it is not clear whether two different types of NLS are required for this two-step mechanism.

As shown by a gene fusion approach in yeast, low-molecular-weight histones (documented for histone H2B) also contain a NLS signal (Moreland *et al.*, 1987). This was unexpected, since histones are small molecules which, according to their size, can passively diffuse through the pores into the nucleus and do not require active transport. Facilitated (probably NLS-mediated) nuclear transport of histone H1 and other small nucleophilic proteins is also true for mammalian cells (Breeuwer and Goldfarb, 1990). It is likely that diffusible small nuclear proteins enter the nucleus much faster when they carry a nuclear-targeting signal. Unexpectedly, mutation or deletion of the histone H2B NLS did not abolish nuclear transport, suggesting that histones H2A and H2B may be cotransported to the nucleus as a heterodimer (Moreland *et al.*, 1987). Examples of this "piggy-back" mechanism of nuclear entry are known from other systems (Loewinger and McKeon, 1988).

In the yeast genetic system, *in vivo* nuclear import can be coupled to complementation assays to prove that NLS-mediated nuclear import is essential for life (Underwood and Fried, 1990). This was shown by first defining and then mutating the NLS within the ribosomal protein L29. When a mutant L29 ribosomal protein with both NLS signals mutagenized was expressed in a haploid yeast strain carrying a disrupted chromosomal L29 gene (L29 is an essential gene in yeast), these cells almost completely stopped cell growth as a consequence of inhibition of ribosome biogenesis. In these mutant cells, L29 was inefficiently imported into the nucleus and therefore the 60S ribosomal subunit could not be assembled inside the nucleolus. This powerful *in vivo* complementation assay coupled to nuclear transport may be applied to select for both intragenic and extragenic suppressor mutations which could help to unravel the mechanism of nuclear transport in yeast.

2. *In Vitro* Assays

In vitro or semi-*in vitro* nuclear import systems which reliably mimic the *in vivo* transport process were until recently not available for yeast. However, a homologous cell-free nuclear localization system consisting only of yeast *S. cerevisiae* components would be desirable, particularly in the study of putative mutants defective in nuclear protein transport. In a homologous *in vitro* system, the primary defect of a putative nuclear import mutant could be directly tested and deficiency could be complemented by extracts and fractions derived from wild-type yeast cells. Furthermore, antibodies against yeast nuclear envelope and NPC components could be tested for their inhibitory effect on nuclear protein accumulation.

Progress was made recently with reports of cell-free transport systems into purified yeast nuclei from two different laboratories (Kalinich and Douglas, 1989; Garcia-Bustos *et al.*, 1991b). Kalinich and Douglas (1989) purified yeast nuclei rapidly and in high yield, free from other subcellular contaminants including cytoskeletal elements, and used these nuclei to import the radiolabeled nuclear proteins SV40 large T antigen and *Xenopus* nucleoplasmin, prepared from the cloned genes by coupled *in vitro* transcription/translation. Both heterologous nuclear proteins were efficiently imported into the purified nuclei since they were protected against exogenously added trypsin–agarose, which could not penetrate through the nuclear pores and thus was excluded from the nuclei. Furthermore, nuclear uptake was ATP-, calcium-, time-, and temperature-dependent and required a functional nuclear localization signal. A mutant form of SV40 large T antigen having a defective NLS or nonnuclear proteins were not imported. This suggests that the yeast nuclear envelope recognizes NLSs from heterologous proteins by a receptor machinery which is similar to the higher eukaryotic nuclear import system. The putative receptors which bind NLS appear to be exposed on the surface of the nuclear envelope (see below); shaving the nuclear surface with protease prior to incubation with the nuclear proteins blocked nuclear transport. Surprisingly, the lectin wheat germ agglutinin, known to block nuclear transport reactions efficiently in the higher eukaryotes (Finlay *et al.*, 1987), failed to inhibit protein import into isolated yeast nuclei. This demonstrates that, although there are many similarities between the two import systems from yeast and higher animal cells, fundamental differences also exist.

Hall and co-workers (Garcia-Bustos *et al.*, 1991b) reported a similar cell-free nuclear import system which confirmed and extended the observations made in the transport system described above. In addition, it was shown that authentic yeast nuclear proteins, the transcription factors MCM1 and STE12, are translocated across the nuclear envelope of purified yeast nuclei, as determined by their inaccessibility to externally added immobilized trypsin protease after the transport.

Unfortunately, the exact location of the *in vitro* imported nuclear proteins could not been shown by direct means (e.g., by indirect immunofluorescence), as is possible in the *in vitro* nuclear transport system reconstituted from *Xenopus* extracts (Newmeyer *et al.,* 1986a), since the radiolabeled proteins were only present in substoichiometric quantities. In yeast, it is also unclear to what extent *in vitro* imported proteins accumulate inside the nucleus and why they become inextractable after association with purified nuclei (only a minor portion is released on salt and detergent treatment of nuclei containing the imported substrate). In other established *in vitro* import assays, such as the mitochondrial transport system, imported soluble proteins can be easily released from the organelle after disrupting the membrane barrier with detergents (Hurt *et al.,* 1984). It is therefore desirable to find new nuclear import substrates which can be released from the nuclear interior after removing the nuclear membrane with detergents.

3. NLS-Binding Proteins

The rate of uptake of nuclear proteins (conjugates between NLS peptides and BSA) is saturable, suggesting the existence of nuclear import receptors (Goldfarb *et al.,* 1986). In order to identify receptors which bind nuclear localization sequences, cell-free binding assays have been applied, both in higher eukaryotic systems (Adam *et al.,* 1989; Li and Thomas, 1989; Yamasaki *et al.,* 1989; Meier and Blobel, 1990) and in yeast (Silver *et al.,* 1989; Lee *et al.,* 1989, 1991; Stochaj *et al.,* 1991). In *S. cerevisiae,* NLS-binding proteins of 70 and 59 kDa (Silver *et al.,* 1989) and 67 kDa (Lee *et al.,* 1989) were found by ligand blotting techniques using yeast nuclear proteins immobilized to nitrocellulose and labeled NLS–peptide conjugates. Both the 70- and 67-kDa NLS-binding proteins are nuclear and the latter is firmly attached to nuclear envelopes. The 67-kDa NLS-binding protein has been cloned and sequenced and an initial molecular characterization of this protein, named NSR1, has been reported (Lee *et al.,* 1991). NSR1 is not essential, but cells with a disrupted gene copy of NSR1 are very sick. It was shown that the NSR1 gene product indeed is responsible for specific *in vitro* NLS-binding and thus could be involved in nuclear transport; however, the location of NSR1 inside the nucleus (most likely within the nucleolus) and the presence of two RNA-recognition motifs and a glycine/arginine-rich repeat domain in the protein sequence make it possible that the protein plays a role in ribosomal rRNA metabolism (Lee *et al.,* 1991). Since nucleolar proteins can shuttle between the cytoplasm and the nucleus (Borer *et al.,* 1989), NSR1 could also play a role in the transport of ribosomal subunits into the cytoplasm. The genetic system of yeast should allow these possibilities to be distinguished.

Stochaj *et al.* (1991) further characterized their 70-kDa NLS-binding protein and presented evidence that it is a component of the nuclear import machinery involved in nuclear protein translocation. At present it is not clear whether this protein is identical to NSR1. The 70-kDa NLS-binding protein was purified by affinity chromatography using column resins with bound NLS from SV40 large T antigen. Antibodies against the 70-kDa NLS-binding protein confirmed that it is associated with nuclei by both subcellar fractionation and indirect immunofluorescence. In comparison to the immunofluorescent localization of the nuclear pore proteins NSP1 and NUP1, a portion of the 70-kDa NLS-binding protein (but not all) may be similarly distributed at the nuclear periphery. A role of the 70-kDa NLS receptor in nuclear transport was measured in an *in vitro* nuclear binding assay (Silver *et al.*, 1989). Binding of ^{125}I-labeled SV40 NLS-HSA (human serum albumin) is reduced in purified yeast nuclei preincubated with anti 70-kDa antibodies or in salt-treated nuclei (salt treatment removes the NLS-binding proteins). The binding activity, however, can be reconstituted to salt-treated nuclei by adding a fraction enriched in NLS-binding proteins. Similarly, antibodies to the NSP1 inhibit binding of the NLS conjugate to purified nuclei. Finally, protease treatment of purified yeast nuclei reduces binding of NLS conjugates, supporting a model in which these receptors are exposed on the nuclear surface; nuclear proteins may first bind to the receptors (e.g., the 70-kDa NLS-binding protein), which can be present in the cytoplasm or on the nuclear surface. The nuclear protein may then be released and imported or cotranslocated with the mobile receptor through the NPC. The cloning of this NLS-binding protein and the generation of yeast mutants should allow testing of this model.

4. Nuclear Transport Mutants

Mutants defective in nuclear protein transport may be identified by the accumulation or appearance of nuclear reporter proteins within the cytoplasm. NLS signals are not removed from the translocated nuclear protein. Therefore, a cytoplasmic appearance of a nuclear protein blocked in translocation can only be detected by immunofluorescence, immuno-EM, or subcellular fractionation methods. Sadler *et al.* (1989) developed an *in vivo* approach to select for nuclear important mutants. A NLS was placed in front of the mitochondrial precursor protein of cytochrome c_1, a mitochondrial inner membrane protein. Within the resulting hybrid protein, the NLS was dominant over the mitochondrial presequence: the fusion protein was no longer imported into mitochondria, but accumulated inside the nucleus, as measured by the indirect immunofluorescence. Accordingly, yeast cells expressing the NLS–cytochrome c_1 in a background of a defective cytochrome c_1 chromosomal gene are respiration deficient due to the missorting of NLS–cytochrome c_1 into the nucleus. This glycerol-negative phenotype was exploited to isolate temperature-sensitive mutants which became respiration competent at the permissive temperature due to a mutation in the nuclear import machinery. Such npl (*n*uclear *p*rotein *l*ocalization) mutants could be obtained which redirected the NLS–cytochrome c_1 hybrid protein into the mitochondrion, giving rise to growth by respiration. The gene for NPL1 was isolated and encodes a protein with homology to DnaJ, an *E. coli* heat-shock protein. Surprisingly, *npl1-1* is allelic to a previously identified yeast mutant, sec63, which is required for protein translocation into the ER lumen (Rothblatt *et al.,* 1989). *npl1* mutants also affect translocation across the ER membrane. It was therefore suggested that NPL1/SEC63 acts in an early common step in translocation of proteins into the nucleus and the ER. Alternatively, the *npl1* mutation first affects the biogenesis of the ER membrane and, as a pleiotropic consequence, disturbs the biogenesis of the NPC or retention of proteins within the nucleus. Finally, *npl1* could directly affect the insertion of an integral membrane protein(s) localized at the NPC into the nuclear membrane, thereby blocking nuclear transport. An integral membrane protein (gp210) as a component of the NPC (Wozniak *et al.,* 1989; Greber *et al.,* 1990) was shown to be required for proper nuclear transport in higher eukaryotic cells (L. Gerace and G. Blobel, personal communication).

Barnes and Rine reported on an elegant nuclear import system which potentially can be exploited to isolate yeast mutants defective in nuclear transport (Barnes and Rine, 1985). The authors expressed functionally active *E. coli* restriction enzyme *Eco*RI under the regulatable *GAL1* promoter in *S. cerevisiae.* Under derepressed conditions, the endonuclease entered the yeast nucleus, cleaved the yeast genome at *Eco*RI-specific recognition sites, and thus killed the cells. Moreover, in *rad52* mutants, which are deficient in double-strand break repair, the toxic effects of *Eco*RI was enhanced. Such a mutant phenotype may be used to select for suppressors deficient in transporting the restriction enzyme (and other authentic nuclear proteins) into the nucleus. To date, no such mutant has been reported.

5. RNA and Ribosome Export from the Nucleus

Nucleocytoplasmic transport of RNA occurs as ribonucleoprotein particles (RNPs) through the NPCs. Ultrastructural studies have shown that RNPs often occupy the center of the NPC during their passage through the nuclear membrane (Maul, 1977). RNA export from the nucleus has been analyzed for U1 snRNAs in *Xenopus* oocytes (Hamm and Mattaj, 1990). A monomethylguanosine cap structure at the 5'-end of snRNA was shown to be a signal for nuclear export. Very little is known about RNA export in yeast. There is evidence that the formation of splicing complexes might prevent export of pre-mRNA, and only spliced mRNA is efficiently transported to the cytoplasm (Legrain and Rosbash, 1989). Unspliced mRNA was only exported if early steps of splicing formation were disturbed, e.g., by mutating the splicing consensus signals or *trans*-acting factors. If later steps in spliceosome formation were blocked, pre-mRNA was retained in the nucleus. Later steps in intron splicing thus appear to be coupled to nuclear export, but how this occurs mechanistically is completely unknown. It is interesting to note that components of the spliceosomes in yeast are located close to the nuclear periphery (see Section IV).

Nucleocytoplasmic transport of ribosomal particles is only poorly understood, EM studies indicate that nuclear pores are the sites from which ribosomal particles are exported, but the mechanism of transport remains unknown (Hadjiolov, 1985).

There is evidence that the maturation of a 43S preribosomal particle to the 40S ribosomal subunit takes place in the yeast cytoplasm coupled to processing of 20S to 18S rRNA. A similar cytoplasmic maturation step is not observed for the 60S subunit (Trapman *et al.,* 1975). Nucleocytoplasmic transport of ribosomes was studied *in vivo* by microinjection of radiolabeled ribosomes from different sources, including yeast, into *Xenopus* oocyte nuclei (Khanna Gupta and Ware, 1989; Bataillé *et al.,* 1990). These mature ribosomes were reexported from the nucleus by an apparent carrier-mediated process which exhibits saturation kinetics (Khanna Gupta and Ware, 1989). Ribosomal export was further inhibited by wheat germ agglutinin, suggesting that nucleoporins on the nucleoplasmic side of the nuclear pore complex are involved in this process.

B. Spindle Pole Body Duplication, Mitotic Spindle Formation, and Nuclear Division

There are a number of genetic approaches in yeast directed toward the identification of components required for nuclear division. A broad genetic approach has been used to isolate many cell division cycle (cdc) mutants (Hartwell *et al.,* 1973) and, among this collection, mutants particular defective in SPB functions, formation of mitotic spindles, and nuclear segregation have been obtained.

Temperature-sensitive (ts) mutations in seven genes *(cdc2, 6, 7, 9, 14,* and *15)* of *S. cerevisiae* confer a defect in nuclear division as determined by fluorescence light microscopy of yeast cells stained for DNA (Culotti and Hartwell, 1971). Depending on the cdc mutation, mutants terminate at the restrictive condition at an early, medial, or later stage of nuclear division.

The cell division cycle mutant *cdc31* is defective in the duplication of the SPB at the restrictive condition, resulting in a single SPB which is abnormally large in size (Byers, 1981; Baum *et al.,* 1986). DNA replication and other cell cycle processes, however, are normal in the *cdc31* mutant. The gene of CDC31 encodes a protein with significant homology to calmodulin and other members of the calcium-binding protein family (Baum *et al.,* 1986). This finding indicates that, in yeast, calcium is important for the duplication of the SPB and, as a consequence, is required for the organization of the mitotic spindle. Nothing is known about the subcellular distribution of the CDC31 gene product within the yeast cell. However, CDC31 shows 50% sequence identity to caltractin, a basal body-associated calcium-binding protein from *Chlamydomonas reinhardtii* (Huang *et al.,* 1988a,b) and thus may be the functional homolog. In analogy to the association of caltractin with the basal body in *Chlamydomonas,* it is likely that the CDC31 protein is associated with the functionally analogous organelle in yeast, the SPB.

Following an elegant genetic approach (the mutant screen was designed to detect ts mutations affecting spindle function by assaying for the formation of diploid cells from haploids), a new gene product, ESP1, was identified in yeast which is required for the regulation of the SPB duplication (Baum *et al.,* 1988). In an esp1 ts mutant, deregulation of SPB duplication is observed. At the non-permissive temperature, mutant cells no longer continue to synthesize DNA and cell division is stopped, but spindle pole duplication still continues for several cycles. As a consequence, the uninucleate mutant cells contain many (up to eight) SPBs within the nuclear envelope and multipolar spindles. In most cdc mutants, nuclear division is generally completed before cells enter a new cell cycle with SPB duplication (Pringle and Hartwell, 1981); in the *esp1* mutant, however, SPB duplication is negatively regulated, which requires a functional ESP1 gene. If this gene is defective, several cycles of SPB duplication can still proceed, whereas cell division stops. The SPB regulation cycle thus can be uncoupled from other cell cycles processes. Nothing is known about the nature of the ESP1 gene product and its subcellular location, but ESP1 might be a regulatory protein of SPB duplication, e.g., by affecting the phosphorylation or the calcium binding of structural components of the SPB. It will be worth further exploiting the above-described genetic approach to select for yeast mutants with increased ploidy in order to find new genes involved in SPB regulation.

A cold-sensitive mutation in the NDC1 gene causes failure of chromosome separation in mitosis at the nonpermissive temperature, but does not block cell division (Thomas and Botstein, 1986). In this mutant, daughter cells are formed by budding, but they become haploid, whereas mother cells become double in ploidy. The SPBs are properly segregated, but chromosomes remain associated with one SPB. NCD1 appears to be required for attachment of chromosomes to the SPB.

C. Nuclear Fusion

A genetic approach exploiting deficiency in nuclear fusion (karyogamy) was successfully used to identify novel components possibly associated with the SPB (Rose and Fink, 1987; Meluh and Rose, 1990). Yeast cells of opposite mating type mate by first initiating cell wall and plasma membrane fusion followed by the fusion of the haploid nuclei. Nuclear fusion mutants can be identified by their failure to form diploids, and mutations in three genes showing a deficiency in karyogamy *(KAR1, KAR2, KAR3)* have been isolated (Conde and Fink, 1976; Polaina and Conde, 1982; Berlin *et al.,* 1990a). From the functional point of view, *KAR3* is best understood from its effect on nuclear fusion. *KAR3* has been cloned by complementation of the karyogamy defect of the *kar3-1* mutation (Meluh and Rose, 1990). The *KAR3* gene is essential for nuclear fusion during the mating process and, accordingly, the expression of *KAR3* is strongly stimulated by the α-factor mating pheromone. By indirect immunofluorescence *KAR3* was localized to the cytoplasmic vicinity of the SPB and to cytoplasmic microtubules (Meluh and Rose, 1990; Alfa and Hyams, 1990). *KAR3* encodes a protein which is homologous to the microtubule motor protein kinesin. The location of *KAR3* and its homology to the force-generating motor domain of kinesin suggest a simple molecular model for how this microtubule-binding motor protein mediates nuclear fusion (Meluh and Rose, 1990): KAR3 could bind via its amino-terminal domain to a component of the two individual SPBs of the as yet unfused nuclei. In this stage the nuclei are interconnected by an array of cytoplasmic microtubules; the carboxy-terminal motor domain of KAR3 would interact with the cytoplasmic microtubules emanating from the SPBs. The force which is generated by the kinesin motor domain then pulls microtubules into the SPB, and, by that movement, nuclei are pulled together. When nuclei are close enough and the cytoplasmic microtubules are shortened due to continuous depolymerization of microtubules at the SPB, the nuclear membranes could fuse to form the diploid nucleus.

The specific role of *KAR1* for karyogamy in *S. cerevisiae* is less clear as compared to *KAR3* (Rose and Fink, 1987). *KAR1* is not only important for nuclear fusion, but is also essential for mitotic cell growth. Temperature-sensitive mutants of *KAR1* arrest with a typical cdc phenotype, and SPB morphology and microtubule organization are altered. SPBs no longer duplicate or separate and their structure is abnormal. Furthermore, cytoplasmic microtubules are long and aberrant. The morphology of the nuclear envelope is altered. In addition, overproduction of *KAR1* leads to cell cycle arrest and an abnormal SBP morphology similar to that found in cdc31 mutants. Taking these abnormalities together, it is likely that the *KAR1* gene product is a component of a multimeric complex, most likely the SPB.

A new group of kem (*k*ar *e*nhancing *m*utation) mutants affecting nuclear fusion in *S. cerevisiae* was identified by exploiting the leaky phenotype of the *kar1-1* mutation which shows a residual nuclear fusion activity (Kim *et al.*, 1990). In a *kar1-1* background, kem mutants were isolated which define three genes, *KEM1*, *KEM2,* and *KEM3,* and reduce or abolish the residual nuclear fusion of *kar1-1* (Kim *et al.*, 1990). The *KEM1* gene has been characterized in more detail. It is not essential, but cells containing a *kem1* null allele grow poorly, and have defects in SPB duplication and/or separation and aberrant nuclear division. There are further pleiotropic effects of *kem1* mutations on microtubule organization and chromosome separation which could all be explained by assuming that KEM1 is a structural component necessary for correct assembly of the mitotic spindle.

A protein which colocalizes with tubulin and the yeast SPB as determined by indirect immunofluorescence is BIK1 (Berlin *et al.,* 1990b). Double mutant strains containing mutations within the BIK1 and the α- or β-tubulin gene yield synthetic lethals consistent with a physical interaction between BIK1 and microtubules. This is further supported by the observation that a repeated sequence motif within BIK1 shows partial sequence homology to the mammalian microtubule-associated protein (MAP) tau. The central domain of BIK1, which reveals a heptad repeat organization characteristic of proteins which form a coiled-coil structure, may be involved in the assembly of higher-order structures and association with other components.

A dynamin-like protein encoded by the yeast sporulation gene *SPO15* appears to be involved in a microtubule-dependent separation of SPBs in meiosis (Yeh *et al.*, 1991). Electron microscopy of *spo15* cells shows a duplicated SPB which cannot be separated. Accordingly, these cells cannot built up the bipolar spindle required for meiotic division. The homology of SPO15 to dynamin, a microtubule-bundling protein, is indicative of the involvement of microtubule-sliding mechanisms for SPB separation and establishment of a mitotic spindle.

IV. Components of the Nuclear Envelope

Nuclear envelope proteins have been identified in yeast using biochemical, immunological, and genetic approaches.

A. Nuclear Pore Complex Proteins

Very little is known about the proteins which build up the NPCs. The molecular mass of 124 MDa for an individual pore complex (Reichelt *et al.*, 1990) suggests that this supramolecular structure is composed of many different proteins (probably more than a hundred), and first estimates support such an assumption (Snow *et al.*, 1987).

Most of the present knowledge of the biochemical composition of the NPC is derived from the mass-biochemical characterization of rat liver nuclei fractions enriched in pore complexes attached to the nuclear lamina (Gerace and Burke, 1988). This classical pore complex/lamina fraction was obtained by treating isolated nuclei with DNase, RNase, and high-salt buffers in order to remove chromatin and other internal contents (Dwyer and Blobel, 1976). The pore complex/lamina fraction from rat liver contains three major polypeptides of 60–70 kDa, which were shown by immuno-EM to be present in the lamina and absent from the pores. Using the combined approach of raising antibodies against nuclear fractions and electron microscopic immunolocalization, a number of pore components were subsequently identified (Krohne *et al.*, 1978; Fisher *et al.*, 1982; Gerace *et al.*, 1982; Berrios *et al.*, 1983; Davis and Blobel, 1986, 1987; Snow *et al.*, 1987; Park *et al.*, 1987; Wozniak *et al.*, 1989). One of these, gp210, is an integral membrane glycoprotein of the nuclear envelope located exclusively at the pores (Gerace *et al.*, 1982; Wozniak *et al.*, 1989; Greber *et al.*, 1990). Another class of NPC polypeptides, called nucleoporins, were identified by monoclonal antibodies and binding to wheat germ agglutinin because of *O*-linked *N*-acetyl-glucosamine (GlcNAc) glycosylation (Starr and Hanover, 1990). The best characterized member of this family has been termed p62 due to its apparent molecular weight (Davis and Blobel, 1986).

NPC proteins have also been found in the yeast *S. cerevisiae* (Aris and Blobel, 1989; Davis and Fink, 1990; Nehrbass *et al.,* 1990). Identification and cloning of NPC components in yeast will now facilitate the analysis of the *in vivo* role of such nuclear envelope proteins in nuclear pore structure and function; furthermore, by exploiting the excellent yeast genetics, new NPC components which interact physically should be identified soon. With the help of *in vitro* systems (e.g., the nuclear binding and transport system; Kalinich and Douglas, 1989; Garcia-Bustos *et al.,* 1991b; Stochaj *et al.,* 1991), the role of these pore proteins for nuclear transport can be extended to a cell-free analysis, and thus should give insight into the mechanism of regulation of nuclear transport by nuclear pores.

1. Yeast Nucleoporins

Aris and Blobel used a monoclonal antibody raised against rat liver NPC proteins to identify cross-reactive proteins in yeast (Aris and Blobel, 1989). In the mammalian system this monoclonal antibody (mAB414) reacts with different nuclear pore proteins, including the prominent nucleoporin p62 (Davis and Blobel, 1986, 1987). Nucleoporins belong to a protein family of NPC proteins which carry *N*-acetylglucosamine sugar modifications at serine/threonine residues (Snow *et al.,* 1987; Holt *et al.,* 1987; Park *et al.,* 1987; Davis and Blobel, 1987), are located in or close to the pore channel (Snow *et al.,* 1987; Park *et al.,* 1987), and play a role in nuclear transport (Featherstone *et al.,* 1988). The monoclonal antibody against rat liver nucleoporins recognizes by immunoblotting two proteins of 110 and 95 kDa in yeast which are bona fide nuclear envelope proteins, as revealed by subcellular and subnuclear fractionation. The monoclonal antibody is also active in indirect immunofluorescence on whole yeast cells and shows a punctate and patchy staining of the nuclear periphery that is indicative of nuclear pore labeling. Postembedding immuno-EM of isolated nuclei reveals the gold label mainly at the nuclear periphery, but whether this corresponds to nuclear pore labeling could not be determined with certainty because pore structures were poorly preserved during specimen preparation.

In summary, antibodies against mammalian NPC components cross-react with NPC constituents in yeast, suggesting that members of the nucleoporin family are evolutionary conserved and, accordingly, may play an essential role in nuclear pore structure and/or function.

2. NUP1

In a similar approach, Davis and Fink (1990) characterized yeast nuclear pore proteins using monoclonal antibodies directed against mammalian nucleoporins. On immunoblots, most of the tested monoclonals recognized a doublet of proteins around 100 kDa which most likely correspond to p110 and p95 initially identified by Aris and Blobel (1989), but one monoclonal antibody, mAB 306, in addition reacted with a further 130-kDa protein. Using the mAB 306 as a probe to visualize the nuclear envelope in yeast, the behavior of the nuclear envelope during the cell cycle could be followed by indirect immunofluorescence.

One anti-nucleoporin monoclonal antibody (mAB 350) was successfully employed in immunoscreening using a yeast genomic λgtll library to isolate the gene encoding the 130-kDa protein which was named NUP1 (*nuc*lear *p*ore). Several other immunopositive clones were obtained in this screen which may correspond to other yeast nucleoporins, but they have not been further characterized. *NUP1* is an essential gene and haploid progeny with a disrupted gene copy can germinate but arrest in the first mitotic division cycle with a large bud. Overproduction of *NUP1* using the strong *GAL* promoter is also lethal and cells stop growth, often in a multibudded stage. This shows that a correct stoichiometry of *NUP1* is important for a normal yeast cell cycle.

From the amino acid sequence, the NUP1 protein was divided into three different domains: an amino-terminal region which is charged and very acidic, a middle portion consisting of 28 degenerate 9-amino acid repeats separated by short and highly charged peptide stretches, and a short, slightly basic, C-terminal domain. NUP1 reveals sequence homology to another yeast nuclear envelope protein, NSP1 (*n*ucleo*s*keletal-like *p*rotein; Hurt, 1988; Nehrbass *et al.*, 1990), which contains a middle repetitive domain similar to that found in NUP1 (see also later). Since NUP1 and NSP1 share common structural motifs within the central repetitive domain, it is likely that this part contains the common immune determinant for the mammalian anti-nucleoporin antibodies. This could be proved by testing a yeast mutant which lacks the repetitive middle domain of NSP1. The truncated NSP1 protein is no longer reactive with the anti-nucleoporin monoclonal antibodies on immunoblots. These observations led to the postulate that yeast NUP1 and NSP1 belong to the nucleoporin protein family and are characterized by having degenerate repeat sequences in common (see also later). Interestingly, the yeast nucleoporins do not appear to be posttranslationally modified by *N*-acetylglucosamine and, accordingly, do not bind the lectin wheat germ agglutinin (Davis and Fink, 1990; Carmo-Fonseca *et al.*, 1991).

3. NSP1

The yeast nuclear pore protein NSP1 was initially identified in the course of raising antibodies against the insoluble fraction of yeast nuclei which can be referred to as the nucleoskeletal fraction or nuclear pore complex/lamina preparation (Hurt, 1988). One antibody out of this collection specifically reacted with a 100-kDa nuclear protein, the NSP1 protein. *NSP1* is an essential gene for cell growth and null mutants of *nsp1* arrest cell growth in a one- to two-cell stage. Strong overexpression of *NSP1* in yeast using the *GAL10* promoter is toxic and causes cell death (Hurt, 1989). In cells with overproduced *NSP1,* the cellular organization drastically changes and the cytoskeleton and nucleoskeleton collapse.

As shown by indirect immunofluorescence, NSP1 is exclusively located at the nuclear periphery. The immunolabeling of the nuclear envelope is heterogeneous, giving a ring-, dot-, and blob-like staining (Fig. 5). Larger blobs could colocalize with the SPB, but this may not be of functional significance: by immuno-EM, NSP1 could be clearly localized to the NPCs (Nehrbass *et al.,* 1990). In frozen thin sections of yeast cells, which gave the best immunolabeling results (Nehrbass *et al.,* 1990), or in purified nuclei (Fig. 6) the gold label is at the NPCs. This distribution resembles the immunogold labeling pattern of mammalian or am-

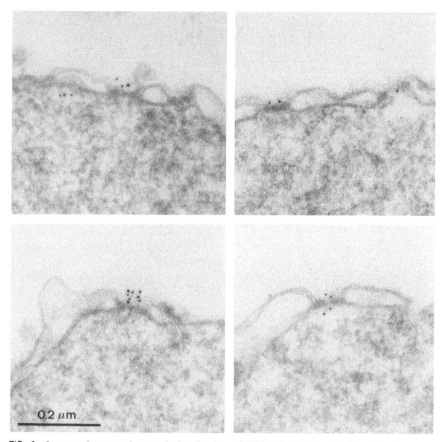

FIG. 6 Immunoelectron microscopic localization of NSP1. Isolated nuclei from *Saccharomyces cerevisiae* were incubated with anti-NSP1 affinity-purified polyclonal antibody and labeled with secondary antibody conjugated to colloidal gold particles (5 nm). NSP1 is located at the nuclear pore complexes.

phibian NPCs using anti-nucleoporin antibodies or wheat germ agglutinin (Snow et al., 1987; Park et al., 1987; Scheer et al., 1988). The location at or in the nuclear pore channel makes NSP1 a very interesting protein and suggests a role in nuclear transport, nuclear pore assembly, or structural organization of the nuclear periphery.

Insight into the possible function of NSP1 has come from the molecular and genetic analysis of NSP1 in living cells. NSP1 is organized into two different functional domains: a carboxy-terminal domain of approximately 220 amino acids which contains the essential function of the protein, and an amino-terminal and middle repetitive domain (the middle domain consists of 22 tandem repeats of a highly conserved 9-amino acid long sequence separated by a 10-residue long charged and hydrophilic sequence) which are dispensable for cell growth (Hurt, 1988; Nehrbass et al., 1990). The carboxy-terminal domain targets the protein to the nuclear periphery and mediates association with the NPCs. A fusion protein consisting of mouse cytosolic dihydrofolate reductase (DHFR) attached to the NSP1 carboxy-terminal domain was expressed in a yeast mutant lacking a functional nsp1 gene (Hurt, 1990). The DHFR-C-NSP1 construct complemented the nsp1 mutant, showing that the fusion protein is functionally active. Furthermore, the hybrid protein was localized at the nuclear periphery, as revealed by indirect immunofluorescence using antibodies against the DHFR moiety. These results support a two-domain model of NSP1: an amino-terminal and middle repetitive part of yet unknown function and an essential carboxy-terminal domain which independently folds and can mediate the association with the nuclear periphery and NPCs.

A molecular explanation for how nuclear pore association is generated is suggested by the structural organization of the carboxy-terminal domain. The entire sequence of the NSP1 carboxy-terminal domain reveals a heptad repeat periodicity, i.e., apolar residues are generally disposed at position 1 and 4 of a 7-residue long repeat sequence (Fig. 7). Heptad repeats of this type are known to occur in intermediate filament proteins such as cytokeratins and lamins, which are required for the formation of α-helical coiled-coil dimers (Steinert and Roop, 1988; Steinert et al., 1985; Cohen and Parry, 1990). Possibly, the NSP1 carboxy-terminal domain is involved in protein/protein interaction and, by this mechanism, can bind to a protein(s) at the NPC exhibiting a similar secondary structure. Candidates for such partner proteins could be NSP2 (see later) or any other yeast nuclear pore protein. Whether NSP1 and other nucleoporins form a filamentous structure at the NPC is not known, but can be tested experimentally, e.g., by in vitro reconstitution assays using purified components.

Experimental evidence for a heptad repeat organization as a functional requirement for nuclear pore association came from a mutational analysis of the NSP1 carboxy-terminal domain (Fig. 7). By site-specific mutagenesis, a glutamic acid (706) and a leucine (707) residue within one heptad repeat of the NSP1 carboxy-terminal domain (for which an α-helical structure is predicted) were changed into a proline and serine. This mutation drastically affects the NSP1 function, causing temperature-sensitive growth and a partial mislocalization of NSP1 within the cytoplasm at the restrictive temperature (Nehrbass *et al.,* 1990). Interestingly, another nuclear marker protein, NOP1 (a nucleolar protein; Schimmang *et al.,* 1989) was found to be also partly mislocalized to the cytoplasm in the *nsp1* mutant at 37°C, suggesting that NSP1 could be involved in nuclear transport. Temperature-sensitive growth at 37°C was caused by the proline mutation, because all intragenic suppressors of the ts *nsp1* that could regrow at 37°C changed the proline into another residue, predominantly a hydrophobic leucine which gave the best complementation at 37°C. Changing to an arginine gave less efficient comple-

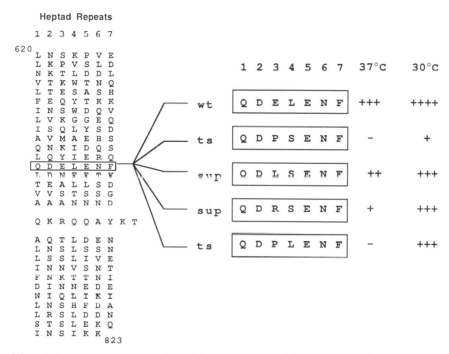

FIG. 7 Intragenic suppressor mutations of the temperature-sensitive *nsp1* mutant. On the left part of the figure, the amino acid sequence of the NSP1 carboxy-terminal domain ranging from residue 620 to 823 is shown. The sequence was aligned in heptad repeats so that hydrophobic amino acids maximally align at positions 1 and 4 of the heptad repeat. Within one heptad repeat (which is boxed), amino acid residues E and L were changed to P and S by site-specific mutagenesis, resulting in a temperature-sensitive (ts) allele of NSP1. Suppressors of this *nsp1* ts mutant, able to grow at 37°C or showing increased growth at 30°C, were selected and analyzed for their generation time (+, growth; −, no growth) at the indicated temperature (right panel). The DNA sequence of the NSP1 carboxy-terminal domain isolated from the corresponding suppressors was determined and the amino acid sequence within the indicated heptad repeat unit is shown. wt, Wild-type NSP1; sup, suppressor of ts *nsp1* able to grow at 37°C; ts, ts mutant of NSP1.

mentation. Furthermore, the hydrophobic leucine (707) at position 4 of the heptad repeat is important for optimal NSP1 function, because replacement of the mutant serine (707) with a leucine restores good growth at an intermediate temperature of 30°C (Fig. 7). In summary, the results derived form the intragenic suppressor analysis of ts *nsp1* are consistent with a model in which the NSP1 carboxy-terminal domain associates with the NPC via hydrophobic α-helical protein/protein interaction.

The function of the second domain, the highly repetitive central part of NSP1, is still not clear. Since this part is not essential for cell growth, its *in vivo* analysis is more difficult to perform. Hints as to how these repeat sequences may function in the eukaryotic cell could come from the observation that this repeated sequence motif is also found in other yeast and mammalian nucleoporins (Davis and Fink, 1990; Nehrbass *et al.,* 1990; Carmo-Fonseca *et al.,* 1991; Starr and Hanover, 1991; Cordes *et al.,* 1991). Accordingly, a role in nuclear transport reactions has been tentatively assigned to the repeat sequences (Carmo-Fonseca *et al.,* 1991; Starr and Hanover, 1991; Robbins *et al.,* 1991). The typical and conserved 9-amino acid repeat sequence within the NSP1 middle domain writes as KPAFSFGAK and similar, but more degenerate repeat sequences are found in NUP1 (Davis and Fink, 1990) and mammalian nucleoporin p62 (Carmo-Fonseca *et al.,* 1991; Starr and Hanover, 1991; Cordes *et al.,* 1991). The similarity between NSP1 and NUP1 is entirely restricted to the central repetitive domains, outside this region (i.e., in the essential NSP1 carboxy-terminal domain) no homology is observed. If one plots the hydrophobic moment for an assumed β-structure for the different nucleoporins whose protein sequences are available, distinct peaks of amphiphilicity are regularly dispersed throughout the whole repetitive domain (Nehrbass *et al.,* 1990; Davis and Fink, 1990; Carmo-Fonseca *et al.,* 1991; Cordes *et al.,* 1991). Alignment of the repetitive domain with the plot for the hydrophobic moment shows that the amphiphilic peaks match to the repetitive nonapeptides, in particular to a central core structure always composed of apolar (phenylalanine)–polar (serine/threonine)–apolar (phenylalanine/leucine) residues flanked by glycines or alanines (consensus motif GFSFG). It is therefore concluded that some nucleoporins share a common repeat sequence element which has similar secondary structure and which has been conserved over 1000 million years of evolution (Carmo-Fonseca *et al.,* 1991; Starr and Hanover, 1991). A simplified model for the structural organization of members of the nucleoporin family is presented in Fig. 8. Different nucleoporins share as a common structural motif a repetitive domain comprising many tandem repeats of the GFSFG type which form a similar secondary structure (zig-zag line) and are recognized by certain anti-nucleoporin monoclonal antibodies. Outside the repetitive domain sequence the homology abrogates. Individual nucleoporins perform their specific function via the non-homologous domains. According to this model, the repetitive domains within nucleoporins are redundant, so their function may be also. Interestingly, the loss of all repeat sequences from the NSP1 does not affect the essential function. Due to the repeat redundancy in the organism, the deletion of repeats from one nucleoporin may be compensated for by similar repetitive sequences found in other NPC proteins. The genetic analysis of different nucleoporin proteins in yeast will allow the testing of such a model.

The function of the GFSFG repeat sequences in nucleoporins is unclear. If they play a role in nuclear transport reactions, they could bind directly or indirectly to NLS-containing proteins and thus act as receptors at the level of the NPC. It is interesting to note that a bipartite pattern was recently recognized in many NLSs

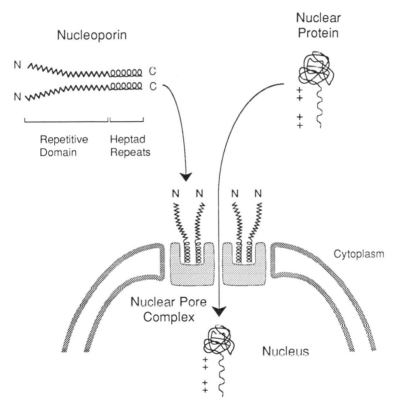

FIG. 8   Model for the structural organization of the nucleoporins at the nuclear periphery. The nuclear pore protein NSP1 and another hypothesized nucleoporin of similar structural organization consist of an amino-terminal repetitive domain (zig-zag line) and a carboxy-terminal domain with the potential to form α-helical interaction (α-helical drawing). The complex is formed by the NSP1 carboxy-terminal domain and, via this part, targeted to the nuclear periphery and assembled into the nuclear pore complex. The structural domain common to nucleoporins is a central repetitive domain (zig-zag line) which is exposed and contains many conserved nonapeptide repeats of the "KPAFSFGAK" type. The repetitive domain of nucleoporins acts and folds independently of the anchorage domain. Whether the repeat sequences play a direct role in nuclear transport needs to be shown.

consisting of two basic domains separated by 10 intervening spacer amino acids (Robbins et al., 1991). This observation prompted the authors to speculate that a cognate receptor for NLS may have a complementary bipartite binding domain. If nucleoporins are members of the NLS receptors, multiple binding to the repeats along the repetitive domain could potentiate receptor recognition and facilitate nuclear transport.

4. NSP2, a Further Nucleoporin in Yeast?

Monoclonal antibodies against mammalian nucleoporins generally recognize on immunoblots of yeast nuclear extracts a 100-kDa doublet of two nuclear proteins (Aris and Blobel, 1989; Davis and Fink, 1990), one of them being NSP1 (Davis and Fink, 1990). Affinity-purified antibodies against NSP1 recognize, besides the predominant 100-kDa NSP1 protein, a second protein with an apparent size of 90 kDa on SDS-PAGE (Nehrbass et al., 1990). We tentatively labeled this protein NSP2 because it shares similar properties with NSP1. NSP2 is not a breakdown product of NSP1 since it is also detected in cell extracts from yeast cells lacking NSP1 or having a truncated version of it. Both proteins have common epitopes typically found within the repetitive middle domain of NSP1 and cofractionate with nuclei and the insoluble nuclear fraction; therefore they may coexist in a higher order structure. It is tempting to speculate that complex formation may involve the carboxy-terminal domain of NSP1. Cloning of NSP2 and further as-yet-unidentified nucleoporins in yeast and sequence analysis will finally reveal the structural and functional relationship between the nucleoporin proteins in general.

5. Nucleoporins and Evolution

NSP1 and NUP1 appear to be key components of the NPC and therefore are expected to have homologs in mammalian cells. The essentiality of both proteins in yeast is suggestive of evolutionary conservation. Affinity-purified antibodies against the NSP1 repetitive middle domain or the carboxy-terminal domain both recognize nucleoporin p62 on immunoblots of HeLa nuclear envelopes and stain the nuclear pores by indirect immunofluorescence (Carmo-Fonseca et al., 1991). Mammalian nucleoporin p62 thus is a good candidate for being the functional homolog of NSP1. Sequence comparison of yeast NSP1 with HeLa nucleoporin p62 (Carmo-Fonseca et al., 1991), rat p62 (Starr and Hanover, 1991), and mouse or Xenopus p62 (Cordes et al., 1991) indeed shows that these proteins are homologous. The homology is mainly restricted to the carboxy-terminal domains, which all reveal a heptad repeat organization, as discussed for the NSP1 protein. It is thus likely that, similar to yeast, human p62 is targeted to the nuclear periphery and nuclear pores via the heptad repeated carboxy-terminal domain.

The amino-terminal half of nucleoporin p62 reveals only limited sequence homology to the NSP1 repetitive middle domain. It shows, however, a similar overall repeat organization of the KPAFSFGAK typical for members of the nucleoporin protein family (Carmo-Fonseca *et al.,* 1991) (see also above). Nucleoporin p62 is modified by *N*-acetylglucosamine and, accordingly, binds the lectin wheat germ agglutinin (Davis and Blobel, 1986, 1987; Starr *et al.,* 1990). The site of modification and binding of anti-nucleoporin monoclonal antibodies was shown to be mainly restricted to the repetitive amino-terminal part of p62 (Carmo-Fonseca *et al.,* 1991; Cordes *et al.,* 1991). Since wheat germ agglutinin and monoclonal antibodies to p62 inhibit nuclear transport (Featherstone *et al.,* 1988; Finlay *et al.,* 1987; Benavente *et al.,* 1989a,b), the repetitive amino-terminal part of nucleoporin and, by analogy, the repetitive middle domain of NSP1 may be involved in nuclear transport (see also above and Fig. 8). How these amphipathic repeat sequences common to different nucleoporins could mediate nucleocytoplasmic transport reactions remains to be shown.

B. Nuclear Lamins and Coiled-Coil Proteins

The predominant constituents of the mammalian nuclear lamina are lamins (polypeptides of 60–70 kDa called lamins A, B, and C) or immunologically related proteins that have been identified in a wide variety of organisms, from man to *Drosophila.* Based on sequence analysis and morphological criteria, the lamins are members of the intermediate filament protein family (Nigg, 1989). In mammalian cells the nuclear lamina appears to associate with specific integral proteins of the inner nuclear membrane (Senior and Gerace, 1988), and a putative lamin B receptor was identified in the avian nuclear envelope (Worman *et al.,* 1988).

To date, no nuclear lamin has been characterized in yeast by molecular means. However, the existence of a nuclear lamina and nuclear lamins in *S. cerevisiae* is suggested by morphological, biochemical, and immunological data (Allen and Douglas, 1989; Georgatos *et al.,* 1989) (see also Section II,B,2) although there is still controversy about this issue (Cardenas *et al.,* 1990). Georgatos *et al.* (1989) identified a lamin B receptor and lamin A and B analogs in yeast by immunological cross-reactivity with antibodies against turkey erythrocyte lamin B receptor and lamins A and B. These analogous yeast proteins have similar molecular weights to their mammalian counterparts, are enriched within the nuclear envelope fraction, and some of their biochemical properties also match. Furthermore, purified turkey erythrocyte lamin B binds to yeast urea-extracted nuclear envelopes, predominantly to the yeast 58-kDa lamin B receptor, which is an integral membrane protein. These data were taken as evidence that typical nuclear lamina components exist in yeast and behave in the same way as their higher eukaryotic analogs. Only cloning of these components in yeast will allow one to test whether they are homologs of bona fide lamins.

Although the existence of authentic lamin proteins in yeast has not been established so far, and therefore molecular and genetic aspects related to lamin location and assembly/disassembly during the cell cycle cannot be studied, Enoch et al. (1991) circumvented this problem by expressing chicken lamin B_2 in the fission yeast *Schizosaccharomyces pombe*. Surprisingly, the chicken lamin assembled into a structure that associates with the nuclear periphery during interphase (it gives a ringlike staining around the nuclear periphery as revealed by indirect immunofluorescence using chicken lamin B_2 antibodies) and becomes dispersed throughout the cytoplasm during mitosis. In higher eukaryotes, the nuclear lamins become hyperphosphorylated during mitosis on mitosis-specific residues (Peter et al., 1990; Ward and Kirschner, 1990), and two of these phosphorylation sites have been mapped and shown to be important for the mitotic disassembly of the nuclear lamina (Heald and McKeon, 1990). Interestingly, chicken lamin B_2 becomes hyperphosphorylated by the cdc2 kinase during mitosis in *Schizosaccharomyces pombe* and this correlates with a change in the intracellular location of chicken lamin B_2. Taking all these data together, the fission yeast *Schizosaccharomyces pombe* contains the machinery to reversibly assemble and disassemble a heterologous nuclear lamin from chicken in a cell cycle-dependent fashion similar to the cycle in higher eukaryotes. It is therefore likely that the fission yeast contains and regulates authentic lamins during the cell cycle in a similar way to that found for chicken lamin B_2.

In *S. cerevisiae,* the protein REP1, which is involved in the segregation of the 2-μm circle plasmid following cell division, exhibits sequence homology to proteins which form coiled-coil structures such as myosin heavy chain, vimentin, and nuclear lamins A and C, indicating a fibrous structure for the REP1 protein (Wu et al., 1987). It was suggested that REP1 regulates correct plasmid partitioning between mother and daughter cells by intercalating into the nuclear lamina and thus providing anchorage sites for attachment of plasmid molecules to the nuclear periphery. REP1 protein is located exclusively in the nucleus and copurifies with the insoluble nuclear fraction that is operationally and morphologically equivalent to the NPC/lamina fraction of higher eukaryotic cells.

The yeast SIR4 protein encoding a protein of 151 kDa is involved in the transcriptional silencing of the mating type loci HMR and HML (Marshall et al., 1987). Interestingly, the SIR4 carboxy-terminal domain is organized into amphiphilic heptad repeats within a long region of predicted α-helicity typically found in coiled-coil intermediate filament proteins. The carboxy-terminus of SIR4 is most closely related to the central rod domain of human nuclear lamins A and C (Diffley and Stillman, 1989). It is suggested that the essential SIR4 carboxy-terminal domain may serve as a dimerization motif, but could be also involved in heterodimerization. This motif could mediate binding to yeast nuclear lamins and thereby allow formation of coiled-coil filamentous structures by SIR4. Given that

SIR4 directly interacts with a nuclear lamina, it would imply that transcriptional silencing involves the interaction of the mating type loci HML and HMR with a yeast nuclear lamina through SIR4 (Diffley and Stillman, 1989).

C. Spindle Pole Body Proteins

The SPBs play a central role in the yeast cell cycle (see also Section II,B,4). SPBs are involved in the organization of cytoplasmic, mitotic, and meiotic microtubules and thus are essential for mitosis, meiosis, and karyogamy (Byers and Goetsch, 1974). There may be further as-yet-unidentified functions of the SPBs for the cell cycle and, accordingly, the molecular organization and function of SPBs is of fundamental scientific interest. It is therefore desirable to identify components of the SPB, and this is particularly challenging in the yeast system because the genetics would allow a rapid advance toward understanding the *in vivo* function of SPB components, by exploiting mutants and suppressors of SPB components (Alfa and Hyams, 1990). A breakthrough in the characterization of the yeast SPB proteins came from Rout and Kilmartin (1990), who successfully applied a biochemical approach and succeeded in purifying yeast SPBs with attached microtubules. Purified SPBs were enriched approximately 600-fold from total cell extracts and the final preparation retained both the microtubule-nucleating characteristics and most of the morphological features seen in thin sections of native yeast nuclear membranes. The enriched fraction of SPBs which contained about 20% tubulin was then used to raise monoclonal antibodies. Fourteen monoclonal antibodies were finally obtained which reacted on immunoblots with the SPB-enriched fraction. These monoclonal antibodies were grouped into three classes, each class recognizing an individual protein of 110, 90, and 80 kDa. Each set of monoclonal antibodies also stained SPBs or attached spindles in whole yeast cells, as visualized by indirect immunofluorescence, giving a one- or two-dot-staining within the nuclear membrane and distinct substructures in isolated SPBs by immunoelectron microscopy. By these immunostaining methods, the 90-kDa component was localized at the outer and inner plaque of the SPB from which cytoplasmic and nuclear microtubules originate, and thus this protein could be involved in microtubule assembly. The 110-kDa protein was localized to the nuclear face of the central plaque and, accordingly, a structural role in the SPB organization was suggested. The 80-kDa component has a somewhat different distribution and is more associated with the spindle microtubules, but always close to the SPB. The specific role of these new SPB components can now be studied in the yeast system and antibodies are currently being used to clone their corresponding genes. From these molecular/genetic studies one can expect interesting insights into how these SPB proteins perform their function at the level of the SPB.

Using an immunological approach and human autoimmune antibodies recognizing the pericentriolar material of mammalian, plant, and insect cells, two SPA (*spindle pole antigen*) antigens, SPA1 and SPA2, were identified in yeast (Snyder and Davis, 1988). SPA1 was subsequently cloned and sequenced and a mutational analysis showed that SPA1 is involved in chromosome segregation and other mitotic functions; spa1 mutants, although viable, grow poorly and missegregate chromosomes and often contain multiple nuclei or no nucleus and deformed spindles. Thus, SPA1 appears to be involved in nuclear migration. It was suggested that the SPA1 protein may be associated with the SPB, and experimental evidence showed that SPA1 indeed cofractionated with the nuclear envelope during subcellular and subnuclear fractionation. Conversely, antibodies prepared against a SPA1::trpE fusion protein react with spindle poles of mammalian cells. SPA1 appears to be a membrane-spanning protein since the DNA sequence predicts a 30-residue-long stretch of hydrophobic amino acids at the amino-terminus, but its specific role can be only determined when its exact location in the yeast cell is known.

A new component of the centrosome was recently found to be a special tubulin, called γ-tubulin (Weil *et al.*, 1986; Oakley *et al.*, 1990; Stearns *et al.*, 1991; Zheng *et al.*, 1991). γ-Tubulin was initially identified by a genetic approach in *Aspergillus nidulans* as a suppressor of a conditional-lethal β-tubulin mutation called mipA (Weil *et al.*, 1986). Strikingly, mipA encoded a tubulin related to α- and β-tubulin, but localized to the SPB and not to microtubule structures (Oakley *et al.*, 1990). γ-Tubulin is now found in a variety of organisms including *Xenopus*, *Schizosaccharomyces pombe*, maize, diatom, a budding yeast (Stearns *et al.*, 1991), *Drosophila*, and human (Zheng *et al.*, 1991). γ-Tubulin is a minor protein, present at less than 1% of the level of α- and β-tubulin, and is only associated with the pericentriolar material. These findings are consistent with the hypothesis that the γ-tubulins are a subfamily within the tubulins and are minus-end nucleators of the microtubule assembly linking centrosome and α- and β-tubulin. The location of γ-tubulin to the centrosome does not depend on intact microtubules, suggesting that this special tubulin form is bound to structural elements of the centrosome by protein/protein interactions not involving α- and β-tubulin. γ-Tubulin has not yet been identified in *S. cerevisiae* by PCR techniques, indicating that this protein is less conserved in this organism, but certainly γ-tubulin should also exist in this budding yeast.

D. Other Nuclear Envelope Proteins

The subnuclear location of the tRNA ligase involved in tRNA splicing in yeast was shown to be the nuclear periphery (Clark and Abelson, 1987). In yeast, a significant portion of the tRNA genes contain a single intron. tRNA splicing is catalyzed by two enzymes, an endonuclease and a ligase. The endonuclease is an integral membrane protein (presumably a nuclear membrane protein), but its exact location has not been determined because of the lack of antibodies. However, the tRNA ligase, which acts in concert with the endonuclease, was purified to homogeneity, its gene was cloned and sequenced, and antibodies were obtained (Phizicky *et al.,* 1986). By indirect immunofluorescence and immuno-EM, the tRNA ligase was shown to be localized primarily at the inner membrane of the nuclear envelope, 40% of the gold label being quite close to visible NPCs. A second location was found to be in a region of the nucleoplasm about 300 nm away from the nuclear envelope. This suggests a functional association of the tRNA splicing apparatus with the inner nuclear membrane, possibly also with the NPC. This location makes it likely that tRNA splicing in yeast is coupled to nuclear export through the NPC (Clark and Abelson, 1987).

The yeast RNA11 component, a protein associated with spliceosomes and essential for pre-mRNA splicing, is localized at the inner periphery of the yeast nucleus (Chang *et al.,* 1988). However, the RNA11 protein appears not to be in direct contact with the nuclear membrane, being frequently found 250 to 300 nm away from the inner side of the nuclear envelope. This distance from the nuclear membrane also occurs with other pre-mRNA splicing components, suggesting that the subnuclear localization of the mRNA splicing machinery in yeast is mainly restricted to the nuclear periphery. Interestingly, mRNA transcription also seems to take place in the vicinity of the nuclear envelope at the same sites where the RNA11 and Sm proteins are located (Chang *et al.,* 1988). How the splicing machinery is kept at a fixed position close to the nuclear envelope, and whether this specific subnuclear location is of importance for nuclear export of mature mRNA through the NPCs remain to be shown.

An abundant nuclear protein which fractionates efficiently with the yeast nuclear scaffold has been identified (Hofmann *et al.,* 1989). This factor, called RAP-1, was shown earlier to be one of the silencer binding proteins, efficiently repressing both mating type loci in yeast. RAP-1 *in vitro* can induce loop formation of DNA which is normally bound to nuclear scaffold sites (Hofman *et al.,* 1989). RAP-1 is located at the nuclear periphery, as shown by indirect immunofluorescence (S. Gasser, personal communication). This location is consistent with RAP-1 also being a telomere binding protein.

Two integral nuclear envelope proteins of as-yet-unknown function have been identified in yeast by raising antibodies against purified nuclear membrane proteins (Hurt *et al.,* 1988; Schimmang and Hurt, 1989). Both a 20- and a 40-kDa protein cofractionate with nuclei during subcellular fractionation and are integral membrane proteins (partition into the Triton X-114 phase and are not extracted by alkaline pH), and antibodies against them stain the nuclear periphery in a ringlike manner, as revealed by indirect immunofluorescence. Using both several cdc mutants and α-factor synchronized yeast cells, the 40-kDa nuclear membrane protein was shown to display cell cycle dependence (Schimmang and Hurt, 1989). The protein was induced around G1/S transition and then changed its location around the nuclear periphery during cell cycle progression. This 40-kDa protein may be associated with a peripheral nucleoskeletal structure displaying cell cycle dependence.

V. Conclusions

In this review, we have demonstrated similarities and differences between nuclear envelopes from yeast and higher eukaryotes. For evolutionarily conserved functions of the nuclear envelope, e.g., nucleocytoplasmic transport, *S. cerevisiae* is without doubt a suitable experimental system, and results obtained from this lower eukaryote are likely to be relevant for other organisms, particularly for those functions which are essential in yeast.

It is the genetic attractability of yeast which adds a further, very powerful, methodological facet to the other approaches existing to study nuclear envelope functions. In addition, the biochemical accessibility of the yeast nuclear envelope has also improved. If biochemical and genetic approaches can be combined in the future, a great potential to study nuclear envelope functions will emerge, especially if mutants can be coupled to both homologous and heterologous *in vitro* systems.

As in other biological disciplines, yeast may once again serve as a model organism with which to unravel the molecular details of how the nuclear envelope fulfills its important role within the eukaryotic cell.

Acknowledgments

We would like to thank Dr. John Aris (Rockefeller University) for kindly providing the monoclonal antibody against NOP1, Dr. Werner Herth and Susanne Diehlmann (University of Heidelberg) for performing freeze-fracture electron microscopic analysis, and Heinz Horstmann (EMBL) for help with electron microscopy. The critical reading of the manuscript by Ralf Jansen, Ulf Nehrbass, Janet Rickard, and Christian Wimmer is gratefully acknowledged.

References

Adam, S. A., Lobl, T. J., Mitchell, M. A., and Gerace, L. (1989). *Nature (London)* **337,** 276–279.
Adam, S. A., Marr, R. S., and Gerace, L. (1990). *J. Cell Biol.* **111,** 807–816.
Aebi, U., Cohn, J., Buhle, L., and Gerace, L. (1986). *Nature (London)* **323,** 560–564.
Akey, C. W. (1989). *J. Cell Biol.* **109,** 955–970.
Alfa, C. E., and Hyams, J. S. (1990). *Nature (London)* **348,** 484.
Allen, J. L., and Douglas, M. G. (1989). *J. Ultrastruct. Mol. Struct. Res.* **102,** 95–108.
Amati, B. B., and Gasser, S. M. (1988). *Cell (Cambridge, Mass.)* **54,** 967–978.
Amati, B. B., and Gasser, S. M. (1990). *Mol. Cell. Biol.* **10,** 5442–5454.

Aris, J. P., and Blobel, G. (1988). *J. Cell Biol.* **107**, 17–31.

Aris, J. P., and Blobel, G. (1989). *J. Cell Biol.* **108**, 2059–2067.

Aris, J. P., and Blobel, G. (1990). *In* "Methods in Enzymology" (C. Guthrie and G. R. Fink, eds.), Vol. 194, pp. 735–748. Academic Press, San Diego.

Barnes, G., and Rine, J. (1985). *Proc. Natl. Acad. Sci. U.S.A.* **82**, 1354–1358.

Bataillé, N., Helser, T., and Fried, H. M. (1990). *J. Cell Biol.* **111**, 1571–1582.

Baum, P., Furlong, C., and Byers, B. (1986). *Proc. Natl. Acad. Sci. U.S.A.* **83**, 5512–5516.

Baum, P., Yip, C., Goetsch, L., and Byers, B. (1988). *Mol. Cell. Biol.* **8**, 5386–5397.

Benavente, R., and Krohne, G. (1986). *J. Cell Biol.* **103**, 1847–1854.

Benavente, R., Dabauvalle, M. C., Scheer, U., and Chaly, N. (1989a). *Chromosoma* **98**, 233–241.

Benavente, R., Scheer, U., and Chaly, N. (1989b). *Eur. J. Cell Biol.* **50**, 209–219.

Berlin, V., Brill, J. A., Trueheart, J., Boeke, J. D., and Fink, G. R. (1990a). *In* "Methods in Enzymology" (C. Guthrie and G. R. Fink, eds.), Vol. 194, pp. 774–794. Academic Press, San Diego.

Berlin, V., Styles, C. A., and Fink, G. R. (1990b). *J. Cell Biol.* **111**, 2573–2586.

Berrios, M., Filson, A. J., Blobel, G., and Fisher, P. A. (1983). *J. Biol. Chem.* **258**, 13384–13390.

Bonner, W. M. (1975a). *J. Cell Biol.* **64**, 421–430.

Bonner, W. M. (1975b). *J. Cell Biol.* **64**, 431–437.

Borer, R. A., Lehner, C. F., Eppenberger, H. M., and Nigg, E. A. (1989). *Cell (Cambridge, Mass.)* **56**, 379–390.

Botstein, D., and Fink, G. R. (1988). *Science* **240**, 1439–1443.

Bourgeois, C. A., and Hubert, J. (1988). *Int. Rev. Cytol.* **111**, 1–52.

Breeuwer, M., and Goldfarb, M. (1990). *Cell (Cambridge, Mass.)* **60**, 999–1008.

Burke, B. (1990). *Curr. Opin. Cell Biol.* **2**, 514–520.

Byers, B. (1981). *In* "The Molecular Biology of the Yeast *Saccharomyces cerevisiae:* Life and Inheritance" (J. N. Strathern, E. W. Jones, and J. R. Broach, eds.), pp. 59–96. Cold Spring Harbor Lab., Cold Spring Harbor, New York.

Byers, B., and Goetsch, L. (1974). *Cold Spring Harbor Symp. Quant. Biol.* **38**, 123–131.

Byers, B., and Goetsch, L. (1975). *J. Bacteriol.* **124**, 511–523.

Byers, B., and Goetsch, L. (1976). *J. Cell Biol.* **69**, 717–721.

Cardenas, M. E., Laroche, T., and Gasser, S. M. (1990). *J. Cell Sci.* **96**, 439–450.

Carmo-Fonseca, M., and David-Ferreira, J. F. (1990). *Electron Microsc. Rev.* **3**, 115–141.

Carmo-Fonseca, M., Cidadao, A. J., and David-Ferreira, J. F. (1987). *Eur. J. Cell Biol.* **45**, 282–290.

Carmo-Fonseca, M., Kern, H., and Hurt, E. C. (1991). *Eur. J. Cell Biol.* **55**, 17–30.

Chang, T.-H., Clark, M. W., Lustig, A. J., Cusick, M. E., and Abelson, J. (1988). *Mol. Cell Biol.* **8**, 2379–2393.

Clark, M. W., and Abelson, J. (1987). *J. Cell Biol.* **105**, 1515–1526.

Cohen, C., and Parry, D. A. D. (1990). *Proteins* **7**, 1–15.

Conde, J., and Fink, G. R. (1976). *Proc. Natl. Acad. Sci. U.S.A.* **73**, 3651–3655.

Cordes, V., Waizenegger, I., and Krohne, G. (1991). *Eur. J. Cell Biol.* **55**, 31–47.

Culotti, J., and Hartwell, L. (1971). *Exp. Cell Res.* **67**, 389–401.

Davis, L. I., and Blobel, G. (1986). *Cell (Cambridge, Mass.)* **45**, 699–709.

Davis, L. I., and Blobel, G. (1987). *Proc. Natl. Acad. Sci. U.S.A.* **84**, 7552–7556.

Davis, L. I., and Fink, G. R. (1990). *Cell (Cambridge, Mass.)* **61**, 965–978.

Diffley, J. F. X., and Stillman, B. (1989). *Nature (London)* **342**, 24.

Dingwall, C. (1985). *Trends Biochem. Sci.* **10**, 64–66.

Dingwall, C., and Laskey, R. A. (1986). *Annu. Rev. Cell Biol.* **2**, 367–390.

Dingwall, C., Robbins, J., Dilworth, S. M., Roberts, B., and Richardson, W. D. (1988). *J. Cell Biol.* **107**, 841–849.

Dwyer, N., and Blobel, G. (1976). *J. Cell Biol.* **70**, 581–591.

Enoch, T., Peter, M., Nurse, P., and Nigg, E. A. (1991). *J. Cell Biol.* **112**, 797–807.

Fahl, W. E., Jefcoate, C. R., and Kasper, C. B. (1978). *J. Biol. Chem.* **253**, 3106–3113.

Featherstone, C., Darby, M. K., and Gerace, L. (1988). *J. Cell Biol.* **107**, 1289–1297.
Finlay, D. R., Newmeyer, D. D., Price, T. M., and Forbes, D. J. (1987). *J. Cell Biol.* **104**, 189–200.
Fisher, P. A., Berrios, M., and Blobel, G. (1982). *J. Cell Biol.* **92**, 674–686.
Franke, W. W. (1974). *Int. Rev. Cytol.* **4**, 72–236.
Franke, W. W., Scheer, U., Krohne, G., and Jarasch, E.-D. (1981). *J. Cell Biol.* **91**, 39–50.
Garcia-Bustos, J. F., Heitman, J., and Hall, M. N. (1991a). *Biochem. Biophys. Acta* **1071**, 83–101.
Garcia-Bustos, J. F., Wagner, P., and Hall, M. N. (1991b). *Exp. Cell Res.* **192**, 213–219.
Gasser, S. M., Amati, B. B., Cardenas, M. E., and Hofmann, J. F. X. (1989). *Int. Rev. Cytol.* **119**, 57–96.
Georgatos, S. D., and Blobel, G. (1987). *J. Cell Biol.* **105**, 117–125.
Georgatos, S. D., Maroulakou, I., and Blobel, G. (1989). *J. Cell Biol.* **108**, 2069–2082.
Gerace, L. (1986). *Trends Biochem. Sci.* **11**, 443–446.
Gerace, L., and Burke, B. (1988). *Annu. Rev. Cell Biol.* **4**, 335–374.
Gerace, L., Ottaviano, Y., and Kondor-Koch, C. (1982). *J. Cell Biol.* **95**, 826–837.
Glass, J. R., and Gerace, L. (1990). *J. Cell Biol.* **111**, 1047–1057.
Goldfarb, D. S., Gariépy, J., Schoolnik, G., and Kornberg, R. D. (1986). *Nature (London)* **322**, 641–644.
Greber, U. F., Senior, A., and Gerace, L. (1990). *EMBO J.* **9**, 1495–1502.
Haarer, B., and Pringle, J. (1987). *Mol. Cell. Biol.* **7**, 3678–3687.
Hadjiolov, A. A. (1985). "The Nucleolus and Ribosome Biogenesis." Springer-Verlag, Vienna.
Hall, M. N., Hereford, L., and Herskowitz, I. (1984). *Cell (Cambridge, Mass.)* **36**, 1057–1065.
Hall, M. N., Craik, C., and Hiraoka, Y. (1990). *Proc. Natl. Acad. Sci. U.S.A.* **87**, 6954–6958.
Hamm, J., and Mattaj, I. W. (1990). *Cell (Cambridge, Mass.)* **63**, 109–118.
Hartwell, L. H., Culotti, J., Pringle, J. R., and Reid, B. J. (1973). *Science* **183**, 46–51.
Heald, R., and McKeon, F. (1990). *Cell (Cambridge, Mass.)* **61**, 579–589.
Henriquez, R., Blobel, G., and Aris, J. P. (1990). *J. Biol. Chem.* **265**, 2209–2215.
Hofmann, J. F.-X., Laroche, T., Brand, A. H., and Gasser, S. M. (1989). *Cell (Cambridge, Mass.)* **57**, 725–737.
Holt, G. D., Snow, C. M., Senior, A., Haltiwanger, R. S., Gerace, L., and Hart, G. W. (1987). *J. Cell Biol.* **104**, 1157–1164.
Huang, B., Watterson, D. M., Lee, V. D., and Schibler, M. J. (1988a). *J. Cell Biol.* **107**, 121–131.
Huang, B., Mengersen, A., and Lee, V. D. (1988b). *J. Cell Biol.* **107**, 133–140.
Hurt, E. C. (1988). *EMBO J.* **7**, 4323–4334.
Hurt, E. C. (1989). *J. Cell Sci.* **12**, 243–252.
Hurt, E. C. (1990). *J. Cell Biol.* **111**, 2829–2837.
Hurt, E. C., Pesold-Hurt, B., and Schatz, G. (1984). *EMBO J.* **3**, 3149–3156.
Hurt, E. C., McDowall, A., and Schimmang, T. (1988). *Eur. J. Cell Biol.* **46**, 554–563.
Jackson, D. A., and Cook, P. R. (1988). *EMBO J.* **7**, 3667–3677.
Jacobs, C. W., Adams, A. E. M., Szaniszlo, P. J., and Pringle, J. R. (1988). *J. Cell Biol.* **107**, 1409–1426.
Jordan, E. G., Severs, N. J., and Williamson, D. H. (1977). *Exp. Cell Res.* **104**, 446–449.
Kalderon, D., Roberts, B. L., Richardson, W. P., and Smith, A. E. (1984). *Cell (Cambridge, Mass.)* **39**, 499–509.
Kalinich, J. F., and Douglas, M. G. (1989). *J. Biol. Chem.* **264**, 17979–17989.
Khanna Gupta, A., and Ware, V. C. (1989). *Proc. Natl. Acad. Sci. U.S.a.* **86**, 1791–1795.
Kilmartin, J. V., and Adams, A. E. M. (1987). *J. Cell Biol.* **98**, 922–933.
Kim, J., Ljungdahl, P. O., and Fink, G. R. (1990). *Genetics* **126**, 799–812.
King, S. M., and Hyams, J. S. (1982). *Macrobiology* **13**, 93–117.
Krohne, G., Franke, W. W., and Scheer, U. (1978). *Exp. Cell Res.* **116**, 85–102.
Lee, W.-C., Xue, Z., and Mélèse, T. (1989). *Proc. Natl. Acad. Sci. U.S.A.* **86**, 8808–8812.
Lee, W.-C., Xue, Z., and Mélèse, T. (1991). *J. Cell Biol.* **113**, 1–12.

Legrain, P., and Rosbash, M. (1989). *Cell (Cambridge, Mass.)* **57**, 573–583.

Li, R., and Thomas, J. O. (1989). *J. Cell Biol.* **109**, 2623–2632.

Loewinger, L., and McKeon, F. (1988). *EMBO J.* **7**, 2301–2309.

Mann, K.-H., and Mecke, D. (1980). *FEBS Lett.* **122**, 95–99.

Mann, K.-H., and Mecke, D. (1982). *Biochim. Biophys. Acta* **687**, 57–62.

Marshall, M., Mahoney, D., Rose, A., Hicks, J. B., and Broach, J. R. (1987). *Mol. Cell. Biol.* **7**, 4441–4452.

Matile, P., Moor, H., and Robinow, C. F. (1969). "Yeast Cytology" Academic Press, New York.

Maul, G. G. (1977). *Int. Rev. Cytol.* **6**, 75–186.

Maul, G. G., and Deaven, L. (1977). *J. Cell Biol.* **73**, 748–760.

McKeon, F. D. (1987). *BioEssays* **7**, 169–173.

Meier, U. T., and Blobel, G. (1990). *J. Cell Biol.* **111**, 2235–2245.

Meluh, P. B., and Rose, M. D. (1990). *Cell (Cambridge, Mass.)* **60**, 1029–1041.

Molenaar, I., Sillevis Smitt, W. W., Rozijn, T. H., and Tonino, G. J. M. (1970). *Exp. Cell Res.* **60**, 148–156.

Moor, H., and Mühlethaler, K. (1963). *J. Cell Biol.* **17**, 609–627.

Moreland, R. B., Nam, H. G., Hereford, L. M., and Fried, H. M. (1985). *Proc. Natl. Acad. Sci. U.S.A.* **82**, 6561–6565.

Moreland, R. B., Langevin, G. L., Singer, R. H., Garcea, R. L., and Hereford, L. M. (1987). *Mol. Cell. Biol.* **7**, 4048–4057.

Nehrbass, U., Kern, H., Mutvei, A., Horstmann, H., Marshallsay, B., and Hurt, E. C. (1990). *Cell (Cambridge, Mass.)* **61**, 979–989.

Nelson, M., and Silver, P. (1989). *Mol. Cell. Biol.* **9**, 384–389.

Newmeyer, D. D., and Forbes, D. J. (1988). *Cell (Cambridge, Mass.)* **52**, 641–653.

Newmeyer, D. D., Finlay, D. R., and Forbes, D. J. (1986a). *J. Cell Biol.* **103**, 2091–2102.

Newmeyer, D. D., Lucocq, J. M., Bürglin, T. R., and De Robertis, E. M. (1986b). *EMBO J.* **5**, 501–510.

Newport, J., and Forbes, D. J. (1987). *Annu. Rev. Biochem.* **56**, 535–563.

Nigg, E. A. (1988). *Int. Rev. Cytol.* **110**, 27–92.

Nigg, E. A. (1989). *Curr. Opin. Cell Biol.* **1**, 435–440.

Nigg, E. A. (1990). *Adv. Cancer Res.* **55**, 271–311.

Oakley, B. R., Oakley, C. E., Yoon, Y., and Jung, M. K. (1990). *Cell (Cambridge, Mass.)* **61**, 1289–1301.

Park, M. K., D'Onofrio, M., Willingham, M. C., and Hanover, J. A. (1987). *Proc. Natl. Acad. Sci. U.S.A.* **84**, 6462–6466.

Pathak, R. K., Luskey, K. L., and Anderson, R. G. W. (1986). *J. Cell Biol.* **102**, 2158–2168.

Peter, M., Nakagawa, J., Doréc, M., Labbé, J. C., and Nigg, E. A. (1990). *Cell (Cambridge, Mass.)* **61**, 591–602.

Peterson, J. B., and Gray, R. H. (1972). *J. Cell Biol.* **53**, 837–841.

Phizicky, E. M., Schwarz, R. C., and Abelson, J. (1986). *J. Biol. Chem.* **261**, 2978–2988.

Pinkham, J. L., Olesen, J. T., and Guarente, L. P. (1987). *Mol. Cell. Biol.* **7**, 578–585.

Polaina, J., and Conde, J. (1982). *Mol. Gen. Genet.* **186**, 253–258.

Potashkin, J. A., Derby, R. J., and Spector, D. L. (1990). *Mol. Cell. Biol.* **10**, 3524–3534.

Pringle, J. R., and Hartwell, L. H. (1981). *In* "The Molecular Biology of the Yeast *Saccharomyces cerevisiae*: Life Cycle and Inheritance" (J. N. Strathern, E. W. Jones, and J. R. Broach, eds.). Cold Spring Harbor Lab., Cold Spring Harbor, New York.

Reichelt, R., Holzenburg, E. L., Buhle, E. L., Jarnik, M., Engel, A., and Aebi, U. (1990). *J. Cell Biol.* **110**, 883–894.

Richardson, W. D., Mills, A. D., Dilworth, S. M., Laskey, R. A., and Dingwall, C. (1988). *Cell (Cambridge, Mass.)* **52**, 655–664.

Robbins, J., Dilworth, S. M., Laskey, R. A., and Dingwall, C. (1991). *Cell (Cambridge, Mass.)* **64**, 615–623.

Robinow, C. F., and Marak, J. (1966). J. Cell Biol. **29,** 129–151.

Rose, M. D., and Fink, G. R. (1987). Cell (Cambridge, Mass). **48,** 1047–1060.

Rothblatt, J. A., Deshaies, S. L., Sanders, S. L., Daum, G., and Schekman, R. (1989). J. Cell Biol. **109,** 2641–2652.

Rout, M. P., and Kilmartin, J. V. (1990). J. Cell Biol. **111,** 1913–1927.

Sadler, I., Chiang, A., Kurihara, T., Rothblatt, J., Way, J., and Silver, P. (1989). J. Cell Biol. **109,** 2665–2675.

Scheer, U., and Benavente, R. (1990). BioEssays **12,** 14–21.

Scheer, U., Dabauvalle, M.-C., Merkert, H., and Benavente, R. (1988). Cell Biol. Int. Rep. **12,** 669–689.

Schimmang, T., and Hurt, E. C. (1989). Eur. J. Cell Biol. **49,** 33–41.

Schimmang, T., Tollervey, D., Kern, H., Frank, R., and Hurt, E. C. (1989). EMBO J. **8,** 4015–4024.

Senior, A., and Gerace, L. (1988). J. Cell Biol. **107,** 2029–2036.

Sillevis Smitt, W. W., Vermeulen, C. A., Vlak, J. M., Rozijn, T. H., and Molenaar, I. (1972). Exp. Cell Res. **70,** 140–144.

Sillevis Smitt, W. W., Vlak, J. M., Molenaar, I., and Rozijn, T. H. (1973). Exp. Cell Res. **80,** 313–321.

Silver, P., and Goodson, H. (1989). CRC Crit. Rev. Biochem. **24,** 419–435.

Silver, P., Sadler, I., and Osborne, M. A. (1989). J. Cell Biol. **109,** 983–989.

Silver, P. A. (1991). Cell (Cambridge, Mass.) **64,** 489–497.

Silver, P. A., Keegan, L. P., and Ptashne, M. (1984). Proc. Natl. Acad. Sci. U.S.A. **81,** 5951–5955.

Snow, C. M., Senior, A., and Gerace, L. (1987). J. Cell Biol. **104,** 1143–1156.

Snyder, M., and Davis, R. W. (1988). Cell (Cambridge, Mass.) **54,** 743–754.

Starr, C. M., and Hanover, J. A. (1990). BioEssays **12,** 323–330.

Starr, C. M., and Hanover, J. A. (1991). BioEssays **13,** 145–146.

Starr, C. M., D'Onofrio, M., Park, M. K., and Hanover, J. A. (1990). J. Cell Biol. **110,** 1861–1871.

Stearns, T., Evans, L., and Kirschner, M. (1991). Cell (Cambridge, Mass.) **65,** 825–836.

Steinert, P. M., and Roop, D. R. (1988). Annu. Rev. Biochem. **57,** 593–625.

Steinert, P. M., Steven, A. C., and Roop, D. R. (1985). Cell (Cambridge, Mass.) **42,** 411–419.

Stochaj, U., Osborne, M., Kurihara, T., and Silver, P. (1991). J. Cell Biol. **113,** 1243–1254.

Thomas, J. H., and Botstein, D. (1986). Cell (Cambridge, Mass.) **44,** 65–76.

Torrisi, M. R., and Bonnatti, S. (1985). J. Cell Biol. **101,** 1300–1306.

Trapman, J., Retèl, J., and Planta, R. J. (1975). Exp. Cell Res. **90,** 95–104.

Underwood, M. R., and Fried, H. M. (1990). EMBO J. **9,** 91–99.

Unwin, P. N., and Milligan, R. A. (1982). J. Cell Biol. **93,** 63–75.

Ward, G. E., and Kirschner, M. W. (1990). Cell (Cambridge, Mass.) **61,** 561–577.

Weil, C. F., Oakley, C. E., and Oakley, B. R. (1986). Mol. Cell Biol. **6,** 2963–2968.

Willison, J. H. M., and Johnson, G. C. (1978). J. Bacteriol. **136,** 318–323.

Worman, H. J., Yuan, J., Blobel, G., and Georgatos, S. D. (1988). Proc. Natl. Acad. Sci. U.S.A. **85,** 8531–8534.

Wozniak, R. K., Bartnik, E., and Blobel, G. (1989). J. Cell Biol. **108,** 2083–2092.

Wright, R., Basson, M., D'Ari, L., and Rine, J. (1988). J. Cell Biol. **107,** 101–114.

Wu. L. C., Fisher, P. A., and Broach, J. R. (1987). J. Biol. Chem. **262,** 883–891.

Yamasaki, L., Kanda, P., and Lanford, R. E. (1989). Mol. Cell. Biol. **9,** 3028–3036.

Yang, C. H., Lambie, E. J., Hardin, J., Craft, J., and Snyder, M. (1989). Chromosoma **98,** 123–128.

Yeh, E., Driscoll, R., Coltrera, M., Olins, A., and Bloom, K. (1991). Nature (London) **349,** 713–715.

Yoneda, Y., Imamoto-Sonobe, N., Yamaizumi, M., and Uchida, T. (1987). Exp. Cell Res. **173,** 586–595.

Zheng, Y., Jung, M. K., and Oakley, B. R. (1991). Cell (Cambridge, Mass.) **65,** 817–823.

The Specialized Junctions of the Lens

G. A. Zampighi,* S. A. Simon,† and J. E. Hall§

*Department of Anatomy and Cell Biology, UCLA School of Medicine, Los Angeles, California 90024; †Department of Neurosciences, Duke University Medical Center, Durham, North Carolina 27710; and §Department of Physiology and Biophysics, University of California, Irvine, California 92712

I. Introduction

The lens is a transparent and moderately deformable organ whose principal function is to focus light on the retina. A long-standing problem in the physiology of the lens is to describe how the vectorial transport of electrolytes and water maintains volume and transparency of the organ. This review proposes that the solution of the problem requires detailed understanding of the relative contributions of the pathways used by ions, nonelectrolytes, and water to diffuse within and out of the lens.

Electrolytes and water move throughout the lens either by crossing plasma membranes of lenticular cells (the "transcellular" pathway) or by diffusing through the extracellular clefts between cells (the "paracellular" pathway). The transcellular pathway is mediated and regulated by ion channels, such as potassium or sodium channels, and the Na,K-ATPase, the enzyme response for the transport of sodium and potassium against their electrochemical gradients. The vectorial movement of sodium, potassium, and water across the lens can be accounted for largely by the asymmetric distribution of the Na,K-ATPase in the lens epithelium (Alvarez et al., 1985; Kobatashi et al., 1982; Palva and Palkama, 1976; Unakar and Tsui, 1980). The transcellular pathway depends also on the presence of specialized regions of plasma membrane contact called "gap junctions" (Revel and Karnovsky, 1967). These junctions are composed of "communicating" channels that mediate and regulate the diffusion of ions and small molecules between the cytoplasm of adjacent cells without allowing them to leak into the extracellular space. In the lens, gap junction channels connect epithelial cells, fiber cells, and also epithelial to fiber cells. This extensive cell-to-cell communication network transforms the lens into a functional syncytium (Mathias et al., 1981).

The most important plasma membrane specialization regulating the permeability of the paracellular pathway to ions and nonelectrolytes is the tight junction. In the lens, tight junctions are found only in the epithelium. Such an asymmetric localization of tight junctions transforms the extracellular clefts in the posterior pole into the pathway of least resistance to the diffusion of molecules within the lens.

In addition to gap and tight junctions, fiber cells are connected by another type of junction composed of a protein designated the main intrinsic polypeptide (MIP). Although MIP junctions have been involved in cell-to-cell communication between lens fibers, recent studies cast doubts about the ability of MIP in making channels of the gap "communicating" variety. This review compares the structure and chemical composition of the different types of junctions in the lens and attempts to distinguish between the possible functional roles of these junctions in ion transport and the maintenance of lens fluid balance.

II. Role of Junctions in Lens Physiology

Although lens metabolism is limited, nutrients must still move within and metabolities diffuse out for the organ to maintain transparency. The assessment of the role played by junctions in these processes requires an understanding of the ion transport pathways, principally those for sodium, potassium, and chloride (Kinsey and Reddy, 1965). The transport of ions drives water fluxes directed from the posterior to anterior poles of the lens. The water fluxes are ultimately responsible for the movement of nutrients and metabolities in and out of the lens. The transport of sodium and potassium requires the establishment of gradients. The gradients are created because the concentration of sodium is higher and the concentration of potassium is lower in the solutions bathing the lens than in the cell cytoplasm; they also must be maintained because the movement of nutrients and metabolities is to be held constant throughout the life of the animal.

The first model attempting to explain the mechanisms maintaining the ionic gradients treated the lens as a giant cell surrounded by a single membrane just below the capsule (Duncan, 1969a,b). This model explained the establishment and maintenance of ionic concentration gradients between the inside and outside of the lens, but it ran into difficulties with the demonstration by AC impedance measurements that the interior of the lens contained many cells and that the extracellular space has a resistivity about 500 times greater than the resistivity of fluids in the anterior or posterior chambers (50,000 ohm-cm to about 100 ohm-cm) and about 5 times greater than the resistivity of the solution in the cytoplasm of fiber cells (Mathias and Rae, 1985a; Mathias *et al.*, 1979, 1981, 1985; Rae *et al.*, 1982a). The large resistivity of the extracellular space predicts that a radial current in the extracellular space would lead to an appreciable radial voltage drop, and in fact the extracellular voltage toward the center of the lens is more negative than at the surface. The converse situation exists for the intracellular space. Although cells in the interior of the lens are coupled by gap junctions, this coupling still has a finite resistance, and thus radial intracellular current flow would also produce an appreciable radial voltage drop. Measurements show that the intracellular potential of the cells near the center of the lens is more positive than that of cells near the surface. Thus, the membrane potential across fiber cells near the center of the lens is lower than that of cells near the surface both because of the more negative extracellular potential and because of the more positive cell interiors.

These and other results are best described by a model which treats the lens as a nonuniform syncytium composed of a single layer of low-resistance epithelial cells on the lens anterior surface coupled via gap junctions to high-resistance fiber cells in the lens interior (Rae *et al.*, 1982a). The nonuniform syncytial model is further supported by AC impedance measurements showing that epithelial cells have specific conductances about 1000 greater than fiber cells (approximately proportional to differences in surface areas) and by patch-clamp evidence which demonstrates there are far fewer channels in fiber cells than epithelial cells (Cooper *et al.*, 1986, 1989, 1990; Duncan *et al.*, 1988).

The nonuniform syncytial model proposes that, to maintain constant volume, sodium, potassium, and chloride move throughout the lens from the posterior to the anterior pole using both the transcellular and the paracellular pathways (Mathias and Rae, 1985a; Mathias *et al.*, 1979). The transcellular pathway consists of the Na,K-ATPase and the ion channels resident in the outside surface of the epithelium and of the gap junctions connecting epithelial cells, fiber cells, and epithelial and fiber cells. The Na,K-ATPase, located principally in the lateral and basolateral membranes of the epithelium, actively transports sodium into the anterior chamber and potassium into the cytoplasm, a process that lowers the concentration of sodium and raises that of potassium inside the cells. The paracellular pathway is composed of the extracellular clefts between epithelia and fiber cells, the tight junctions between epithelial cells, the MIP junctions between fiber cells, and the suture lines formed by the ends of the fiber cells (see below). A key problem in understanding the physiology of the lens is to determine the relative contributions of these pathways to the movement of ions and molecules into and out of the organ.

Although the pathways followed by sodium, potassium, and chloride throughout the lens are complex, their location and directions can be at least approximately inferred using a combination of electrophysiological and transport measurements (Patterson, 1981; Rae, 1979; Robinson and Patterson, 1983; Wind *et al.*, 1988). The cells must satisfy the principle of electroneutrality that requires equality between the number of negative and positive charges in a cell. Thus, the concentration of negative charges contributed by crystallines, which in the lens is unusually high, must be taken into account. Duncan estimated that crystalline proteins (and to a small extent ATP) contribute about 100 mM of negative charges to fiber cells at physiological pH (Duncan and Croghan, 1969). This produces a large chloride concentration gradient, $(Cl^-)_i < (Cl^-)_o$. In frog lens, the average chloride equilibrium potential (E_{Cl}) is about -55 mV (Mathias and Rae, 1985a,b). If E_{Cl} is constant throughout the lens, chloride would enter fiber cells at potentials more depolarized than -55 mV (toward 0 mV) and would exit at more hyperpolarized potentials. Mathias and Rae (1985b) showed that, as a consequence of the radially dependent voltage drops in intra- and extracellular space, the membrane potential is lower than the E_{Cl} toward the lens nucleus (i.e., Cl^- enters) and greater than E_{Cl} in the lens cortex (i.e., Cl^- exits). In fiber cells where chloride enters, sodium passively follows (to preserve electroneutrality) and water follows salt (to reduce the osmotic pressure). If this process is not balanced by sodium and chloride efflux, occurring principally in the epithelium, fiber cells would swell and rupture. Indeed, when fiber cells are removed form their supporting epithelial layer, or the ability of the epithelium to pump salt is compromised, this is readily observed.

The high concentration of Na,K-ATPase in the epithelium and the coupling of epithelial and fiber cells predicts an outward transcellular sodium efflux directed from posterior to anterior. Commencing from the posterior chamber, sodium diffuses first into the lens extracellular space, then radially inward toward the nucleus of the lens. As it moves inward, sodium continually crosses fiber cell plasma membranes and diffuses radially outward through their cytoplasm via gap junction channels until it reaches the epithelium, where it is actively transported into the anterior chamber of the eye. The presence of a large voltage drop in the extracellular space may also modulate the sodium influx throughout the lens. The depolarization of fiber cells results in smaller electrochemical gradients for sodium entry and in smaller sodium influx. Thus, fiber cells in the lens nucleus should be maximally depolarized.

Evidence collected with a vibrating probe electrometer suggests that the route through the lens for sodium may be even more complex than the pathway described above. Robinson and Patterson (1983) used the vibrating probe to measure the outward and inward currents surrounding the lens of rats and found outward sodium currents at the equatorial surface and inward currents at the anterior and posterior poles. An inward current at the posterior pole is expected because the posterior surface lacks epithelial cells containing a high concentration of Na,K-ATPase. The inward sodium current at the anterior pole is somewhat unexpected. Because the Na,K-ATPase in the epithelium must generate an outward sodium current, the inward sodium current at the anterior pole probably represents the difference between the outward pump current and the inward current moving by all other pathways.

The circulation of potassium through the lens is also complex. Because the inward potassium current measured by the vibrating probe is abolished by ouabain, it is probably generated principally by the Na,K-ATPase (Patterson, 1981). Once potassium is in the epithelium, it diffuses either into the extracellular space between the epithelium and the first layer of fiber cells or via gap junction channels into the cytoplasm of fiber cells adjoining the epithelium. A recirculating current occurs when potassium diffuses out of the cells first into the lens extracellular space and later into the anterior and posterior chambers to reiterate the cycle. Because potassium efflux is enhanced by cell depolarization, a voltage drop in the extracellular space would lead to a greater efflux of potassium from nuclear fiber cells than from cortical fiber cells.

Thus, gap junction channels provide a substantial component of movement of cations, anions, and low-molecular-weight molecules in the recirculating current loops. Involvement of the paracellular pathway in the transport processes implies that the permeability of the extracellular clefts between epithelial cells and between fiber cells to ions and small molecules must also be regulated. In this manner, pockets of solution that would scatter light are not created in the extracellular space of the lens. In most epithelia, the permeability of the paracellular pathway is regulated solely by tight junctions, but, in the lens, tight junctions are found only in the epithelium (Lo and Harding, 1983). Below, we discuss structural information suggesting that fiber cells are connected by distinctive "tight junction-like" structures made of MIP. The principal characteristic of MIP junctions is an extremely narrow extracellular gap (0.5–0.7 nm) that should increase the resistance of the extracellular space. The extensive networks of junctions composed of MIP may account for the voltage drop measured from the anterior and posterior chambers to the nucleus of the lens.

III. Morphology of Lens Junctions

The lens is composed of two types of cells (Fig. 1A–C). A simple cuboidal epithelium underlies the anterior and equatorial capsules. A compact mass of highly elongated cells ("fiber" cells) arising from differentiation of the epithelium (Fig. 1C) comprises the bulk of the lens (Cohen, 1965; Leeson, 1971; Maisel *et al.*, 1981; Rafferty and Esson, 1974; Wanko and Gavin, 1959). These cell types differ in critical morphological and physiological properties. Fiber cells contain large concentrations of specialized proteins called "crystallines." The crystalline concentration increases markedly toward the lens nucleus and thus provides a gradient in the index of refraction that reduces spherical aberration. Fiber cells lack most cellular organelles, as such organelles would scatter light. They also appear to possess relatively few proteins capable of ion transport, such as pumps and channels. In contrast, epithelial cells contain the normal complement of cellular organelles and are responsible for most of the active transport of sodium and potassium in the lens. Also, epithelial and fiber cells differ in the structure, protein composition, and function of the junctions connecting their plasma membranes.

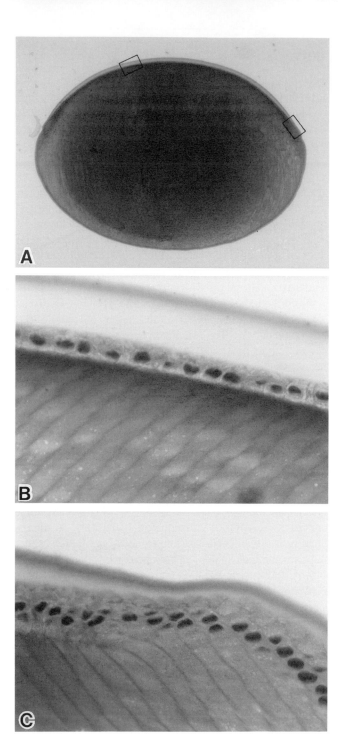

A. Junctions in the Epithelium

The epithelium is polarized with its apical surface oriented toward the lens interior and its basal surface toward the lens capsule.[1] Tight and gap junctions are located in the lateral surfaces of the cells in a region closer to the apical surface (i.e., the surface closer to fiber cells; Fig. 2A–C). Tight junctions are specialized regions of the cell surface that, in thin sections, appear as punctated regions of extremely close plasma membrane apposition. In freeze-fracture, tight junctions appear as ridges composed of 5-nm-diameter particles on the P fracture face (arrow in Fig. 2C) and complementary furrows on the E face (arrow in Fig. 2B). In the lens, tight junctions are found only between cells of the anterior epithelium (Goodenough *et al.*, 1980; Lo and Harding, 1983). In freeze-fracture, these epithelial tight junctions are formed by three to four strands encircling the entire cell perimeter. Electron-dense markers, such as lanthanum and horseradish peroxidase (HRP), indicate that tight junctions form a barrier to diffusion (Lo and Harding, 1983) (Fig. 2A). In contrast, the paracellular space in the lens posterior surface is freely permeable to extracellular tracers such as lanthanum (Fig. 3A,B). Thus, the asymmetric location of tight junctions in the lens influences permeability of ions and nonelectrolytes by making the extracellular space in the posterior pole the pathway of least resistance (Fig. 3).

Epithelial cells of human, bovine, rat, and rabbit lens are connected by gap junctions (Benedetti *et al.*, 1976; Kuszak *et al.*, 1978, 1982, 1985b; Lo and Harding, 1986). In thin sections, these junctions appear as two closely apposed plasma membrane of 16–18 nm in overall thickness (Revel and Karnovsky, 1967; Zampighi *et al.*, 1980). In freeze-fracture, gap junctions contain plaques of particles on the P face and complementary pits on the E face (Fig. 2C). Particles and pits are arranged either hexagonally (i.e., when the particle locations are related by symmetry operations of the plane group P6) or randomly (i.e., when the particles are clustered in plaques and their locations are unrelated by symmetry operations) (Broekhuyse *et al.*, 1978; Lo and Harding, 1986; Peracchia and Dulhunty, 1976).

Physiological studies show that epithelial cells of bovine, mouse, chick, frog, rat, and rabbit lenses are coupled to the passage of current and fluorescent dyes of

[1]This arrangement reflects the polarity found in the original lens vesicle where apical ("outside") corresponded to the lumen of the vesicle. At later stages of development, the epithelial cells of the posterior surface (i.e., those closer to the developing retina) differentia' into fiber cells and fill the lumen of the original vesicle.

FIG. 1 (A) Mouse lens sectioned in the anterior–posterior direction. The anterior surface (located upward) contains the layer of simple, cuboidal epithelium, whereas the bulk of the lens is composed of fiber cells. (B and C) Higher magnification views of the areas boxed in (A). Note that cells in the epithelium elongate at a region close to the lens equator and differentiate into fiber cells. Magnifications: A, 24×; B and C, 940×.

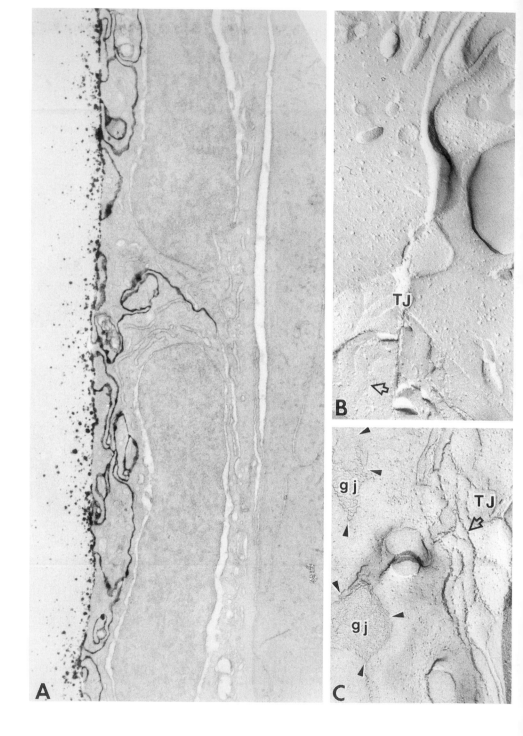

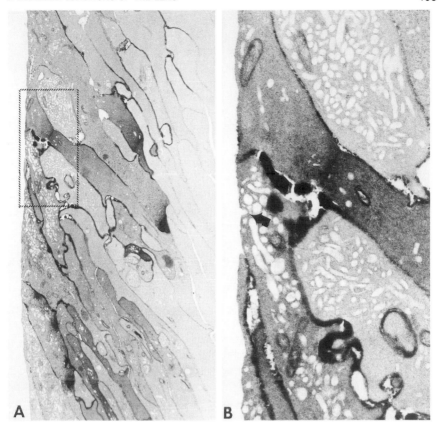

FIG. 3 The posterior surface of the lens is covered by fiber cells. The ends of fiber cells meet at regions called suture lines. Lanthanum infiltrates the extracellular space in the posterior surface, suggesting that fiber cells are not connected by tight junctions. The cytoplasm of fiber cells located in the most superficial layers contains numerous vesicles and elongated tubules with clear lumens that, on occasions, appear continuous with the plasma membrane (black arrow in B). Magnifications: A, 8700×; B, 43,500×.

FIG. 2 Cells in the epithelium are connected by tight and gap junctions. (A) A low-magnification view of the lens anterior surface. The extracellular space was labeled with lanthanum that appears as a black precipitate between epithelial cells. The basolateral membranes are highly invaginated, with numerous projections that interdigitate with projections from neighboring cells. Tight and gap junctions are located in the lateral membranes in a region closer to the apical surface (i.e., the surface of the cell toward the lens interior). Tight junctions (TJ) appear as strands of particles on the P face and complementary linear depressions on the E fracture face in freeze-fracture replicas (B and C). Gap junctions (gj; regions indicated by arrowheads in C) appears as plaques of particles on the P face and complementary pits on the E face. The plaques are usually located in close proximity and sometimes surrounded by tight junctional strands. Magnifications: A, 11,700×; B and C, 58,500×.

low molecular weight such as Procian Yellow (Bernardini et al., 1981; Rae, 1991; Rae et al., 1989). Whole-cell patch-clamp of cultured human, rat, and chicken epithelial cell pairs identifies cell-to-cell channels of 100–200 pS unitary conductance (Jacob, 1986; Mathias and Rae, 1989; Miller et al., 1992; Rae and Levis, 1984; Stewart et al., 1988). These are similar to channels recorded from pairs of cultured cells from lacrimal glands that are also coupled via gap junctions (Neyton and Trautmann, 1985). In addition, high-stringency Northern blot analysis performed in epithelial cells shows the expression of an mRNA coding for connexin 43, the heart myocyte gap junction protein (Beyer et al., 1988). Thus, morphological, physiological, and biochemical data indicate that the epithelium is coupled via gap junction channels made from a protein of the connexin family.

The nonuniform syncytium model predicts that the low-resistance epithelial cells are coupled to the high-resistance fiber cells (Mathias et al., 1981). Gap junctions between these cells (the "epithelium–fiber cell gap junction") have been identified in rat, chick, and mouse lenses (Goodenough et al., 1980; Lo and Harding, 1986). In thin sections, epithelium–fiber cell gap junctions are composed of closely apposed plasma membranes of 16–18 nm in overall thickness. In freeze-fracture, they are composed of plaques of intramembrane particles on the P fracture face and complementary pits on the E fracture face (Goodenough et al., 1980). The epithelium–fiber cell gap junctions, however, differ from gap junctions between epithelial cells in two important aspects. First, the intramembrane particles of epithelium–fiber cell gap junctions are arranged only in random packings instead of the random and hexagonal packings described between epithelial cell gap junctions (Goodenough et al., 1980; Lo and Harding, 1986). Second, epithelium–fiber cell gap junctions are insensitive to uncoupling by cytoplasmic acidification, an experimental manipulation that uncouples epithelial cells electrically and blocks the passage of fluorescent dyes (Miller and Goodenough, 1986). The protein composition of these gap junctions is presently unknown but the differences in structure and in sensitivity to uncoupling by acidification suggest these junctions are formed, at least in part, from a different protein than that forming the junctions between epithelial cells.

Thus, the distribution of gap junctions proteins in the epithelium appears to be more complex than previously anticipated; two types of gap junctions having different sensitivity to cytoplasmic acidification and structure are recognized. The gap junction in the lateral surfaces couples epithelial cells to each other while the other, located in the apical surface, couples epithelial to fiber cells. It seems plausible that different gap junctions may provide distinctive regulatory mechanisms for the communication between epithelial cells and epithelial and fiber cells.

B. Junctions between Fiber Cells

In human lens, fiber cells measure 8–12 mm in length, 7 μm in width, and 4 μm in thickness. Because their length so greatly exceeds the other dimensions, these cells have been called "fiber cells" or simply "fibers." The lens grows inward; the oldest fiber cells form the nucleus and the youngest cells form the cortex. Each fiber cell has six surfaces and two ends. The surfaces articulate through extensive junctional networks with adjacent fiber cells from the same layer and adjacent layers. The ends of fibers cells meet at highly specialized regions in the lens called suture lines (Fowlks, 1965). A large percentage of the limited extracellular space in the lens (only 1–3% of total volume excluding the capsule) is located at the suture lines. Suture lines are permeable to tracers such as lanthanum (Fig. 3A,B), suggesting that their extracellular space is accessible to the diffusion of ions and small molecules. At the suture lines, fiber cells have elaborate plasma membrane infoldings and even some membrane-bound organelles such as mitochondria (Gorthy and Anderson, 1980). Also, numerous membrane vesicles and tubules with clear lumens are found in the sutural ends of the most externally located fiber cells at the posterior pole (Fig. 3A,B). Gorthy and Anderson (1980) speculated that the infoldings contained intrinsic membrane proteins different from those in fiber cell surfaces, such as glycoproteins, ion channels, and pumps. If this is the case, the ends of fiber cells at suture lines may participate in ion transport and the regulation of fluid balance in the lens.

The surfaces contacting adjacent fiber cells contain elaborate interdigitations. One type, "ball-and-socket" interdigitations, consists of short cytoplasmic stalks and larger spherical heads invaginating into the cytoplasm of adjacent cells; they are most frequent between cortical fiber cells. Another type, "tongue-and-groove" interdigitations, is formed by interlocking surfaces with a wavelength of 0.3–0.4 µm; they are found with increasing frequency toward the nucleus (Lo and Harding, 1984). Both types of interdigitations contain extensive networks of close plasma membrane contacts named "occluded" or "fiber cell" junctions (Fig. 4A) (Dickson and Crock, 1972; Leeson, 1971; Nonaka *et al.*, 1976; Wanko and Gavin, 1959). Earlier studies using thin-section electron microscopy showed that fiber cell junctions exhibit pentalamellar structures similar to gap junctions between liver hepatocytes or heart myocytes. This observation suggested that the entire junctional network connecting fiber cell interdigitations belongs to the gap "communicating" variety. However, other studies showed that this network is formed by junctions differing in structure, protein composition, and perhaps also function. By comparing the structure of gap junctions to the structure of junctions in the interdigitations it is now possible to show that only a small percentage of the junctional network conforms to the structure of gap junctions.

C. Comparison of the Structure of Fiber Cell Junctions and Gap Junctions

Gap junctions in epithelia and connective, muscular, and nerve tissues exhibit a common structure. In transverse sections, they appear as closely apposed plasma membranes of 15–18 nm in overall thickness separated by a small 2- to 3-nm-wide

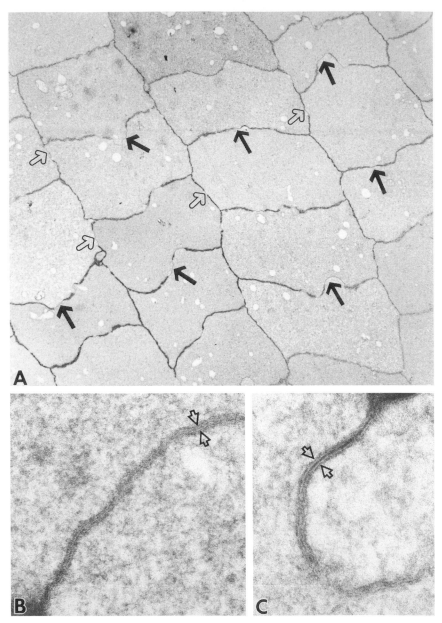

FIG. 4 Fiber cells are connected by an extensive network of junctions. (A) A low-magnification view of cortical fiber cells of a mouse lens sectioned perpendicular to their long axis. The extracellular space of the lens was infiltrated with lanthanum to outline regions of close plasma membrane apposition between fiber cells (black and open arrows). The junctions are located on both the flat surfaces that connect fiber cells from different layers (black arrows) and on the edges connecting fiber cells in same layer (open arrows). (B and C) Higher magnification views of the different types of junctions found connecting fiber cells. These junctions differ in their overall thickness and permeability of their extracellular gap to lanthanum. B shows a junction of 11–13 nm in overall thickness that is impermeable to lanthanum. C shows a junction of 18–20 nm in overall thickness that is permeable to lanthanum. Magnifications: A, 10,650×; B and C, 106,500×.

extracellular gap which is permeable to lanthanum (Philipson *et al.,* 1975). The plane of the junction is composed of two-dimensional plaques of annular units usually arranged randomly or with hexagonal symmetry (Fig. 5A,B). Each annular unit is a cell-to-cell channel that spans the two plasma membranes and the intervening extracellular gap. Each channel is constructed of two coaxially aligned hexamers (hemichannels) joined through their external domains (Unwin and Ennis, 1984; Unwin and Zampighi, 1980). The geometric center of this 12-subunit assembly (dodecamer) contains an approximately 2-nm-diameter hydrophilic pore directly connecting the cytoplasmic spaces of adjacent cells (Unwin and Ennis, 1984; Unwin and Zampighi, 1980).

The junctions located in the interdigitations of fiber cell junctions are formed by closely apposed plasma membranes and, in this regard, they resemble gap junctions (Fig. 4A–C). However, they differ from gap junctions in their overall thickness and ability to be infiltrated by lanthanum. Based on these characteristics, fiber cell junctions have been classified into two types. One type, representing the majority of fiber cell junctions, is impermeable to lanthanum and measures 11–13 nm in overall thickness (Fig. 4B). The other type is permeable to lanthanum and measures 18–20 nm in overall thickness (Fig. 4C).

Additional differences between fiber cell junctions and gap junctions are revealed by comparing their overall thickness. Hepatocyte gap junctions are significantly thicker than fiber cell junctions, which are impermeable to lanthanum (15–16 nm against 11–13 nm). About half of this difference in thickness arises from the smaller width of the extracellular gap (2–3 nm in gap junctions versus 0.5–0.7 nm in fiber cell junctions). Hepatocyte gap junctions are, on the other hand, significantly thinner than fiber cell junctions permeable to lanthanum (18–20 nm); most of this difference, however, arises from the larger thickness of the dense layers on the cytoplasmic surfaces of the junction. A prediction derived from this observation is that proteins with different molecular weights should form these distinctive junctions.

Freeze-fracture replicas show that the two types of fiber cell junctions differ among themselves and also differ when compared to gap junctions between myocytes and hepatocytes. Fiber cell junctions permeable to lanthanum contain particles arranged randomly, whereas fiber cell junctions impermeable to lanthanum contain particles usually crystallized in tetragonal arrays (Fig. 5D) (Lo and Harding, 1984; Okinami, 1978). For comparison, in Fig. 5, plaques of particles and pits from hepatocyte gap junctions arranged randomly (Fig. 5A) and hexagonally (Fig. 5B) are shown.

The tetragonal symmetry observed in some fiber cell junctions indicates that each junctional particle (i.e., the hemichannel) is formed by four subunits rather than the six subunits determined for the structure of the hemichannels of hepatocyte gap junctions. Another difference between these junctions and gap junctions is found in tongue-and-groove interdigitations in the nucleus of human lens. The junctions are composed of extensive tetragonal arrays that adhere to apposed, particle-free membranes (Lo and Harding, 1984). These observations raised the possibility that fiber cell junctions composed of tetragonal crystals are not of the gap "communicating" variety (Kistler and Bullivant, 1980a; Zampighi *et al.*, 1982).

At one time, a great deal of attention was focused on the packing of intramembrane particles as determined in freeze-fracture replicas (Fig. 5A–C). It was proposed that plaques of junctional particles arranged randomly implied "communicating" channels in the open state, whereas plaques of particles arranged with crystalline packings represented "communicating" channels in the closed

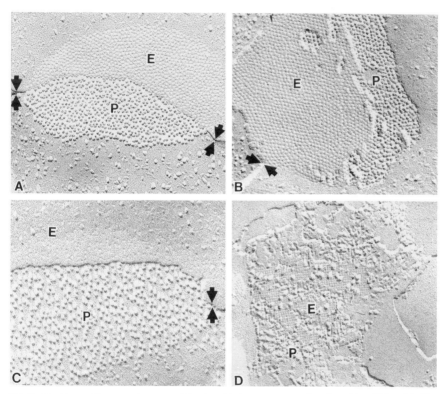

FIG. 5   Packings of intramembrane particles found in hepatocyte gap junctions (A and B) and in fiber cell junctions (C and D). In the four plaques, the fracture plane passes from the external (E) to the protoplasmic (P) fracture face. This type of fracture shows the E and P faces and the small fracture step (black arrows) that includes the narrow extracellular gap separating the junctional membranes. In hepatocyte gap junctions, the P face particles and complementary E face pits are arranged either randomly (A) or hexagonally (B). In contrast, fiber cell junctions have particles and pits arranged either randomly or tetragonally (C and D). Note that the "random" arrangements of particles and pits in the junctions between hepatocytes (A) and fiber cells (C) differ substantially in the center-to-center spacing (9–11 nm versus 13–20 nm). Magnifications: A–D, 97,500×.

state (Peracchia, 1978; Peracchia and Dulhunty, 1976). This hypothesis dominated the interpretation of data reported in early studies of junctions because it rationalized, at least partially, a large body of evidence from fiber cell junctions in whole lenses and in isolation.

The hypothesis correlating the packing of particles to gap junction channel conductance was first proposed in studies performed in gap junctions between the lateral giant axons of the crayfish abdominal cord (Peracchia and Dulhunty, 1976). The large diameter of these axons greatly simplifies the measurement of both gap junctional conductance and packing of intramembrane particles in the same junctions of the same axons. These studies showed that axons coupled by gap junctions fixed in the high-conductance state contain plaques of particles arranged *randomly,* whereas nerve cords fixed with gap junctions in the low-conductance state contain plaques of particles arranged *hexagonally.* These results suggested the notion that the transformation of random into hexagonal packing implied the transition of gap junction channels from the open to the closed state.

The extension of this notion to fiber cell junctions relies mostly on studies of the packing of junctional particles shown by freeze-fracture methods, rather than on the simultaneous measurement of conductance and particle packing density as determined in gap junctions of the crayfish nerve cord (Bernardini *et al.,* 1981; Peracchia and Girsch, 1985a; Peracchia and Peracchia, 1980a,b). The principal experimental observation used to extend this notion to fiber cell junctions was that an increase in free calcium concentration or a decrease of intracellular pH crystallized the random packing of junctional particles into *tetragonal* arrays. In isolated junction fractions, the transformation takes place when the free calcium concentration is increased from 5×10^{-7} to $5 \times 10^{-5} M$ or the pH decreased from 7.0 to 6.0 (Peracchia and Girsch, 1985a; Peracchia and Peracchia, 1980a,b). Because gap junctions in most tissues are coupled at free calcium concentrations of $1 \times 10^{-7} M$ and pH 7.0 and uncoupled when free calcium concentrations are increased or the intracellular pH is acidified (Bassnet and Duncan, 1988; Schuetze and Goodenough, 1982), the random and tetragonal packings appear to be correlated to the open and closed states of the channels.

Doubts about the validity of the hypothesis arose when several investigators failed to reproduce the transformation from random to tetragonal arrays by increasing the concentrations of calcium and hydrogen ions. In junctions isolated from bovine lenses, calcium failed to transform random arrays into tetragonal arrays when its concentration was increased up to 10 mM (Goodenough, 1979). Goodenough (1979), however, proposed that the communicating channels in fiber cell junctions have adapted to the limited metabolism of the lens by locking permanently into the "open" state. According to Goodenough, the inability of the channels to close predicted that the lens might be extremely vulnerable to injury because damage to a single cell will result in widespread depolarization. However, this prediction has not been supported by several physiological studies. First, impedance measurements of frog lens indicate that cortical fiber cells can be uncoupled reversibly by treatment with 2,4-dinitrophenol (DNP), a chemical that uncouples gap junctions in many tissues by lowering the intracellular pH (Rae et al., 1982b). Second, studies where normal and mechanically injured rat lenses are preincubated with ^{42}K and Procian Yellow, a fluorescent dye that diffuses through gap junction channels, also showed that fiber cells can uncouple (Bernardini et al., 1981). In these experiments, lenses are loaded with ^{42}K and then mechanically damaged. The rate at which the isotope diffuses out of the lens correlates with the degree of fiber cell coupling. ^{42}K slowly diffuses out of lenses, suggesting that the damaged regions of the lens are not connected to the undamaged regions via open gap junction channels. This conclusion is further supported by the observation that Procian Yellow loaded into mechanically damaged fiber cells does not diffuse into undamaged regions of the lens (Bernardini et al., 1981). Thus, fiber cells can uncouple and avoid the consequence that injuries to a single cell spread to other regions of the lens.

Studies involving whole lens and isolated junctions have failed to establish the correlation between particle packing and free calcium concentration. In developing chicken, a quantitative analysis of rapidly frozen lenses found no correlation between packing of junctional particles and coupling state of fiber cells (Miller and Goodenough, 1985). In isolated junctions from sheep lens, free calcium concentrations of up to 1 mM resulted only in partial transformation (Kistler and Bullivant, 1980b). In isolated junctions from bovine lens, free calcium concentration of 1×10^{-8} M or lower resulted in extensive tetragonal arrays, as determined in freeze-fracture replicas and in low angle X-ray diffraction patterns (Zampighi et al., 1982). The simplest explanation of the data is that the free calcium concentration does not affect directly the in-plane packing of junctional particles. Thus, in fiber cell junctions, the demonstration that the packing of junctional particles reflects the conductance state of the gap "communicating" channel is still lacking.

IV. Chemical Composition of Lens Junctions

A. Protein Composition of Fiber Cell Junctions

Lens proteins separate into water-soluble (the crystallines) and water-insoluble fractions (Lasser and Balazs, 1972). Junctional membranes are isolated by extraction of the water-insoluble fraction with high concentrations of urea or guanidine hydrochloride that solubilize most extrinsic proteins adhering to the plasma membranes. The isolation of fiber cell junctions is simpler than the isolation of gap junctions because treatment with detergents and/or digestion with exogenous proteases is not required (Broekhuyse and Kuhlmann, 1974, 1978; Broekhuyse et al., 1976; Dunia et al., 1974; Goodenough, 1979; Zampighi et al., 1982). Sodium dodecyl sulfate–polyacrylamide gel electrophoresis (SDS-PAGE) analysis shows that purified lens fiber junctions contain intrinsic polypeptides with apparent masses ranging from 18 to 140 kDa (Alcala *et al.*, 1975; Benedetti *et al.*, 1976; Bloemendal *et al.*, 1972; Dunia *et al.*, 1974; Harding and Dilley, 1976). A 26-kDa polypeptide represents over 50% of the total amount of intrinsic protein in lens plasma membranes (Broekhuyse *et al.*, 1976). It has been named MP26 by some investigators, due to its migration in SDS-PAGE, and by others, the main intrinsic protein, MIP (Broekhuyse *et al.*, 1976). The complete sequence of MIP indicates a mass of 28 kDa (Gorin *et al.*, 1984), and some investigators have also referred to this protein as MP28. In this report, we refer to this protein as MIP. The polypeptide with an apparent molecular mass of 17–18 kDa has been found to have a true molecular mass of 19.6 kDa by cDNA cloning and is called MP19 (Gutekunst *et al.*, 1990).

An issue in lens cell-to-cell communication is the determination of the protein composition and function of the different types of junctions connecting fiber cells. The problem has been investigated by raising antibodies against particular plasma membrane proteins and then using the antibodies to localize the proteins in isolated fractions and whole lens (Bouman and Broekhuyse, 1981; Broekhuyse *et al.*, 1979; Vallon *et al.*, 1985; Waggoner and Maisel, 1978; Zigler and Horwitz, 1981).

The first studies of this type reported that MIP is contained in fiber cell junctions (Bok *et al.*, 1982; Fitzgerald *et al.*, 1985). Figure 6A–C shows thin sections of fiber cells from a mouse lens labeled with antibodies against MIP visualized with gold particles by the postembedding method. At low magnifications, MIP is located throughout the entire surface of fiber cells, including the interdigitations (Fig. 6B). At higher magnifications, MIP is found in both single plasma membrane and pentalamellar structures of 11–13 nm in thickness (Fig. 6A–C). Thus, in fiber cells, MIP is found in 11- to 13-nm junctions and in single plasma membranes. The presence of MIP in junctions supports the notion that MIP may be involved in cell-to-cell communication. However, the large concentrations of MIP in single plasma membranes (Bok *et al.*, 1982; Fitzgerald *et al.*, 1985) are puzzling because this is not observed in gap junction-forming proteins of other tissues. Other studies have located a 70-kDa protein (MP70) and MP19 in fiber cell junctions (Gruijters *et al.*, 1987; Kistler *et al.*, 1985; Zampighi *et al.*, 1989).

A question addressed in several studies was to find out which of these proteins exhibit homology to hepatocyte (connexin 32) and myocyte gap junction (connexin 43) proteins. Analysis of the amino acid sequences of MIP and MP19 shows that the two proteins are not related to each other and, more importantly, lack significant homology to gap junction proteins of the "connexin" family (Beyer *et al.*, 1987; Ebihara *et al.*, 1989; Gimlich *et al.*, 1988; Gorin *et al.*, 1984; Kumar and Gilula, 1986; Paul, 1986; Zhang and Nicholson, 1989). Although the complete amino acid sequence of MP70 is not available, this protein has been classified as a connexin because the sequence of its N-terminal exhibits homology to that of connexin 32 and connexin 43 from liver and heart, respectively. In addition, high-stringency Northern blots have identified an mRNA that codes for a protein of 46 kDa (Beyer *et al.*, 1988). The predicted amino acid sequence of the protein is homologous to proteins of the connexin family (connexin 46). At present, it is unknown whether connexin 46 is expressed in fiber cells and, more importantly, whether it forms "communicating" channels.

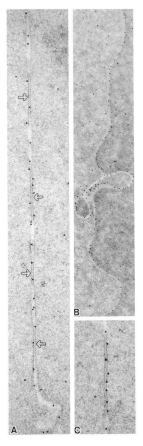

FIG. 6   Immunolocalization of MIP in fiber cells of mouse lens by postembedding staining. (A) A region of the surface of two fiber cells containing four junctional segments characterized by penta-lamellar structures (arrows). The antibody labeled both junctional regions (arrows) and also nonjunctional plasma membrane seen between the junctions. (B) A lower magnification view of the surface between two fiber cells that contains "ball-and-socket" interdigitations. The labeling pattern suggests that MIP is located throughout the cell surface, including the interdigitations. (C) A pentalamellar structure labeled by the anti-MIP antibody at higher magnification. The junctions resemble the 11- to 13-nm junctions whose extracellular gap is not permeable to lanthanum. The primary anti-MIP was prepared by Bok *et al.* (1982). In this experiment, it was used at a dilution of 1 : 1000 and visualized with protein A complexed with gold particles. Magnifications: A, 41,400×; B, 23,000×; C, 69,000×.

The presence of several proteins forming fiber cell junctions raises the question of which protein(s) forms communicating channels and what is the function of proteins that form junctions but do not form channels of the "communicating" variety. It is possible that fiber cell junctions are composed of mixtures of all of these proteins but that only one of them forms the "communicating" channel (i.e., MIP, MP70, or connexin 46). Alternatively, more than one protein may be able to form "communicating" channels. For example, MP70 and connexin 46 could both mediate coupling but in different regions of the lens (e.g., cortex and nucleus).

To resolve these issues, proteins suspected to form fiber cell junctions are being characterized biochemically, with great attention focused on MIP because it comprises over 50% of the total intrinsic protein of fiber cells. Immunological studies have shown that MIP is highly conserved in different species (Takemoto et al., 1985). MIP is an intrinsic membrane protein because it is soluble in detergents and most apolar organic solvents and insoluble in solutions containing high concentrations of urea or guanidine hydrochloride, or with alkaline pH. MIP has been solubilized with the nonionic detergent octylglucopyranoside and purified to homogeneity by chromatographic methods (Aerts et al., 1990; Ehring et al., 1990; Johnson et al., 1991; Manenti et al., 1988; Zampighi et al., 1985). MIP binds calmodulin in a calcium-independent manner (Welsh et al., 1982). Studies using circular dichroism show that 50% of the solubilized protein is probably α-helical and 20% β-sheet (Horwitz and Bok, 1987). Glutaraldehyde cross-linking and Western blots showed that detergent-solubilized MIP is monomeric with about 10 mol of phospholipids per mole of 28 MIP (Ehring et al., 1991). However, other studies using similar solubilization protocols but different methods of protein purification suggest that the detergent-solubilized MIP is most likely a tetramer (Aerts et al., 1990). At present, it is unclear whether the MIP tetramers were formed during purification of the soluble protein or during the solubilization process.

In intact lens, MIP is endogenously proteolyzed to a 21- to 22-kDa (MP21 or MP22) protein with a half-life of about 40 years (Broekhuyse and Kuhlmann, 1980; Horwitz and Wong, 1980; Horwitz et al., 1979; Ngoc et al., 1985). "In vitro" digestion of isolated membrane fractions with α-chymotrypsin, papain, trypsin, and Staphylococcus aureus proteases also reduces MIP to a 21- to 22-kDa polypeptide (Keeling et al., 1983; Nicholson et al., 1983). N-Terminal sequence analysis and cyanogen bromide cleavage show that most of the cleaved peptide corresponds to the C-terminal exposed on the cytoplasmic surface and that proteases remove only five residues at the N-terminal (Keeling et al., 1983; Nicholson et al., 1983). Thus, it appears that both amino- and carboxyl-terminals are exposed on the cytoplasmic surface and that about 80% of MIP is embedded in the plasma membrane (Keeling et al., 1983).

Phosphorylation of MIP has been observed in whole-lens homogenates incubated with [γ-³²P]ATP and also in fragments of fetal and bovine lenses incubated with [³²P]orthophosphate (Garland and Russell, 1985; K. R. Johnson et al., 1985,

1986; R. Johnson et al., 1985; Lampe and Johnson, 1989, 1991; Lampe et al., 1986; Takats et al., 1978). In these studies, phosphorylation is mediated by protein kinase C, is calcium-dependent, and is responsive to forskolin, cAMP, and 12-O-tetradecanoylphorbol-13-acetate (TPA). In HPLC purified MIP, the protein was phosphorylated with a stoichiometry of 1:1 by both protein A and protein C kinases at serine-243, located in the sequence of the C-terminal.

MP70 is a 70-kDa membrane-bound protein that is also involved in the formation of fiber cell junctions. In intact lens, MP70 is most abundant in cortical fiber cells (Gruijters et al., 1987; Kistler et al., 1985). MP70, like MIP, is endogenously proteolyzed in an age-dependent manner (Kistler and Bullivant, 1987). N-Terminal analysis shows that MP70 and two other proteins (MP64 and MP38) have identical amino acid sequences, demonstrating that the low-molecular-weight proteins arise from proteolytic cleavage of MP70 (Kistler et al., 1988). MP70 and MP64 (but not MP38) are phosphorylated *"in vitro"* with cAMP-dependent protein kinase A (Voorter and Kistler, 1989). Immunofluorescence shows that MP70 and MP64 are located in cortex whereas MP38 is exclusively located in the nucleus. The N-terminal sequence of MP70, MP64, and MP38 is homologous to the N-terminal sequence of connexin 32 and connexin 43 (Kistler et al., 1988). Thus, MP70 appears to be a junction-forming protein of the connexin family.

MP19 contributes approximately 10% of the total membrane protein and is the second most abundant membrane-bound protein in fiber cell plasma membrane after MIP (Galvan et al., 1989; Louis et al., 1989; Voorter et al., 1989). Immunocytochemical studies performed in fractions of isolated fiber cell plasma membranes suggest that MP19 is located in junctions (Louis et al., 1989). However, other immunocytochemical studies using different antibodies show that MP19 appears to be located in single membranes (Voorter et al., 1989). At present, there is no explanation for this apparent controversy between these two studies. Amino acid sequence analysis shows that MP19 is not related to any previously sequenced protein in the lens or to proteins of the connexin family (Louis et al., 1989). Hydrophobicity plots calculated from the sequence suggest four intramembrane segments, probably α-helices, with N- and C-terminals both located on the cytoplasmic surface (Gutekunst et al., 1990). This model is similar to the accepted topological model of the connexins (Milks et al., 1988). It has been suggested that MP19 might contribute to the assembly of fiber cell junctional channels rather than itself composing the channels (Louis et al., 1989), but data supporting this notion are lacking.

B. Lipid Composition of Fiber Cell Junctions

The lipid-to-protein weight ratio and the total lipid composition of fiber cell plasma membranes have several unique characteristics (Broekhuyse, 1973, 1981).

First, they both depend on the age of the lens. In the lens of a 30-year-old human, the protein–lipid ratio between the cortex and the nucleus increases from 1.5 to 2.4 while in the lenses of a 70-year-old human, the ratio between the cortex and nucleus increases form 1.6 to 10.5. Second, the cholesterol:phospholipid molar ratio is unusually large and also increases with aging (Broekhuyse, 1973, 1981). In the lens of a 38-year-old human, the ratio increases from 2.4 to 3.2 (Broekhuyse, 1973, 1981). Such high ratios are surprising because the known solubility limit of cholesterol in phospholipids is about 1:1. Thus, it is possible that cholesterol might form a separate phase in fiber cell plasma membranes. Third, the concentration of the phospholipid sphingomyelin (SPH) is unusually large (Broekhuyse, 1973, 1981). In plasma membranes of brain tissues, SPH represents about 30% by weight of the total phospholipid content. In contrast, SPH forms approximately 80% of the total phospholipids in the nucleus of the lens of a 70-year-old human (Broekhuyse, 1973, 1981). The structure of SPH differs from that of glycerol-based phospholipids in that the long aliphatic chain of the sphingosine backbone forms one of the two hydrocarbon chains of the lipid. In the lens, the acyl chains of SPH are asymmetric. One acyl chain is sphingosine and the other a fatty acid with an amide linkage to the sphingosine backbone. As a fiber cell elongates, the fatty acid chain changes its length. The content of oleic acid (18:1 cis in position 9) decreases from 16 to 5% and the content of nervonic acid (24:1 cis in position 15) increases from 17 to 39%. The six-carbon increase in chain length means that SPH should interdigitate to adopt the bilayer configuration (Broekhuyse, 1973, 1981). The prediction that SPH forms interdigitated phases is important because these phases are crystalline at 37°C and also may tend to phase-separate. Perhaps the interaction of SPH and cholesterol fluidizes the crystalline bilayers and solves, at least partially, the packing of membranes of such unusual lipid composition. Nevertheless, the high concentration of these two lipids, with their concomitant packing difficulties, suggests that the structure of the membrane of the nuclear fiber cells may not be that of a fluid bilayer.

The large concentrations of SPH and cholesterol predict that apposing membranes composed of mixtures of SPH and cholesterol may form junction-like structures. These "lipidic" junctions should have a predicted overall thickness of approximately 12.5 nm, including the fluid layer in the extracellular space. This dimension is very similar to the overall thickness of MIP junctions (11–13 nm), and thus it is likely that some of the "thin" junctions described in isolated fractions may correspond to "lipidic" junctions.

A fourth unique aspect of the lipid composition of fiber cell plasma membranes is that the concentrations of lysophosphatidylcholine (lysoPC) and lysophosphatidylethanolamine (lysoPE) are larger than in other membranes. These lyso lipids contain only one acyl chain and, by themselves, can destabilize the bilayer structure of plasma membranes. However, cholesterol that exits at high concentration in fiber cell plasma membranes may combine with these lyso lipids and prevent the disintegration of the fiber cell plasma membrane. The lysophospholipids may also induce fusion between plasma membranes of adjacent fiber cells. Extensive regions of plasma membrane fusion are indistinguishable from gap junctions.

Thus, in the lens, both protein and lipid are degraded by oxidation and endogenous proteases and lipases to fatty acids and smaller polypeptides. Only stable proteins such as MIP and stable lipids such as SPH and cholesterol are likely to remain in the fiber cells in the lens nucleus. The functional significance of such a massive posttranslational modification of membrane components could be large in animals of long life span, such as humans.

C. Physiological Properties of Major Intrinsic Protein

The physiological properties of MIP have been studied in lentoids, the oocyte expression system, and by reconstituting the purified protein into liposomes and planar bilayers. Lentoids are clusters of cultured lens cells. They have been used as *"in vitro"* models of fiber cells because they have the ability to synthesize crystallines and are coupled electrically and to the passage of Procian Yellow (Fitzgerald and Goodenough, 1986; Johnson *et al.*, 1988; Menko *et al.*, 1984). In one study, the ability of MIP form cell-to-cell channels was studied by injecting a single cell in the lentoid with solutions containing antibodies against MIP (Johnson *et al.*, 1988). Procian Yellow injected in other cells of the same lentoid diffused into all cells in the lentoid *except* into the cell injected with the antibody, thus creating a "black hole." The simplest interpretation of the formation of the "black hole" is that MIP forms "communicating" channels which close when the antibody binds to their cytoplasmic domains. If correct, this interpretation indicates that proteins other than connexins may form "communicating" channels in the lens. Yet, the possibility still exists that the lack of diffusion of the dye into the cell injected with the anti-MIP may result from mechanical cell damage rather than antibody-mediated closure of cell-to-cell channels.

Xenopus oocytes have the ability to express and process exogenous proteins when their mRNAs are injected into the cytoplasm. For the expression of gap junction-forming proteins, the injected oocytes are manipulated into contact and the plasma membranes of each oocyte voltage-clamped independently (Dahl *et al.*, 1987). Studies of oocytes injected with mRNAs coding for connexin 32 and connexin 43 show that these proteins form communicating channels when brought into contact (Dahl *et al.*, 1987; Swenson *et al.*, 1989).

In contrast, injections of mRNA coding for MIP do not induce conductances in single oocytes nor communicating channels when the two cells are manipulated mechanically into contact (Swenson *et al.*, 1989). Immunocytochemistry shows that MIP was synthesized by the injected oocytes, a fact that suggests that perhaps MIP lacks channel-forming capabilities. Negative results, however, should be interpreted with caution because it is possible that the MIP may have been inserted incorrectly, if indeed it is inserted at all, into the plasma membrane of the oocyte.

Evidence suggesting that MIP forms channels, at least in single membranes, has been obtained by reconstituting the protein into liposomes and planar lipid bilayers (Gooden et al., 1985a,b; Peracchia and Girsch, 1985b; Scaglione and Rintoul, 1989). In contrast to the negative results obtained in oocytes, liposomes reconstituted with MIP isolated from bovine and human normal and cataractous lenses are permeable to ions and sugars as large as maltoheptose (1.15 kDa) (Nikaido and Rosenberg, 1985). The permeability decreases on addition of calcium-activated calmodulin to the MIP-containing liposomes (Girsch and Peracchia, 1985a,b; Peracchia and Girsch, 1985a,b). These observations are significant because physiological experiments show that fiber cells became uncoupled by calcium-activated calmodulin (Gandolfi et al., 1990). As with the permeabilities measured in liposomes, the uncoupling of fiber cells is blocked by calmodulin inhibitors such as calmidazolium and compounds of the W series (Gandolfi et al., 1990). Also, the ability of calmodulin to affect liposome permeability depends on the integrity of the C-terminal of MIP (Peracchia and Girsch, 1985a).

Peracchia and Girsch (1985a,b) argue that the ability of MIP to form junctions between fiber cells and to increase the permeability of ions and small molecules in liposomes strongly suggests that MIP forms channels of the "communicating" variety. Although this conclusion is premature (since MIP is reconstituted in single-walled liposomes and not in double membranes), these observations are significant because they predict that MIP possesses the ability to form channels of large conductance in artificial systems.

Direct information about the channel-forming properties of MIP was obtained by reconstituting purified MIP into planar lipid bilayers (Ehring and Hall, 1988; Ehring et al., 1990; Zampighi et al., 1985). In these experiments, MIP is first reconstituted in liposomes which are then fused with preformed planar lipid bilayers. These studies characterized the electrical properties of the MIP channels. The number of MIP monomers forming the channels was estimated from experiments using liposomes reconstituted with high protein/lipid ratios. These liposomes contain MIP crystallized in tetragonal arrays, an observation suggesting that the each channel is composed of four copies of the 28-kDa polypeptide.

The principal characteristics of the reconstituted MIP channel are large single-channel conductance, limited anionic selectivity, symmetrical voltage dependance, rapid opening and slow closing rates, and insensitivity to calcium and hydrogen ion concentration (Ehring and Hall, 1988; Ehring et al., 1990; Zampighi et al., 1985). The voltage dependence is symmetric (depending on the magnitude but not the sign of the voltages) and is modified by phosphorylation (Ehring et al., 1991). Channels induced by the 28-kDa polypeptide isolated from hepatocyte gap junctions (connexin 32) and reconstituted into planar bilayers exhibit similar single-channel conductance and voltage dependance and are also insensitive to calcium and pH (Young et al., 1987). The similarity in the properties of channels induced by connexin 32 and MIP in planar bilayers is puzzling in view of the negative results obtained by injecting oocytes with mRNA coding for MIP.

The voltage dependence of the MIP channels is modulated by phosphorylation. Dephosphorylated MIP forms channels that are largely voltage-independent, while phosphorylated MIP forms channels that are significantly more voltage-dependent (Ehring et al., 1992). Similarly, the voltage-independent channels can be converted into voltage-dependent channels by adding directly the phosphorylating mixture into the bilayer chamber. Analysis of steady-state conductance–voltage data and channel kinetics demonstrates that the effect is all-or-none; i.e., channels are either voltage dependent or voltage independent. In addition, the sign of the phosphorylation-induced gating charge is consistent with the addition of a negatively charged phosphate to the gating mechanism. Single-channel data showed that the effect on the channel voltage dependence results from an alteration in channel kinetics rather than a change in single-channel conductance.

Direct comparison of the reconstituted MIP channel to channels in the lens is presently not possible because of a lack of detailed characterization of channels in intact fiber cells. However, the conductance in the order of 10^{-6} S/cm^2 of fiber cells suggests that very few, if any, open MIP channels with the conductance of the reconstituted MIP channel can exist in these cells (Mathias et al., 1981). This poses an intriguing problem because fiber cell plasma membranes contain extensive regions with MIP either forming junctions or restricted in single membranes. If only a small percentage of the protein forms channels with the conductance measured for the reconstituted MIP channel, the cell would rupture due to osmotic swelling. Clearly, the physiological role of MIP requires a more complete assessment of the ion channels present in fiber cells and perhaps also the study of cell lines transfected with the gene coding for this protein.

D. Connexins of the Lens

Gap junction proteins from heart, liver, Xenopus embryo, and oocytes share substantial amino acid sequence homology (Beyer et al., 1987; Ebihara et al., 1989; Gimlich et al., 1988; Kumar and Gilula, 1986; Paul, 1986; Zhang and Nicholson, 1989). Presently, the connexin family contains 12 proteins, but the total number is expected to increase. Topological models depicting the path followed by the polypeptide in the membrane have been obtained by combining hydrophobicity plots calculated from the amino acid sequence with studies using antibodies raised against synthetic peptides having the sequence of specific regions of the proteins (Milks et al., 1988; Zimmer et al., 1987). These studies show that all connexins share a common motif composed of four transmembrane domains with both N- and C-terminals exposed on the cytoplasmic surface. The regions of highest homology among connexins are located at the N-terminal and the external loops. The region of lowest homology is located at the C-terminal. One intramembranous domain (M3) exhibits a stretch of amphiphilic amino acids that is thought to form, in association with the other 11 monomers in the dodecamer, the hydrophilic pore of the "communicating" channel (Milks *et al.*, 1988).

In the lens, the epithelium and fiber cells express different connexins. The epithelium expresses connexin 43, the heart myocyte gap junction protein (Beyer *et al.*, 1988), whereas fiber cells express connexin 46 and MP70 (Beyer *et al.*, 1988; Gruijters *et al.*, 1987; Kistler *et al.*, 1985). The complete amino acid sequence of MP70 is not available, but its N-terminal is homologous to the N-terminals of connexin 32 and connexin 43 (40% identical with liver and 50% with heart) (Kistler *et al.*, 1988). In addition, MP70 is located in fiber junctions of the lens cortex but is less abundant in the fiber cells of the nucleus. Although the demonstration is still elusive, it seems likely that MP70 is one of the proteins involved in the formation of the lens "communicating" channels. In contrast, there are not enough data to determine whether connexin 46 is located in junctions or if it forms channels of the gap "communicating" variety. The only information available is that its mRNA is expressed in fiber cells and that its amino acid sequence is homologous to the connexins. In the oocyte expression system, injection of mRNA for connexin 46 induces large conductances but the formation of "communicating" channels spanning the plasma membrane of two oocytes is still lacking.

In contrast, the amino acid sequence of MIP exhibits no significant homology with connexins. Hydrophobicity plots predict six transmembranous domains for MIP instead of the four predicted for connexins (Gorin *et al.*, 1984). One of these domains (no. 6) is amphiphilic, which is consistent with an oligomer of MIP forming a hydrophilic pore. Analysis of the amino acid sequence shows that MIP belongs to a new family of membrane proteins found in plants, insects, and bacteria. Among these are nodulin, the protein found in the root of nitrogen-fixing plants (Baker and Saier, 1990); glycerol facilitator found in the external membrane of in *Escherichia coli* (Lin, 1986); tonoplast intrinsic protein (TIP) (Johnson *et al.*, 1990); and the protein in the locus of the mutant "big brain" from *Drosophila* (Rao *et al.*, 1990). These proteins are located only in *single membranes*, an important difference with MIP, which in fiber cells is present in both single and junctional membranes. Although the function of the proteins in this new family is unknown, it appears that they increase the permeability of plasma membranes to low-molecular-weight molecules.

V. Structure of Fiber Cell Junctions

The principal aim of this survey has been to show that the extensive networks of plasma membrane that contact connecting fiber cells are composed of different types of junctions. We have reviewed evidence indicating that these junctions differ in their permeability to lanthanum (Fig. 4B,C), their arrangement of intramembrane particles (Fig. 5C,D), and their overall thickness (Zampighi *et al.*,

1982, 1989). In this section, we review information regarding their structure and protein composition. In an attempt to simplify the description of the data, the difference in overall thickness is used to classify fiber cell junctions arbitrarily into a 11- to 13-nm junction and an 18- to 20-nm junction.

A. The 11- to 13-nm Junction

The 11- to 13-nm junctions are composed principally (perhaps, in some regions of the lens, exclusively) of MIP. They measure 11–13 nm in overall thickness and in freeze-fracture replicas they are composed of plaques of intramembrane particles arranged either randomly or tetragonally (Fig. 5C,D). Their surface topology can be either wavy or flat. Wavy junctions have undulating shapes with wavelengths of 0.2–0.3 μm and are formed by closely apposed convex and concave membranes (Fig. 7A). A striking feature of wavy junctions is that the convex membrane is approximately 2 nm thicker than the concave membrane (Zampighi et al., 1982, 1989). In the nucleus, wavy junctions are found principally in tongue-and-groove interdigitations (Lo and Harding, 1984), while in the cortex, flat junctions are most abundant.

Immunocytochemical studies show that 11- to 13-nm junctions of either the wavy or flat variety are labeled by anti-MIP, but with strikingly different patterns (Zampighi et al., 1989). Wavy junctions are labeled asymmetrically on the convex but not on the apposing concave membranes (Fig. 7A,B and Fig. 8, black arrow). In contrast, flat junctions are labeled symmetrically on the cytoplasmic surfaces of both junctional membranes, but with one membrane of the pair usually more heavily labeled than the other (Fig. 7C and Fig. 8, open arrow). This unusual situation, where the same protein adopts different labeling patterns depending on the type of junction under study, has resulted in conflicting interpretations (Paul and Goodenough, 1983; Sas et al., 1985). The more symmetrical patterns have been interpreted as strong support for the notion that MIP is involved in com-

---

FIG. 7   Immunolocalization of MIP in membranes isolated from bovine lenses and labeled with a polyclonal anti-MIP [prepared by Bok et al. (1982)] and with protein A–gold (A and B) and a monoclonal anti-MIP [prepared by Paul and Goodenough (1983)] visualized by the peroxidase reaction (C). (A) A low-magnification view of an extensive profile containing junctions and single membranes. MIP is located in both junctional and nonjunctional membranes (black arrows in A and the region where the junction splits in the lower right corner of C). (B) A higher magnification view of a junction in A. The anti-MIP labeled these junctions asymmetrically in only one membrane of the pari. B also contains a short junctional segment of 18–20 nm in overall thickness located between two regions of 11- to 13-nm junctions (open arrow). This thicker type of junction is not labeled with the anti-MIP and presumably does not contain the protein. (C) A junction labeled with a monoclonal anti-MIP. Both junctional and single plasma membranes are labeled with a monoclonal anti-MIP. In contrast to the asymmetrical labeling seen in experiments using the polyclonal anti-MIP, labeling with the monoclonal shows reaction products on both membranes. Magnifications: A, 24,300×; B, 202,500×; C, 243,000×. A and B are from Zampighi et al. (1989).

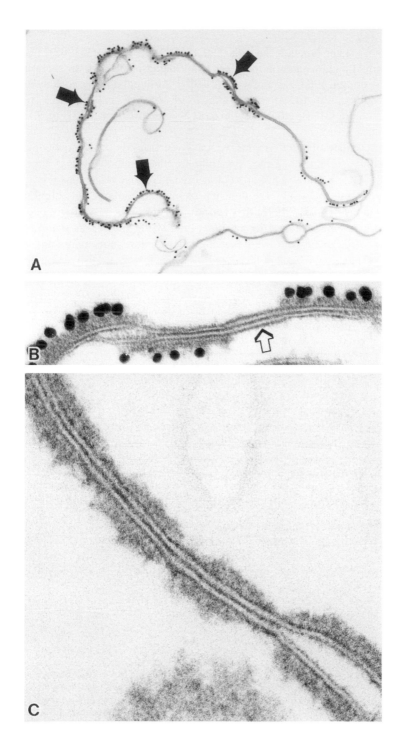

munication via gap junction channels (Bok *et al.,* 1982; Sas *et al.,* 1985). The more asymmetrical patterns are, on the other hand, inconsistent with this notion (Paul and Goodenough, 1983; Zampighi *et al.,* 1989). Although the controversy is by no means resolved, a partial solution is suggested from the analysis of the structure of wavy junctions, those exhibiting the most asymmetrical labeling (Fig. 7A,B and Fig. 8, black arrow).

The determination of the structure of wavy junctions requires knowledge of the protein composition of the concave membrane (the membrane that is not labeled by the anti-MIPs). The absence of labeling on the concave membrane is consistent with three possibilities: (1) The concave membrane contains proteolytic break-down products of MIP (MP22) that are not recognized by the antibody. (2) The concave membrane contains proteins other than MIP. (3) The concave membrane is a protein-free membrane. Studies using label-fracture, a technique that allows labeling of junctional fracture faces, permit investigators to distinguish between these possibilities (Zampighi *et al.,* 1989).

Label-fracture replicas show that anti-MIPs label tetragonal arrays on P fracture

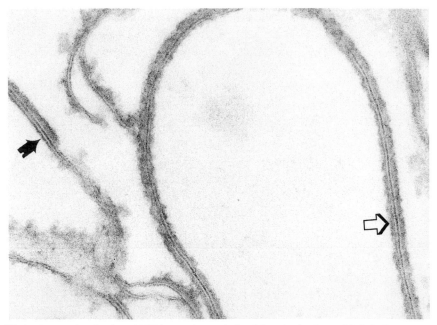

FIG. 8   Immunolocalization of MIP in membranes isolated from bovine lenses labeled with a mono-clonal anti-MIP. The antibody labeled some junctions asymmetrically on only one membrane of the pair (black arrow). Other junctions, however, are labeled more symmetrically (open arrow). But, even in these more symmetrically stained junctions, one membrane of the junction is usually more heavily stained than the other. Magnification: 140,000×. From Zampighi *et al.* (1989).

faces and smooth particle-free E fracture faces (Fig. 9). The interpretation of these labeled fracture faces is straightforward because, after staining, the junctions are adhered to a glass substrate and then fractured. This provides a definitive reference plane that distinguish junctions from single membranes because in junctions E and P faces are separated by small fracture steps (Fig. 9). Since junctions can only adhere to the glass via their cytoplasmic surfaces, all P faces arise from the membrane that adhered to the glass and all E faces from the apposing junctional membrane. The labeling of P faces demonstrates that tetragonal arrays are composed partially, or perhaps exclusively, of MIP. The labeling on smooth E faces and the absence of labeling on complementary plaques of pits arranged tetragonally demonstrates that the apposing membrane is particle free (Zampighi *et al.*, 1989). Thus, the MIP junction is formed of one membrane composed of MIP arranged tetragonally adhering to an apposing protein-free membrane (Fig. 10) (Zampighi *et al.*, 1989).

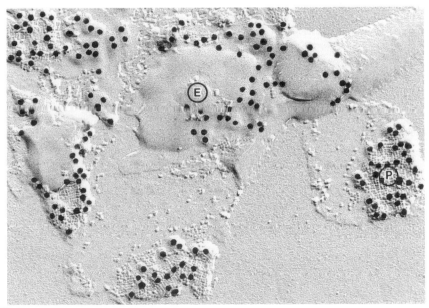

FIG. 9   Immunolocalization of MIP in membranes isolated from bovine lens by the label-fracture method. The junctions were first labeled with primary and secondary antibodies and then adhered onto polylysine-coated glass slides. The junctions were covered with copper hats and frozen without fixation or glycerinization by immersion in liquid propane (Zampighi *et al.*, 1989). Gold particles do not cast shadows because they are located between the membrane and the glass substratum. Label-fracture replicas are labeled on P faces composed of tetragonal arrays 6–7 nm center-to-center apart and also on smooth E fracture faces. The labeling on the smooth E faces arises from MIP located in the apposing membrane. These observations indicate that MIP junctions are formed by tetramers of MIP in one membrane abutted against protein-free apposing membranes. Magnification: 128,250×.

The different labeling patterns observed in thin-section immunocytochemistry (Figs. 7 and 8) can now be interpreted as arising from junctions exhibiting the same structure but having crystals of MIP of different sizes. MIP arranged in extensive crystals adheres against equally large protein-free apposing membranes, an arrangement that results in a highly asymmetrical labeling pattern (Fig. 7A,B). MIP arranged randomly, or in small tetragonal crystals, displays more symmetrical labeling patterns because the size of the apposing protein-free membrane is smaller (Figs. 7C and 8). The distribution of MIP in aggregates of different size may explain why the protein can form junctions with wavy and flat topologies. The precise radius of curvature of wavy junctions resides in the intrinsic shape (or in a particular grouping of charged residues) of MIP and in its strong tendency to crystallize. The flat topology arises when MIP crystallizes in small plaques because opposing plaques have a tendency to bend in opposite directions, thus canceling their bending movements. If 11- to 13-nm junctions are constructed of plaques of MIP arranged randomly, no bending movement is generated and the resulting junctions are flat (Zampighi *et al.*, 1989).

A model where tetramers of MIP abut against a protein-free, apposing membrane provides an explanation for a number of observations regarding the structure of the junctions made of MIP. For example, the presence of MIP in only one of the junctional membranes explains why the convex membrane in wavy junctions is always about 2 nm thicker than the concave membrane (Simon *et al.*, 1982; Zampighi *et al.*, 1982). The close apposition between a MIP tetramer and a protein-free membrane can also explain the small width of the fluid layer that does not permit the diffusion of lanthanum.

The model of the structure of the MIP junctions suggests a function for MIP in

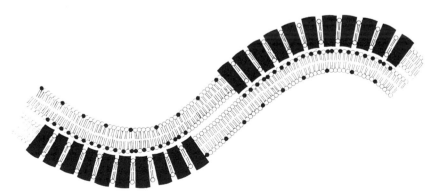

FIG. 10 Model of junctions made of MIP deduced form morphological, immunocytochemical, and biochemical methods. The model depicts the junction formed of MIP units (tetramers) arranged in crystals. These MIP crystals are present in only one of the membranes of the junction. The junction is formed when the MIP crystal abuts on the apposing particle-free membrane. From Zampighi *et al.* (1989).

increasing the resistance of the extracellular space. Costello *et al.* (1985, 1989) proposed that the two membranes are held together by electrostatic attractive forces between positively charged amino acids in the external loops of MIP and negatively charged lipids in the external monolayers of the apposing membrane. However, given the paucity of negatively charged lipids in these plasma membranes (Aster *et al.*, 1986; Heslip *et al.*, 1986), the attractive interaction holding the two membranes so close together may not be entirely due to salt bridges but may also involve other short-range forces such as hydrogen bonds. Reduction of the average width of the fluid layer may result from the reduction in the repulsive undulation pressure from thermally induced bending moments in fluid membranes. This pressure is inversely proportional to the membrane compressibility modulus. In fiber cell membranes, the pressure is further reduced because of their high cholesterol and SPH content and the presence of crystalline plaques of MIP. The low concentration of glycolipids and glycoproteins in fiber cells ensures that steric repulsive pressure between fiber cells is kept to a minimum. All these factors reduce the width of the extracellular fluid space to a small percentage of the total volume of the lens.

Thus, the 11- to 13-nm junctions are composed principally of MIP. They cover a large percentage of the fiber cell surface and are a factor in reducing the volume of the extracellular space of the lens. Perhaps the MIP junctions are the structures that account in part for the radially dependent voltage drop measured in the lens extracellular space.

B. The 18- to 20-nm Junction

The 18- to 20-nm junctions are most abundant in the lens cortex and markedly decrease toward the nucleus. These junctions are formed of closely apposed plasma membranes separated by a 4- to 5-nm extracellular gap permeable to lanthanum (Fig. 4C). They are composed of extensive plaques of intramembrane particles that are about 7–8 nm in diameter and arranged randomly (Gruijters *et al.*, 1987; Zampighi *et al.*, 1989). On occasion, the particles are arranged in hexagonal lattices of spaced approximately 9 nm apart, similar to those in gap junctions (regions encircled by arrowheads in Fig. 11).

The determination of the structure of the 18- to 20-nm junctions has not progressed to the state where the subunit composition of the 7- to 8-nm diameter units and position in the junction can be resolved in detail. The best known aspect of this type of junction is that it contains MP70 on both plasma membranes (Fig. 12A–D). In contrast to the asymmetrical labeling found in 11- to 13-nm junctions, anti-MP70 antibodies label 18- to 20-nm junctions symmetrically on both plasma membranes. No asymmetry was detected when the primary antibodies were visualized with gold particles (Fig. 12A–C) or by the peroxidase reaction (Fig. 12D).

The 18- to 20-nm junction contains unlabeled regions (arrowheads in Fig. 12A).

Double-labeling these junctions with anti-MP70 and anti-MIP antibodies shows that the isolated junctions are formed of segments of 11- to 13-nm junctions composed of MIP and segments of 18- to 20-nm junctions composed of MP70 (Fig. 13). The transition from one type of junction to the other is abrupt, suggesting that the two proteins intermingle with each other to form the extensive junctional profiles in fiber cells.

The symmetrical arrangement of MP70 on both junctional membranes is a strong suggestion that 18- to 20-nm junctions are formed of coaxially aligned "communicating" channels. The possibility that MP70 forms "communicating" channels is also supported by the observation that nonionic detergents solubilize the 18- to 20-nm junction into particles with a sedimentation coefficient of 16S (Kistler and Bullivant, 1988). By negative staining, these particles appear to be formed of two domains. Thus, it appears that MP70 forms coaxially aligned channels spanning the entire thickness of the 18- to 20-nm junction.

The presence of at least four different proteins in the 18- to 20-nm junctions suggests at least two plausible subunit arrangements for the putative communicating channel. One depicts the channel to be constructed by copies of the four different polypeptides (as in acetylcholine receptors). An alternative arrangement depicts the channels as constructed of subunits of either MP70 or connexin 46. In this second possibility, the channels are made only of MP70 or connexin 46, although plaques of these channels may coexist in the same junction. The observation that isolated 18- to 20-nm junctions contain plaques of particles arranged in hexagonal arrays suggests that the channels are composed of identical subunits (Fig. 11). Clearly, three-dimensional reconstructions of crystalline 18- to 20-nm junctions are necessary to determine the subunit composition of channels and the presence and extension of the hydrophilic pore. The unraveling of the functional role of these channels will probably be an equally challenging task.

VI. Conclusions

Morphological, biochemical, and physiological information indicates that cells in the lens are connected by at least four types of plasma membrane contacts: tight junctions, gap junctions, 18- to 20-nm junctions, and 11- to 13-nm junctions.

FIG. 11   The packing of particles and complementary pits in isolated junctions from bovine lens. The junctions were adhered on glass coated with polylysine, covered with a copper hat, and then frozen without prior fixation or glycerinization (Zampighi et al., 1989). The P and E fracture faces are composed mostly of particles and pits arranged randomly. However, two small regions of the junction E face exhibit pits arranged hexagonally (regions encircled by arrowheads) approximately 8–9 nm in unit cell dimension. This arrangement is similar in appearance and dimensions to E faces of hepatocyte gap junctions. Magnification; 150,000×.

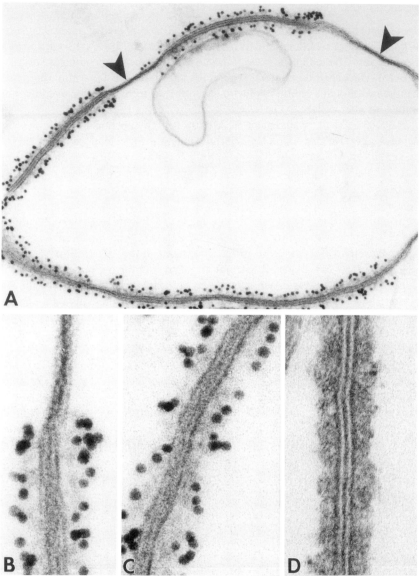

FIG. 12   Immunolocalization of MP70 using a monoclonal antibody [prepared by Kistler *et al.* (1985)] visualized with anti-mouse complexed with gold particles (A–C) and by the peroxidase reaction (D). (A) A low-magnification view of an extensive junctional profile. Most of the junction was labeled symmetrically with gold particles on both membranes of the junctional pair. This pattern contrast sharply with the asymmetrical labeling pattern obtained with anti-MIP antibodies in Figs. 6A,B and 7. The arrowheads point to regions composed of 11- to 13-nm junctions that are not labeled with the anti-MP70. (B) A region of transition between a labeled and an unlabeled junction at higher magnification. (C and D) Higher magnification views to demonstrate that the 18- to 20-nm junctions contains MP70. Magnifications: A, 72,000×; B–D, 216,000×.

These junctions appear to mediate and control different pathways for the diffusion of ions and small nonelectrolytes within and out of the lens and thus play fundamental roles in lens fluid balance and transparency.

The tight junctions are found only between epithelial cells in the anterior pole of the lens. Tight junctions should function as barriers to the diffusion of large molecules from the anterior chamber of the eye into the extracellular space of the lens. Their function is crucial to lens physiology because, if osmotically active electrolytes (other than salts) or nonelectrolytes are allowed to enter into the lens extracellular space, they may spend considerable time in this space because of its large diffusional resistance. These osmotically active molecules would drive water into the extracellular space and increase its volume, a situation that might lead to lens opacification and cataract formation. Tight junctions in the epithelium might also provide a regulated pathway for the diffusion of cations such as sodium and anions such as chloride within the extracellular space of the lens. The magnitude of the contribution of this pathway to the recirculating currents in the lens is unknown.

Gap junction channels mediate and regulate the diffusion of ions and small molecules between the cytoplasm of adjacent cells. In the lens, gap junctions exist between epithelial cells, fiber cells, and epithelium and the first layer of fiber cells. Gap junctions between epithelial cells contain connexin 43, the heart myocyte gap junction protein. The proteins forming gap junctions between fiber cells appear to be two new members of the connexin family, MP70 and connexin 46. These connexins might form gap junction channels either by themselves or in association with each other or with nonconnexin proteins such as MIP and/or MP19. The protein forming the gap junction coupling epithelial to fiber cells has not been identified yet.

The major intrinsic protein (MIP) comprises over 50% of the total membrane protein of the lens. MIP is expressed exclusively in fiber cells and is located in both single membrane and in junctions. The junctions are formed of tetramers of MIP abutted against an apposing protein-free plasma membrane. The structure of the MIP junctions contrasts sharply with the structure of gap junctions that are formed of hexamers joined through their external domains. Analysis of the amino acid sequence shows that MIP is not a connexin but exhibits sequence homology to protein found in plants, *E. coli,* and the brain of *Drosophila.* When reconstituted in planar bilayers, MIP forms voltage-dependent channels of large unitary conductance. Although the function of MIP in the lens is still unresolved, evidence is accumulating that the protein might have a role in regulating the paracellular resistance of the lens.

The unusual high concentrations of cholesterol and SPH predict that these lipids might form separate phases in the plasma membrane of fiber cells, particularly in the nucleus. This prediction is significant because phases made of cholesterol or SPH do not exhibit the fluid bilayer organization of normal plasma membranes. In addition, membranes composed of combinations of SPH and cholesterol may

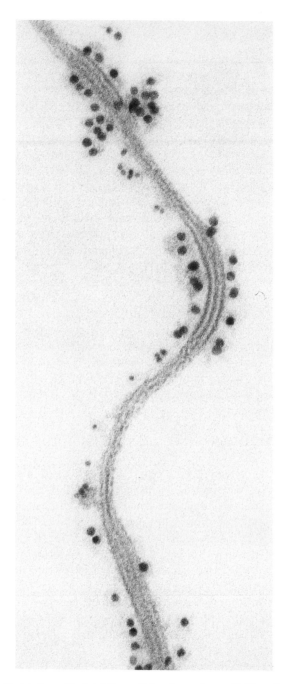

FIG. 13 Double-labeling of isolated fiber cell junctions using anti-MIP antibodies (visualized with 5-nm-diameter gold particles) and with anti-MP70 antibodies (visualized with 15-nm-diameter particles). The 18- to 20-nm junction contains MP70, while the 11- to 13-nm junction contains MIP. Thus, it appears that the extensive network of close plasma membrane contacts in fiber cells is formed by a mixture of junctional membranes composed of different proteins. Magnification: 172,000×. From Zampighi *et al.* (1989).

form junctional structures that should have a predicted overall thickness of 12.5 nm. The low concentration of glycolipids and glycoproteins toward the lens nucleus will help the formation of junctions composed exclusively of lipids. These predicted "lipidic" junctions will also have a small fluid layer and might also increase the resistance of the lens paracellular pathway.

The MIP junctions and perhaps also the predicted "lipidic" junctions can provide at least a partial explanation of the potential drop across the lens that is thought to be the underlying mechanism driving the vectorial water flow that carries nutrients into and metabolites out of the lens. The high-resistance pathway provided by MIP junctions is in series with a lower resistance extracellular space contained in the suture lines. Ions and small molecules entering or leaving the lens via the extracellular space should diffuse much faster through the network formed by the sutures than through the clefts between fiber cells.

Acknowledgments

We are indebted to Dr. L. Kruger for his comments. We are also thankful for the excellent technical assistance of Mr. M. Kreman. This work was supported by National Institutes of Health grants EY4110 and EY05661.

References

Aerts, T., Xia, J.-Z., Slegers, H., de Block, J., and Clauwaert, J. (1990). *J. Biol. Chem.* **265**, 8675–8680.
Alcala, J., Lieska, N., and Maisel, H. (1975). *Exp. Eye Res.* **21**, 591–595.
Alvarez, L. J., Candia, O. A., and Grillone, L. R. (1985). *Curr. Eye Res.* **4**, 143–152.
Aster, J. C., Brewer, G. J., and Maisel, H. (1986). *J. Cell Biol.* **103**, 115–122.
Baker, M. E., and Saier, M. H. (1990). *Cell (Cambridge, Mass.)* **60**, 185–186.
Bassnett, S., and Duncan, G. (1988). *J. Physiol. (London)* **398**, 507–521.
Benedetti, E. L., Dunia, I., Bentzel, C. J., Vermorken, A. J. M., Kibbelaar, M., and Bloemendal, H. (1976). *Biochim. Biophys. Acta* **457**, 353–384.
Bernardini, G., Peracchia, C., and Venosa, R. A. (1981). *J. Physiol. (London)* **320**, 187–192.
Beyer, E. C., Paul, D. L., and Goodenough, D. A. (1987). *J. Cell Biol.* **105**, 2621–2629.
Beyer, E. C., Goodenough, D. A., and Paul, D. L. (1988). *In* "Gap Junctions" E. L. Hertzberg and R. G. Johnson, eds.), pp. 167–175. Alan R. Liss, New York.
Bloemendal, H., Zweers, A., Vermorken, F., Dunia, I., and Benedetti, E. L. (1972). *Cell Differ.* **1**, 91–106.
Bok, D., Dockstader, J., and Horwitz, J. (1982). *J. Cell Biol.* **92**, 213–220.
Bouman, A. A., and Broekhuyse, R. M. (1981). *Exp. Eye Res.* **33**, 299–308.
Broekhuyse, R. M. (1973). *Ciba Found. Symp.* [N.S.] **19**, 135–149.
Broekhuyse, R. M. (1981). "Mechanisms of Cataract Formation in the Human Lens." Academic Press, New York.
Broekhuyse, R. M., and Kuhlmann, E. D. (1974). *Exp. Eye Res.* **19**, 297–302.
Broekhuyse, R. M., and Kuhlmann, E. D. (1978). *Exp. Eye Res.* **26**, 305–320.
Broekhuyse, R. M., and Kuhlmann, E. D. (1980). *Exp. Eye Res.* **30**, 305–310.
Broekhuyse, R. M., Kuhlmann, E. D., and Stols, A. L. (1976). *Exp. Eye Res.* **23**, 365–371.

Broekhuyse, R. M., Kuhlmann, E. D., Bijvelt, J., Verkleij, A. J., and Ververgaert, P. H. J. T. (1978). *Exp. Eye Res.* **26,** 147–156.

Broekhuyse, R. M., Kuhlmann, E. D., and Winkens, H. J. (1979). *Exp. Eye Res.* **29,** 303–313.

Cohen, A. I. (1965). *Invest. Ophthalmol. Visual Sci.* **79,** 189.

Cooper, K. E., Rae, J. L., and Gates, P. (1989). *J. Membr. Biol.* **111,** 215–227.

Cooper, K. E., Tang, J. M., Rae, J. L., and Eisenberg, R. S. (1986). *J. Membr. Biol.* **93,** 259–269.

Cooper, K. E., Gates, P., Rae, J. L., and Dewey, J. (1990). *J. Membr. Biol.* **117,** 285–298.

Costello, M. J., McIntosh, T. J., and Robertson, J. D. (1985). *Curr. Eye Res.* **4,** 1183–1201.

Costello, M. J., McIntosh, T. J., and Robertson, J. D. (1989). *Invest. Ophthalmol. Visual Sci.* **30,** 975–989.

Dahl, G., Miller, T. M., Paul, D. L., Voellmy, R., and Werner, R. (1987). *Science* **236,** 1290–1293.

Dickson, D. H., and Crock, G. W. (1972). *Invest. Ophthalmol.* **11,** 809–815.

Duncan, G. (1969a). *Exp. Eye Res.* **8,** 315–325.

Duncan, G. (1969b). *Exp. Eye Res.* **8,** 406–412.

Duncan, G., and Croghan, P. C. (1969). *Exp. Eye Res.* **8,** 421–430.

Duncan, G., Stewart, S., Prescott, A. R., and Warn, R. M. (1988). *J. Membr. Biol.* **102,** 195–204.

Dunia, I., Ghosh, C. S., Benedetti, E. L., Zweers, A., and Bloemendal, H. (1974). *FEBS Lett.* **45,** 139–144.

Ebihara, K., Beyer, E. C., Swenson, K. I., Paul, D. L., and Goodenough, D. A. (1989). *Science* **243,** 1194–1195.

Ehring, G. R., and Hall, J. E. (1988). *Proc. West. Pharmacol. Soc.* **31,** 251–253.

Ehring, G. R., Zampighi, G. A., Horwitz, J., Bok, D., and Hall, J. E. (1990). *J. Gen. Physiol.* **96,** 631–664.

Ehring, G. R., Lagos, N., Zampighi, G. A., and Hall, J. E. (1992). *J. Membr.* (in press).

Fitzgerald, P. G., and Goodenough, D. A. (1986). *Invest. Ophthalmol. Visual Sci.* **27,** 755–771.

Fitzgerald, P. G., Bok, D., and Horwitz, J. (1985). *Curr. Eye Res.* **4,** 1203–1217.

Fowlks, W. L. (1965). *Invest Ophthalmol.* **4,** 367–368.

Galvan, A. C., Lampe, P. D., Hur, K. C., *et al.* (1989). *J. Biol. Chem.* **264,** 19974–19978.

Gandolfi, S. A., Duncan, G., Tomlinson, J., and Maraini, G. (1990). *Curr. Eye Res.* 533–541.

Garland, D., and Russell, P. (1985). *Proc. Natl. Acad. Sci. U.S.A.* **82,** 653–657.

Gimlich, R. L., Kumar, N. M., and Gilula, N. B. (1988). *J. Cell Biol.* **107,** 1065–1073.

Girsch, S. J., and Peracchia, C. (1985a). *J. Membr. Biol.* **83,** 217–225.

Girsch, S. J., and Peracchia, C. (1985b). *J. Membr. Biol.* **83,** 227–233.

Gooden, M. M., Rintoul, D. A., Takehana, M., and Takemoto, L. J. (1985a). *Biochem. Biophys. Res. Commun.* **128,** 993–999.

Gooden, M. M., Takemoto, L. J., and Rintoul, D. A. (1985b). *Curr. Eye Res.* **4,** 1107–1115.

Goodenough, D. A. (1979). *Invest. Ophthalmol. Visual Sci.* **18,** 1104–1122.

Goodenough, D. A., Dick, John S. B., and Lyons, J. E. (1980). *J. Cell Biol.* **86,** 576–589.

Gorin, M. B., Yancey, S. B., Cline, J., Revel, J. P., and Horwitz, J. (1984). *Cell (Cambridge, Mass.)* **39,** 49–59.

Gorthy, W. C., and Anderson, J. W. (1980). *Visual Sci.* **19,** 1038–1052.

Gruijters, W. T., Kistler, J., Bullivant, S., and Goodenough, D. A. (1987). *J. Cell Biol.* **104,** 565–572.

Gutekunst, K. A., Rao, G. N., and Church, R. L. (1990). *Curr. Eye Res.* **9,** 955–961.

Harding, J. J., and Dilley, K. J. (1976). *Exp. Eye Res.* **22,** 1–73.

Heslip, J., Bagchi, M., Zhang, S., Aloust, S., and Maisel, H. (1986). *No Journal Found* **5,** 949–958.

Horwitz, J., and Bok, D. (1987). *Biochemistry* **26,** 8092–8098.

Horwitz, J., and Wong, M. M. (1980). *Biochim. Biophys. Acta* **622,** 134–143.

Horwitz, J., Robertson, N. P., Wong, M. M., Zigler, J. S., and Kinoshita, J. H. (1979). *Exp. Eye Res.* **28,** 359–365.

Jacob, T. J. C. (1986). *J. Physiol. (London)* **377,** 37P.

Johnson, K. D., Hofte, H., and Chrispeels, M. J. (1990). *Plant Cell* **2,** 525–532.

Johnson, K. R., Panter, S. S., and Johnson, R. G. (1985). *Biochim. Biophys. Acta* **844**, 367–376.

Johnson, K. R., Lampe, P. D., Hur, K. C., Louis, C. F., and Johnson, R. G. (1986). *J. Cell Biol.* **102**, 1334–1343.

Johnson, K. R., Sas, D. F., and Johnson, R. G. (1991). *Exp. Eye Res.* **52**, 129–139.

Johnson, R., Frenzel, E., Johnson, K., *et al.* (1985). *In* "Gap Junctions" (M. V. L. Bennett and D. C. Spray, eds.), pp. 91–105. Cold Spring Harbor Lab., Cold Spring Harbor, New York.

Johnson, R. G., Klukas, K. A., Hong, L. T., and Spray, D. C. (1988). *In* "Gap Junctions" (E. L. Hertzberg and R. G. Johnson, eds.), pp. 81–98. Alan R. Liss, New York.

Keeling, P., Johnson, K. R., Sas, D. F., Klukas, K. A., Donahue, P., and Johnson, R. G. (1983). *J. Membr. Biol* **74**, 217–228.

Kinsey, E., and Reddy, D. V. N. (1965). *Invest Ophthalmol.* **4**, 104–116.

Kistler, J., and Bullivant, S. (1980a). *FEBS Lett.* **111**, 73–78.

Kistler, J., and Bullivant, S. (1980b). *J. Ultrastruct. Res.* **72**, 27–38.

Kistler, J., and Bullivant, S. (1987). *Invest. Ophthalmol. Visual Sci.* **28**, 1687–1692.

Kistler, J., and Bullivant, S. (1988). *J. Cell Sci.* **91**, 415–421.

Kistler, J., Kirkland, B., and Bullivant S. (1985). *J. Cell Biol.* **101**, 28–35.

Kistler, J., Christie, D., and Bullivant, S. (1988). *Nature (London)* **331**, 721–73 .

Kobatashi, S., Roy, D., and Spector, A. (1982). *Curr. Eye Res.* **2**, 327–334.

Kumar, N. M., and Gilula, N. B. (1986). *J. Cell Biol.* **103**, 767–776.

Kuszak, J. R., Maisel, H., and Harding, C. V. (1978). *Exp. Eye Res.* **27**, 495–498.

Kuszak, J. R., Rae, J. L., Pauli, B. U., and Weinstein, R. S., (1982). *J. Ultrastruct. Res.* **81**, 249–256.

Kuszak, J. R., Macsai, M. S., Bloom, K. J., Rae, J. L., and Weinstein, R. S. (1985a). *J. Ultrastruct. Res.* **93**, 144–160.

Kuszak, J. R., Shek, Y. H., Carney, K. C., and Rae, J. L. (1985b). *Curr. Eye Res.* 1145–1153.

Lampe, P. D., and Johnson, R. G. (1989). *J. Membr. Biol.* **107**, 145–155.

Lampe, P. D., and Johnson, R. G. (1991). *Eur. J. Biochem.* **194**, 541–547.

Lampe, P. D., Bazzi, M. D., Nelsestuen, G. L., and Johnson, R. G. (1986). *Eur. J. Biochem.* **156**, 351–357.

Lasser, A., and Balazs, E. A. (1972). *Exp. Eye Res.* **13**, 292–308.

Leeson, T. S. (1971). *Exp. Eye Res.* **11**, 78–82.

Lin, E. C. (1986). *In* "Methods in Enzymology" (S. Fleischer and B. Fleischer, eds.), Vol. 125, pp. 467–473. Academic Press, Orlando, Florida.

Lo, W.-K., and Harding, C. V. (1983). *Invest. Ophthalmol. Visual Sci.* **24**, 396–402.

Lo, W.-K., and Harding, C. V. (1984). *J. Ultrastruct. Res.* 228–245.

Lo, W.-K., and Harding, C. V. (1986). *Cell Tissue Res.* **214**, 253–263.

Louis, C. F., Hur, K. C., Galvan, A., *et al.* (1989). *J. Biol. Chem.* 19967–19973.

Maisel, H., Harding, C. V., Alcala, J. R., Kuszak, J. R., and Bradley, R. (1981). *In* "Molecular and Cellular Biology of the Eye Lens" (H. Bloemendal, ed.). Wiley, New York.

Manenti, S., Dunia, I., Le Marie, M., and Benedetti, E. L. (1988). *FEBS Lett.* **233**, 148–152.

Mathias, R. T., and Rae, J. L. (1985a). *Am. J. Physiol.* **249**, C181–C190.

Mathias, R. T., and Rae, J. L. (1985b). *Curr. Eye Res.* **4**, 421–430.

Mathias, R. T., and Rae, J. L. (1989). *In* "Cell Interactions and Gap Junctions" (N. Sperelakis and W. C. Cole, eds.), Vol. 2. CRC Press, Boca Raton, Florida.

Mathias, R. T., Rae, J. L., and Eisenberg, R. S., (1979). *Biophys. J.* **25**, 181–201.

Mathias, R. T., Rae, J. L., and Eisenberg, R. S. (1981). *Biophys. J.* **34**, 61–83.

Mathias, R. T., Rae, J. L., Ebihara, L., and McCarthy, R. T. (1985). *Biophys. J.* **48**, 423–434.

Menko, A. S., Klukas, K. A., and Johnson, R. G. (1984). *Dev. Biol.* **103**, 129–141.

Milks, L. C., Kumar, N. M., Houghten, R., Unwin, P. N., and Gilula, N. B. (1988). *EMBO J.* **7**, 2967–2975.

Miller, A. G., Zampighi, G. A., and Hall, J. E. (1992). *J. Membr. Biol.* (in press).

Miller, T. M., and Goodenough, D. A. (1985). *J. Cell Biol.* **101**, 1741–1748.

Miller, T. M., and Goodenough, D. A. (1986). *J. Cell Biol.* 194–199.

Neyton, J., and Trautmann, A. (1985). *Nature (London)* **317,** 331–335.
Ngoc, L. D., Paroutaud, P., Dunia, I., Benedetti, E. L., and Hoebeke, J. (1985). *FEBS Lett.* **181,** 74–78.
Nicholson, B. J., Takemoto, L. J., Hunkapiller, M. W., Hood, L. E., and Revel, J. P. (1983). *Cell (Cambridge, Mass.)* **32,** 967–978.
Nikaido, H., and Rosenberg, E. Y. (1985). *J. Membr. Biol.* **85,** 87–92.
Nonaka, T., Nishiura, M., and Ohkuma, M. (1976). *J. Electron Microsc.* 35–36.
Okinami, S. (1978). *Albrecht von Graefes Arch. Klin. Exp. Opththalmol.* **209,** 51–58.
Palva, M., and Palkama, A. (1976). *Exp. Eye Res.* **22,** 220–236.
Patterson, J. W. (1981). *Invest. Ophthalmol. Visual Sci.* **20,** 40–46.
Paul, D. L. (1986). *J. Cell Biol.* **103,** 123–134.
Paul, D. L., and Goodenough, D. A. (1983) *J. Cell Biol.* **96,** 625–632.
Peracchia, C. (1978). *Nature (London)* **271,** 669–671.
Peracchia, C., and Dulhunty, A. F. (1976). *J. Cell Biol.* **70,** 419–439.
Peracchia, C., and Girsch, S. J. (1985a). *Am. J. Physiol.* **248,** H765–H782.
Peracchia, C., and Girsch, S. J. (1985b). *Curr. Eye Res.* **4,** 431–439.
Peracchia, C., and Girsch, S. J. (1985c). *Biochem. Biophys. Res. Commun.* **133,** 688–695.
Peracchia, C., and Peracchia, L. L. (1980a). *J. Cell Biol.* **87,** 708–718.
Peracchia, C., and Peracchia, L. L. (1980b). *J. Cell Biol.* **87,** 719–727.
Philipson, B. T., Hanninen, L., and Balzas, E. A. (1975). *Exp. Eye Res.* **21,** 205–219.
Rae, J. L. (1979). *Curr. Top. Eye Res.* **1,** 37–90.
Rae, J. L. (1991). *Invest. Ophthalmol. Visual Sci.* **13,** 147–150.
Rae, J. L., and Levis, R. A. (1984). *Mol. Physiol.* **6,** 115–162.
Rae, J. L., Mathias, R. T., and Eisenberg, R. S. (1982a). *Exp. Eye Res.* **35,** 471–489.
Rae, J. L., Thomson, R. D., and Eisenberg, R. S. (1982b). *Exp. Eye Res.* **35,** 597–609.
Rae, J. L., Lewno, A. W., Cooper, K. E., and Gates, P. (1989). *Curr. Eye Res.* **8,** 859–869.
Rafferty, N. S., and Esson, E. A. (1974). *J. Ultrastruct. Res.* **46,** 239–253.
Rao, Y., Yan, L. Y., and Jan, Y. N. (1990). *Nature (London)* **345,** 239–253.
Revel, J. P., and Karnovsky, M. J. (1967). *J. Cell Biol.* **33,** C7–C12.
Robinson, K. R., and Patterson, J. W. (1983). *Curr. Eye Res.* **2,** 843–847.
Sas, D. F., Sas, M. J., Johnson, K. R., Menko, A. S., and Johnson, R. G. (1985). *J. Cell Biol.* **100,** 216–225.
Scaglione, B. A., and Rintoul, D. A. (1989). *Invest. Ophthalmol. Visual Sci.* **30,** 961–966.
Schuetze, S. M., and Goodenough, D. A. (1982). *J. Cell Biol.* **92,** 694–705.
Simon, S. A., Zampighi, G., McIntosh, T. J., Costello, M. J., Ting Beall, H. P., and Robertson, J. D. (1982). *Biosci. Rep.* **2,** 333–341.
Stewart, S., Duncan, G., Marcantonio, J. M., and Prescott, A. R. (1988). *Invest. Ophthalmol. Visual Sci.* **29,** 1713–1725.
Swenson, K. I., Jordan, J. R., Beyer, E. C., and Paul, D. L. (1989). *Cell (Cambridge, Mass.)* **57,** 145–155.
Takats, A., Antoni, F., Farago, A., and Kertesz, P. (1978). *Exp. Eye Res.* **26,** 389–397.
Takemoto, L. J., Hansen, J. S., and Horwitz, J. (1985). *Exp. Eye Res.* **41,** 415–422.
Unakar, N. J., and Tsui, J. Y. (1980). *Invest. Ophthalmol. Visual Sci.* **19,** 630–641.
Unwin, P. N. T., and Ennis, P. D. (1984). *Nature (London)* **307,** 609–613.
Unwin, P. N. T., and Zampighi, G. (1980). *Nature (London)* 545–549.
Vallon, O., Dunia, I., Favard-Sereno, C., Hoebeke, J., and Benedetti, E. L. (1985). *Biol. Cell* **53,** 85–88.
Voorter, C. E. M., and Kistler, J. (1989). *Biochim. Biophys. Acta* **986,** 8–10.
Voorter, C. E. M., Kistler, J., Gruijters, W. T. M., Mulders, J. W. M., Chritie, D., and de Jong, W. W. (1989). *No Journal Found* 697–706.
Waggoner, P. R., and Maisel, H. (1978). *Exp. Eye Res.* **27,** 151–157.
Wanko, T., and Gavin, M. A. (1959). *J. Biophys. Biochem. Cytol.* **6,** 97–99.
Welsh, M. J., Aster, J. C., Ireland, M., Alcala, J., and Maisel, H. (1982). *Science* **216,** 642–644.
Wind, B. E., Walsh, S., and Patterson, J. W. (1988). *Exp. Eye Res.* **46,** 117–130.

Young, J. D., Cohn, Z. A., and Gilula, N. B. (1987). *Cell (Cambridge, Mass.)* **48,** 733–743.

Zampighi, G. A., Corless, J. M. and Robertson, J. D. (1980). *J. Cell Biol.* **86,** 190–198.

Zampighi, G. A., Simon, S. A., Robertson, J. D., McIntosh, T. J., and Costello, M. J. (1982). *J. Cell Biol.* **93,** 175–189.

Zampighi, G. A., Hall, J. E., and Kreman, M. (1985). *Proc. Natl. Acad. Sci. U.S.A.* **82,** 8468–8472.

Zampighi, G. A., Hall, J. E., Ehring, G. R., and Simon, S. A. (1989). *J. Cell Biol.* **108,** 2255–2276.

Zhang, J. T., and Nicholson, B. J. (1989). *J. Cell Biol.* 3391.

Zigler, J. S., and Horwitz, J. (1981). *Invest. Ophthalmol. Visual Sci.* **21,** 46–51.

Zimmer, D. B., Green, C. R., Evans, W. H., and Gilula, N. B. (1987). *J. Biol. Chem.* **262,** 7751–7763.

Colloidal Gold and Its Application in Cell Biology

Marc Horisberger

Nestec Ltd, CH-1800 Vevey, Switzerland

I. Introduction

A. Historical Perspective

1. The Early Days

Colloidal gold prepared from gold and plant juice has been known in China and India for many centuries as a drug for longevity and is said to be an antecedent of Western alchemy (Mahdihassan, 1984). Colloidal gold also has a long tradition as a staining reagent, dating from the second half of the seventeenth century. It was used for the production of red or ruby glass (purple of Cassius) and a range of enamel colours from pink to maroon (Hunt, 1981). Although colloidal gold was known to the alchemists of the seventeenth century, it was Michael Faraday (1791–1867) who first made a scientific study of its preparation and properties. Some of Faraday's original gold dispersions in water, prepared in 1857, are still preserved at the Royal Institution in London. Faraday also discovered that adding small amounts of electrolytes turns the color of gold sols from ruby to blue and coagulates them. More importantly, he demonstrated that these effects can be prevented by the addition of gelatin and other macromolecules (Faraday, 1857). The true nature of purple of Cassius, a colloidal gold preparation, was only elucidated at the turn of the century by the famous Viennese chemist, Richard Zsigmondy (1869–1925).

2. The Pioneers

Over the years, colloidal gold has drawn the attention of many other inquisitive minds. As noted by Roth (1986), it is rather surprising that colloidal gold was used in one of the very first applications of electron microscopy in biology, namely, the study of viruses (Kausche and Ruska, 1939; Kausche *et al.*, 1939). However, the enormous development in immunocytochemistry started with a short paper by Faulk and Taylor (1971). These authors prepared an immunogold marker by stabilizing a gold sol against coagulation by ions through the absorption of an antiserum. They successfully stained *Salmonella* surface antigens at the transmission electron microscopy (TEM) level by a rabbit anti-*Salmonella* antiserum conjugated to 5-nm gold particles. The following year, the group of Horisberger reported the first application of colloidal gold labeled with a well-defined molecule, i.e., a lectin (Bauer *et al.*, 1972). In this publication, the presence of mannan in bud scars isolated from a yeast was detected at the TEM level by colloidal gold conjugated to concanavalin A. Subsequently, Romano *et al.* (1974) applied the method to stain red blood cells using an antiglobulin gold marker. They described an indirect method in which colloidal gold was labeled with affinity-purified horse antibodies to human IgG, and the resulting complex was used to map the distribution of rhesus antigen sites on human erythrocyte ghosts.

The gold method was introduced for scanning electron microscopy (SEM) in 1975 by Horisberger *et al.* (1975). The use of a protein A–gold complex was first described by Romano and Romano (1977) in a preembedding technique for the localization of antigens present at the cell surface. Roth *et al.* (1978) introduced the postembedding technique with such complexes. Subsequently, the group of Roth contributed to a large extent to the development of the gold method both at the light and TEM level. The first comprehensive publication appeared in 1977 (Horisberger and Rosset, 1977a), in which the preparation of a variety of specific gold complexes was described with application in both SEM and TEM. Shortly after, a second comprehensive publication was published, but the applications were limited to TEM (Geoghegan and Ackerman, 1977). Historical aspects of the immunogold labeling for electron microscopy have been reviewed by Romano and Romano (1984).

Since 1977, the method has become the most widely used tool for the cytochemical detection of specific ligands, thanks to the initiative of a few groups. This rapid development is not surprising, considering its potential:

Advantages of Colloidal Gold as a Marker in EM Immunocytochemistry[a]

Monodisperse gold sols with gold particles of uniform size and shape can be rapidly, reproducibly, and inexpensively prepared in a particle size range of 2–150 nm mean diameter. Gold sols remain stable for many months under appropriate storage conditions at 4°C

The particulate nature of colloidal gold allows fine localization of marked sites; the size range guarantees high flexibility in lateral resolution

Gold particles are negatively charged and can be complexed by noncovalent electrostatic adsorption with various macromolecules (e.g., staphyloccocal

---

[a]From Hodges *et al.* (1987; Table 1, p. 3).

protein A, immunoglobulins, lectins, toxins, glycoproteins, enzymes, streptavidin, hormones, peptide antigens conjugated to bovine serum albumin),
forming stable and bioactive gold–ligand complexes termed gold probes.
Under appropriate storage conditions these will remain stable and retain
much of their bioactivity for many months at 4°C; high labeling flexibility is
guaranteed through a wide choice of reagents

Gold particles demonstrate high electron density because of the high atomic
number of gold and are capable of strong emission of secondary and backscattered electrons; these physical characteristics make gold particles excellent markers for TEM and SEM

The high electron backscattering coefficient of gold suggests that the enhanced
contrast of backscattered electron imaging could provide a superior alternative for visual or computer-aided quantitative analysis of target molecules

Also, the characteristic X-ray signals emitted by gold could be used to image
and quantify cell-bound gold markers by application of X-ray microanalyti-
cal techniques and appropriate computer programs

Double- or multiple-labeling of different target sites is possible by the application of monodisperse gold probes of various sizes

Quantification can be achieved by direct counting of gold particles

Due to the low degree of nonspecific adsorption of gold probes to specimen
surfaces, the signal-to-noise ratio is very high

Because of high binding constants of gold probes to specimen surfaces, biological samples can be processed with minimal loss of gold particles

Because gold particles absorb or reflect light and can be amplified by silver
enhancement procedures, they are applicable for a variety of light microscope
marking techniques as well as nonmicroscopical procedures (including immunoblotting and immunoprecipitation), thereby providing a range of cor-
relative methodologies

First applied in TEM, gold markers have now found uses in light microscopy as
well as in SEM, where the secondary electron and backscattered electron modes
can be combined.

Among the suppliers of gold-labeled products, Janssens Life Sciences has
contributed most, since 1982, to the dissemination of knowledge on the technol-
ogy and some of the applications of the colloidal gold method.

3. Actual Development and Applications

Cytochemists now have at their disposal monodisperse gold sols, with particle
sizes varying from 1 to 80 nm and even larger, for various applications, including
multilabeling and quantitative evaluations. The introduction of ultrasmall gold
particles with increased penetration capabilities has enlarged the potential of
colloidal gold reagents, particularly in the field of light microscopy. The signal

produced by the ultrasmall colloidal markers must be visualized by silver enhancement (Holgate *et al.*, 1983).

Colloidal gold was initially used as an "immuno-colloid" with either primary antibodies (Faulk and Taylor, 1971) or secondary antibodies (Romano *et al.*, 1974). The method rapidly broadened to include lectins (Bauer *et al.*, 1972; Horisberger and Rosset, 1977a), protein A (Romano and Romano, 1977; Roth *et al.*, 1978), protein G (Bendayan, 1987), and a number of other macromolecules (for review, see Horisberger, 1981a) in order to study receptor–ligand interactions and internalization of molecules (Roth, 1983a; Handley and Chien, 1987).

The diversity of probes has led to a corresponding diversity of applications. High-resolution cytochemical techniques based on colloidal gold are now available to cell biologists for the subcellular and suborganelle localization of most types of biological molecules: peptides and proteins as antigenic sites, carbohydrate residues as antigenic sites or through their affinity for various lectins and enzymes, nucleic acids, and phospholipids (Coulombe *et al.*, 1988) for their affinity for enzymes. The latter method is not specific but could be useful for studying the transverse and lateral distribution of phospholipids in cell membranes. Other developments, such as surface replica and low-temperature embedding procedures, have certainly also contributed to the success of the technique, as they improve the biologically relevant resolution. Colloidal gold has been also introduced as a sensitive marker for the detection of antigens in Western blot analysis (Teradaira *et al.*, 1983; Hsu, 1984) and in blot overlays on nitrocellulose strips. The sensitivity of the method in these applications was improved by the silver enhancement technique, with a detection level in the picogram range (Moeremans *et al.*, 1984, 1989). While it was first applied to localize statically surface and intracellular ligands, the gold method can now be applied to study dynamic processes such as endocytosis and cellular processing of molecules.

A number of reviews and technical manuals are presently available that confirm that colloidal gold can be truly described as a universal cytochemical tool. The preparation of gold markers has now evolved into a series of standardized methods. However, the scope of the applications is constantly broadening.

B. Current Trends

In the 1980s, procedures were elaborated to produce gold particles of more uniform size using controlled growth procedures (Slot and Geuze, 1985; van Bergen en Henegouwen and Leunissen, 1986). Procedures are also being developed for producing ultrasmall gold particles (1 nm) in order to improve resolution and/or penetration of gold probes in preembedding methods. Newer improvements of instruments include the use of dark-field microscopy to visualize colloidal gold labeling directly (De Mey, 1983; De Waele *et al.*, 1983), photoelectron microscopy (Birrell *et al.*, 1985), and backscattered electron imaging with secondary electron signals (De Harven *et al.*, 1984). Two major developments have improved the detection of colloidal gold probes in SEM. First, the introduction of a new high-resolution-field SEM with its application to biological samples (Nagatani and Saito, 1986; Pawley and Albrecht, 1988) has permitted the detection of 4- to 20-nm gold particles on cell surfaces by secondary electron emission in samples coated with carbon. Second, the development of an improved single-crystal detector for backscatter electron imaging (Autrata, 1989) has permitted the detection of 1-nm gold particles. The versatility of the use of colloidal gold is also demonstrated by the development of gold probes for *in situ* hybridization at the ultrastructural level (Binder *et al.*, 1986; Wolber *et al.*, 1988).

II. Preparation of Gold Probes

A. Preparation of Colloidal Gold

There are a number of review articles covering the various aspects of colloidal gold preparation, mostly for application in TEM. Gold particles, with sizes varying from 1 to 150 nm, can be prepared by different methods using reducing chloroauric acid ($HAuCl_4$) with number of reducing agents (Frens, 1973; Roth, 1983a). These include the white phosphorus method published by Faraday (1857) and used by Faulk and Taylor (1971), the citrate method of Frens (1973) employed by Horisberger and Rosset (1977a), and the modified tannic acid method of Slot and Geuze (1985). The type of reducing agent and the concentration of the reacting components largely determine the ratio between nucleation and growth and, in turn, the final particle size and size distribution (Handley, 1989; Leunissen and De Mey, 1989).

More recently, a method has been published for the controlled growth of existing gold particles using reduction by white phosphorus (van Bergen en Henegouwen and Leunissen, 1986). The particle diameter, initially 5.5 nm, can be gradually increased to approximately 12 nm by repetitive reduction cycles, thus affording high particle density and a low diameter spreading. A narrow size distribution can also be obtained by the controlled growth procedure elaborated by Slot and Geuze (1985) as a modification of the method published by Mühlpfordt (1982) using tannic acid and sodium citrate. Unlike larger particles, the ultrasmall gold particles (1 nm or even smaller) are not stable as such and will grow with time. However, after conjugation to proteins, the particles are stabilized and maintain their original size. The final diameter of the colloidal gold particles is determined by the number of icosahedral nuclei formed at the beginning of the reaction compared with the subsequent rate of shell condensation. Thus, gold colloids are composed of an internal core of pure gold that is surrounded by a surface layer of adsorbed $AuCl_2^-$. It is these ions that confer the negative charge to colloidal gold and prevent particle aggregation by electrostatic repulsion (Weiser, 1933; Pauli, 1949).

All gold colloids display a single adsorption peak (λ_{max}) in the visible range between 510 and 550 nm. With increasing particle diameter, the λ_{max} shifts to larger wavelength (Horisberger, 1985). The width of the adsorption spectra relates to the size range. Smaller particles are basically spherical, while larger particles (30–80 nm) show increased particle eccentricities related to the ratio of the major to minor axis (Handley, 1989).

B. Labeling of Colloidal Gold with Macromolecules

1. General Considerations

The addition of polymers is a well-known method for controlling the stability of colloidal dispersions. Nonionic macromolecules such as polyethylene glycol and polysaccharides as well as charged macromolecules can be adsorbed onto a metal surface. The specific property of gold markers arises from the adsorption of macromolecules onto the surface of the particles. For practical uses, the adsorbed macromolecules must not only keep their activity and specificity, but also be stable in physiological buffers (Horisberger, 1989a).

Polymers do not appear to be adsorbed over their entire length on any interface. It is thought that, at low concentration, the polymer coils attach themselves extensively in the form of partly flattened (random) coils. Further deposition causes more interpolymer contact and a reduction in surface contact per macromolecule. In this adsorption phenomenon, flexibility of the polymer is important for keeping their specific ligand properties. Indeed, bioactivity of gold markers can be influenced by conformational changes of the protein during adsorption, but most macromolecules adsorbed onto gold particles have shown no change in their apparent bioactivity (Horisberger, 1981a, 1989a). However, catalase is inactivated through adsorption onto colloidal gold (Horisberger and Rosset, 1977a; Horisberger, 1978). This is attributed to the instability of catalase in the presence of H_2O_2, as reported for catalase immobilized on organic support (Ferrier et al., 1972). A second exception was reported by Clerc et al. (1988) with IgE–gold complexes. They are inactive toward EgE receptors due to the fact that adsorbed IgE molecules are prevented from bending to interact with their receptors. Elliott and Dennison (1990) have also encountered unexpected difficulties in the preparation of active probes for tissue proteinases, using relatively small proteinase inhibitors (M_r ~12,000) as ligands. The proteinase inhibitors stabilized the colloids against an electrolyte challenge, but the authors were unable to prepare functional, specific probes. Although it is believed that macromolecules bind at random to a metal surface, Baschong and Wrigley (1990) have presented some evidence indicating that Fab fragments bind preferentially around their papain cleavage site, and not near their antigen-binding site. For gold particles of a size greater than 5 nm, more than one protein molecule is generally attacked to each particle. As a consequence, the valency of the gold–ligand complex is greater than that of the ligand alone by several orders of magnitude. This was quantatively demonstrated with lectins (Horisberger and Rosset, 1977b; Horisberger and Tacchini-Vonlan-then, 1983a) and protein A (Horisberger and Clerc, 1985).

Gold particles have been labeled with toxins, hormones conjugated to proteins, glycoproteins, protein A and G, low-density lipoprotein, enzymes, immuno-globulins, lectins, etc. (reviewed by Horisberger, 1981a) and, more recently, with streptavidin, nucleic acids, and monoclonal immunoglobulins. Colloidal gold probes have also been obtained with polyelectrolytes such as polylysine (Skutel-sky and Roth, 1986) and chitosan, a polymer of β-$(1\rightarrow4)$-linked D-glucosamine (Horisberger and Clerc, 1988a), and with neutral macromolecules such as poly-saccharides (Horisberger and Rosset, 1977a; Molday and Laird, 1989).

The adsorption of macromolecules onto colloidal gold depends on a number of factors, such as stability of the colloid itself; the concentration, shape, con-figuration, and isoelectric point of macromolecules; and the ionic strength, pH, and temperature of the suspending medium. Understanding these factors deter-mines success or failure in the preparation and utilization of colloidal gold mark-ers.

Colloidal gold particles of different sizes have been labeled with a variety of proteins and glycoproteins. Data on the number of macromolecules adsorbed per gold particle have occasionally been reported. In many instances, the number of molecules adsorbed per particle was vastly superior to the number a particle can accommodate as a monolayer (Horisberger and Vauthey, 1984), especially when determined with chemically radiolabeled macromolecules. This prompted Horis-berger and Vauthey (1984) to undertake detailed studies on protein adsorption using β-lactoglobulin as a model. In the presence of a saturating amount of β-lactoglobulin, a maximum of 13–14 molecules was adsorbed per 12.4-nm particle, compared to a theoretical maximum of 20. The molecules were irrevers-ibly adsorbed on the particles. Only markers labeled with more than five molecules per particle could be totally bound by immobilized anti-β-lactoglobulin anti-bodies. In addition, their was no evidence for the formation of a multilamellar shell. The lack of reactivity of particles at low coverage was attributed to a number of factors (Horisberger and Vauthey, 1984; Horisberger, 1990).

Only a few reports have appeared on the quantitative aspect of labeling colloidal gold (reviewed by Horisberger, 1989a, 1990). Studies on quantitative labeling of colloidal gold have been reported using proteins differing widely in size and shape, i.e., protein A (single polypeptide chain, 42,000 Da), β-lactoglobulin (di-meric protein, 36,000 Da), and monoclonal IgE (four polypeptides chains, 203,000 Da) (Horisberger and Clerc, 1985; Horisberger and Vauthey, 1984; Clerc *et al.*, 1988; Ghitescu and Bendayan, 1990).

In all cases thoroughly studied, the adsorption isotherms could be divided into two regions, i.e., regions of low and high coverage (Figs. 1 and 2). At low protein concentration, practically all molecules are adsorbed and the isotherm is largely pH independent. At the higher concentration, an equilibrium is established between free and adsorbed molecules. Adsorption is maximal at the isoelectric point of the molecules where they assume the most compact form. In all of these quantitative studies, no evidence for the formation of a multilamellar shell was observed. The markers are stable on storage when prepared at low coverage. However, in the case of a single polypeptide chain such as protein A, when the protein concentration is increased, the molecule becomes attached by a shorter segment, with a concomitant decrease in its affinity for the gold particle.

2. Protein A

Gold particles were first coated with protein A for preembedding staining by Romano and Romano (1977) and subsequently by Roth et al. (1978) for postembedding ultrastructural localization of tissue antigens. Protein A–gold complexes have become very popular and are now widely used as a secondary reagent for the localization of both surface and intracellular antigens (Roth, 1982).

Protein A is a single polypeptide chain (42,000 Da; pI 5.1) and has a very extended shape. It is functionally bivalent and reacts with the F_c region of IgG (Langone, 1982). The adsorption isotherms of protein A on 11-nm gold particles buffered at different pHs were studied by Horisberger and Clerc (1985) (Fig. 1). Maximum coverage occurred at the isoelectric point of protein A. At low coverage, desorption of protein A was extremely slow. However, at very high coverage (25 molecules/particle), the particles were progressively depleted of protein A during storage. This indicates that protein A molecules are attached by short segments at high coverage. A Scatchard plot analysis indicated that the association constant at low coverage ($1.5 \times 10^8 \, M^{-1}$ at pH 6.1) was reduced to $7.4 \times 10^6 \, M^{-1}$ at high coverage. Some of these results were confirmed by Ghitescu and Bendayan (1990).

Like staphylococcal protein A, streptococcal protein G has a high affinity for IgGs from various mammalian species. In contrast to protein A, protein G reacts only with IgG and possesses an avidity for a broader spectrum of animal IgGs (Åkerström et al., 1985). Protein G was introduced by Bendayan (1987) as a tool in immunocytochemistry. While Taatjes et al. (1987b) found no superiority of protein G–gold complexes for immunolabeling, Bendayan and Garzon (1988) reported that such complexes appear to be a better probe because of enhanced reactivity with monoclonal antibodies and broader affinity for polyclonal antibodies. Bendayan (1989) has reviewed the potential of protein A and protein G–gold complexes for high-resolution quantitative immunocytochemistry.

3. Immunoglobulins

In the early days of developments in immunocytochemistry, Faulk and Taylor (1971) obtained a functional marker for use at the TEM level by labeling colloidal

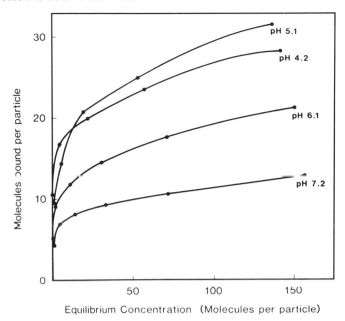

FIG. 1   Adsorption isotherms of ^{125}I-labeled protein A adsorbed onto 11.2-nm gold particles as a function of pH and equilibrium concentration. From Horisberger and Clerc (1985; Fig. 2, p. 220).

gold particles (5 nm) with a dialyzed antiserum. A functional probe was also obtained by Horisberger *et al.* (1975) by labeling much larger gold particles (60 nm) with a total anti-*Candida utilis* antiserum dialyzed overnight against 5 mM NaCl. However, the most "active" and specific markers are produced by coating colloidal gold particles of all sizes with affinity-purified antibodies. We have observed (M. Horisberger and M. F. Clerc, unpublished) that, when gold particles (50 nm) are labeled with ammonium sulfate-precipitated IgG from an anti-mannan antiserum, only 70% of the particles are bound by *Saccharomyces cerevisiae* cells. However, with affinity-purified antibodies, all the particles are bound onto the yeast cell surface mannan. This indicates that the presence of unspecific macromolecules on the gold particle surface hinders the binding of specific IgG to the antigen. With smaller particles (5 nm or less), the problem is less serious since each particle can accommodate only one macromolecule.

The adsorption isotherm of IgE (203,000 Da) as a function of pH has been determined with gold particles of 27.5 nm (Fig. 2) (Clerc *et al.*, 1988). At low surface coverage, the isotherm was independent of pH (6.1–8.8). In the presence of a large excess of IgE, the highest coverage was obtained at pH 6.1, near the pI of IgE (5.2–5.8). The isotherms were examined at low and high surface coverage by Scatchard analysis (Fig. 3 and Table I). At low coverage, the affinity constant to the particle (many contact points, side-on adsorption) was higher than

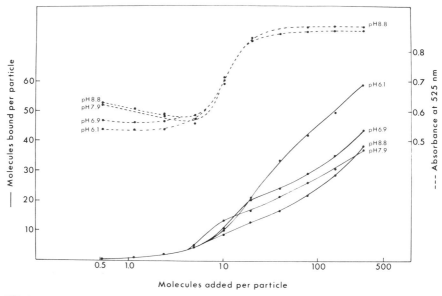

FIG. 2 Coagulation (dashed line) and adsorption (solid line) isotherms of monocloncal IgE adsorbed on gold particles (27.5 nm in diameter). The salt-induced coagulation test (Horisberger and Rosset, 1977) indicated that the colloid buffered at different pHs was stabilized with an identical amount of IgE added per particle (25 molecules per particle). At pH 6.1, 6.9, 7.9, and 8.8, this amounts to 25, 22, 18, and 14 molecules adsorbed per gold particle respectively. From Clerc *et al.* (1988; Fig. 2, p. 345).

at high coverage (fewer contact points). Such IgE–gold complexes did not bind to surface receptors of mast cells. This was attributed to the fact that IgE must bend in order to bind to its receptor through its F_c portion (Holowka *et al.*, 1985). Such bending would be mostly prevented when IgE is adsorbed on a solid surface. To map IgE receptors successfully, a three-step procedure was used (see Section V,C,4,c).

4. Lectins

Lectins possess the ability to recognize classes of complex glycoconjugates and have been extensively used to investigate properties of cell-surface glycoconjugates. Lectins were among the first purified macromolecules to be conjugated to colloidal gold (Bauer *et al.*, 1972; Horisberger *et al.*, 1975; Horisberger and Rosset, 1977a; Roth and Wagner, 1977). Several reviews are available on various aspects of the preparation and application of lectin–gold complexes (Horisberger, 1985; Roth, 1987; Benhamou, 1989).

The optimal pH for preparing lectin-labeled gold complexes has been reported for most common lectins (Roth, 1983b, 1987). Much evidence indicates that lectins, adsorbed onto gold particles as a monolayer, remain firmly attached and

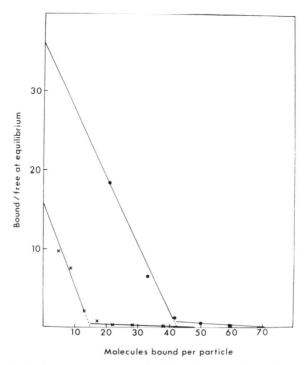

FIG. 3   Scatchard plots for IgE adsorbed on gold particles (27.5 nm in diameter) at pH 6.1 (•) and 8.8
(x). From Clerc *et al.* (1988; Fig. 3, p. 346).

TABLE I

Number (N) of IgE Molecules Adsorbed per Gold Particle (28 Nm in Diameter) at
Saturation, and Binding Constants (K)[a]

pH of adsorption	N	Amount adsorbed [μ/cm²]	K [M^{-1}]
6.1	42	0.15	2.2×10^9
	68	0.24	6.3×10^7
6.9	27	0.01	2.6×10^9
	56	0.20	3.6×10^7
8.8	15	0.05	2.9×10^9
	48	0.17	3.9×10^7

[a]From Clerc *et al.* (1988; Table 1, p. 346).

keep their carbohydrate-binding capacity for months or even years (Horisberger and Tacchini-Vonlanthen, 1983a). The affinity constants of lectin-labeled gold particles to cell surface glycoconjugates (10^9–10^{10} M^{-1}) are several orders of magnitude higher than those of lectins to monosaccharides (Horisberger and Tacchini-Vonlanthen, 1983a). They result from multivalent interactions and, possibly, from secondary interactions such as hydrophobic bonds which add to the strength of binding. This property enables marked specimens to be poststained or processed for embedding with minimal loss of bound particles (Horisberger, 1985).

Since 1977, several authors have mentioned that difficulties could be encountered in the adsorption of some lectins to colloidal gold. Horisberger and Rosset (1977a) reported first that gold particles with diameters greater than 20 nm could not be stabilized with wheat germ lectin. The lectin was therefore cross-linked to bovine serum albumin to increase its size and modify its isoelectric point. Under these conditions, a functional and stable gold complex was obtained. Lectins considered "identical" in terms of monosaccharide specificity possess the ability to recognize fine differences in more complex structures (Debray *et al.*, 1981). It is, therefore, imperative to interpret cytochemical observations very carefully with respect to lectin specificity. Many lectin–gold complexes can be used with the new-generation acrylic plastics with equal success (i.e., Lowicryl K4M and LR White). However, comparisons of lectin-binding studies using similar but not identical methods should be interpreted with caution (Frisch and Phillips, 1990). Frisch and Phillips have also shown that, for goblet cells stained prior to embedding, biotinylated lectins (visualized by the avidin–biotin–peroxidase technique) were more efficient than the lectin–gold complex.

5. Enzymes

The possibility of detecting enzyme substrates by the use of enzyme–gold complexes was first demonstrated by Bendayan (1981). The enzyme–gold technique, initially applied to the localization of nucleic acids, has been rapidly extended to other cellular or tissue substrates (reviewed by Bendayan, 1985). Coupling with colloidal gold is performed at a pH near the pI of the enzyme. For labeling thin sections, the pH for optimum labeling is generally related to the pH of the enzyme optimal activity. Thus, for each enzyme–gold system, conditions of coupling and labeling have to be determined since they differ (Bendayan, 1985). More recently, hyaluronidase (Chardin *et al.*, 1990), glucanase4, and β-galactosidase (Vian *et al.*, 1991) have been conjugated to colloidal gold, forming functional probes. The number of enzyme molecules adsorbed per gold particle has not been determined, with the exception of catalase (Horisberger, 1978), which was inactivated by adsorption (Horisberger and Rosset, 1977a). The number of catalase molecules adsorbed per gold particles of different sizes was proportional to the surface of the particles.

6. Cationic Colloidal Gold

Several methods have been developed for the visualization of cell surface anionic sites by TEM (for references, see Horisberger and Clerc, 1988b). Poly-L-lysine–colloidal gold complexes (5–14 nm in diameter) which can be used at both physiological and acid pHs were developed for TEM (Skutelsky and Roth, 1986). They have been applied to the localization of anionic sites on tissues embedded in hydrophilic media (Vorbrodt, 1987). Flocculation can easily occur during the preparation of these complexes (Vorbrodt, 1987). More recently, Horisberger and Clerc (1988b) introduced chitosan–gold complexes as polycationic probes for uses in TEM and SEM. Chitosan is a polymer of $\beta(1\rightarrow4)$-linked D-glucosamine obtained by deacetylation of chitin. These complexes were shown to be suitable for the detection of anionic sites on cell surfaces of various origins both in TEM and SEM under a wide range of pH and ionic strength.

Because of the relatively nonspecific nature of charge-based interactions, it is often difficult to specify the biochemical identity of the entities binding polycationic probes. Bertolatus (1990) has developed a probe-specific method based on specific covalent binding of biotin to sialic acid residues or –COOH groups followed by identification of biotinylated sites by colloidal gold conjugates. Ultimately, the identity of the entities bearing anionic sites should be confirmed by immunocytochemistry, as demonstrated by Lin (1990). In this study, heparan sulfate proteoglycan (HSPG), a common component of all basal laminae, was localized with the cationic dye ruthenium red and by immunogold cytochemistry using an antibody to the core protein of HSPG. The results obtained with thin sections of brain and retinal capillaries of rat were compared. Gold particles appeared randomly distributed in the basal lamina of capillaries, in contrast to discrete ruthenium red-staining sites near the endothelial plasma membranes, as shown in previous studies of the retinal vessel.

7. Dextran Conjugates

Alternative probes have been proposed, such as the protein–gold–dextran conjugates. In this procedure, well-defined hydrophilic polysaccharide derivatives are initially adsorbed to the gold particles. This coating stabilizes the particle against salt-induced aggregation. Proteins such as protein A, streptavidin, wheat germ lectin, goat anti-mouse immunoglobulin, and various monoclonal antibodies can then be covalently linked to the reactive groups of the polysaccharide derivatives (Molday and Laird, 1989). The advantage of the method is the stability of the probes in the presence of buffers and proteins and their ability to conjugate a variety of proteins which otherwise do not form stable protein–gold reagents. However, the procedure is time consuming. These probes (immuno-gold-dextrans) have been exclusively used by Molday and co-workers to study the organization of rhodopsin and several minor proteins in the membranes of rod and cone photoreceptor cells (see Molday and Laird, 1989).

C. Stabilization of Gold Probes

When gold particles are labeled with macromolecules (particularly with proteins), full stabilization against coagulation by electrolytes is not always observed during storage, especially with the larger size markers (>20 nm in diameter). Full stabilization in the suspending buffer can be achieved by the addition of macromolecules such as polyvinylpyrrolidone, bovine serum albumin, or, more effectively, by Carbowax 20 M (steric stabilization). The term "steric stabilization" denotes the fact that the adsorption of flexible polymers of sufficiently high molecular weight leads to polymer chains protruding from the particle surface. The slight interpenetration of these chains protruding from two colliding particles keeps them at a distance too large to give a van der Waals interaction sufficient for coherence. This effect has been discussed by Horisberger (1979, 1981a, 1989a).

III. Properties of Gold Probes

A. Stability of Gold Probes

When properly prepared under standard conditions, the stability of gold markers stored in buffers is generally excellent with respect to dispersion of the particles and activity (Horisberger, 1989a).

A major factor in the rapid adoption of immunogold methods for cytochemistry has been the commercial availability of stable gold complexes with antibodies, lectins, protein A, and streptavidin. In some commercial preparations of immunoglobulin complexes, Kramarcy and Sealock (1991) have detected traces of free antibody molecules which will compete with the gold complexes for binding sites and thus cause a drop in labeling intensity, rendering quantitative comparisons meaningless.

A native gold sol is most likely charged due to the adsorption of $H(AuCl_4)$ and partially reduced $H(AuCl_2)$ or $H(AuOHCl)$ and $H(AuOH_2)$ to the surface of the individual Au_0 spheres. The nature of the charge is the result of dipole orientation (Pauli, 1949). As observed by Baschong and Wrigley (1990), conjugation to the negatively charged colloid might polarize the protein at least locally, which would result in a mutual influence of the dipole of the gold sphere and the local isoelectric value of the protein, i.e., affect the conjugate stability.

B. Steric Hindrance

As relatively large size gold markers are still used for multiple marking experiments and, particularly, in SEM cytochemistry, steric hindrance can be a major problem, especially in direct marking procedures. As a result of steric hindrance, the number of particles bound to the cell surface decreases when the particle size is increased. The causes have been discussed by Horisberger (1981a). Most of the problems can be overcome by indirect procedures, which generally lead to denser marking when the different interacting species have a high affinity.

Another consequence of steric hindrance is that labeling studies will preclude a one-to-one correspondence of marker to target molecule. Therefore, there is no direct relationship between the number of bound particles and the number of cell surface ligand sites.

C. Nonspecific Interactions

Nonspecific staining of thin sections by immunogold reagents occasionally may cause some difficulties due to the lack of understanding of the exact causes. Many researchers have shown that, in certain cases, the presence of an anionic group can lead to staining, irrespective of the stabilizing agent of the colloid (Behnke et al., 1986). In studies of whole-mount cytoskeletal preparations of cultured cells, Birrell et al. (1987) encountered a problem with nonspecific binding of small colloidal gold particles (particularly those made using the citrate–tannic acid procedure). They proposed a number of modifications of experimental conditions to limit this problem. More recently, staining of thin sections of human kidney showed that positivity could not be abolished in most cells by omitting the primary antibody (anti-MF VIII-RA antibodies) in the protein A procedure (Onetti-Muda et al., 1990). It has been hypothesized that exposed residual binding sites on the colloidal surface may interact nonspecifically with the material under investigation. Such interactions could also explain the observed nonspecific interaction of small gold probes with cytoskeletal entities (Birrell et al., 1987). Although the gold method generally provides a low background of nonspecific interactions, techniques for reducing nonspecific labeling have been described by Roth et al. (1989), with recommendation on the correct concentration of the protein A gold complex to be used.

IV. Technical Applications of Gold Probes

A. Light Microscopy

Although colloidal gold was used initially in immunoelectron microscopy, it soon became apparent that gold particles were also suitable for light microscopy. Geoghegan et al. (1978) were the first to report the use of colloidal gold as a marker for light microscopy. Surface immunoglobulins (IgD and IgM) on human

B lymphocytes were detected by a two-step technique. The technique was further developed by De Mey *et al.* (1981a) and De Waele *et al.* (1983) for the enumeration of T lymphocytes and their subclasses. Postembedding techniques for staining frozen and paraffin sections were developed by Gu *et al.* (1981) and Roth and co-workers (reviewed by Lucocq and Roth, 1985) (Fig. 4a,b).

In the detection of cell surface antigens, positive cells are characterized by the presence of dark, purplish granules at their surface. As such, the method is not widely applied because the signal generated is of low intensity and adequate only for high concentration of antigens. The staining signal can be amplified when the preparations are examined with dark-field microscopy or polarized light epi-illumination. In sections, positive structures appear light to dark red when examined by bright-field illumination (Lucocq and Roth, 1985; De Mey, 1983). The method became a routine tool in histology when staining intensity was enhanced by the silver method of Danscher (1981). Silver enhancement not only enhances the visible gold staining but also increases the detection limit on paraffin sections (Holgate *et al.*, 1983) (Fig. 4a) and on semithin sections (Taatjes *et al.*, 1987a) (Fig. 4b).

The immunogold silver staining (IGSS) method offers many advantages: it produces an intense black reaction product which is insoluble in commonly used dehydrating reagents, it permits the use of a number of counter stains, and it produces a signal that can be enhanced by epipolarization. The IGSS method allows the use of smaller gold particles, which give a higher labeling efficiency due to reduced steric hindrance and better penetration. In an attempt to improve labeling efficiency further, 1-nm particles have been developed. However, in a recent study, De Valck *et al.* (1991) have shown that 1-nm probes do not offer a major advantage in comparison with 5-nm probes for the study of cell surface antigens in light microscopy.

The potential of reflection contrast microscopy has been investigated by Cornelese-ten Velde and Prins (1990), resulting in the successful detection of colloidal gold (15-nm particles) or silver-enhanced gold particles on ultrathin sections. These authors observed an increased detection sensitivity of reflection contrast microscopy compared with bright-field microscopy.

Video microscopy has been used to study the dynamic reorganization of individual cell components on living cells (Sheetz *et al.*, 1989). Nanovid microscopy, which is the application of video-enhanced light microscopy to specific colloidal gold probes, can detect the mobility of individual markers on the cell

---

FIG. 4 Light microscopical lectin–gold staining technique with silver intensification. (a) Demonstration of the sialic acid α-2,3-*N*-acetylglucosamine sequence with the *Maackia amurensis* lectin in a paraffin section of human colon. (b) Detection of sialic acid residues with the *Limax flavus* lectin–fetuin–gold technique in a semithin section of low-temperature Lowicryl K4M-embedded rat colon. Courtesy of J. Roth, University of Zürich, Switzerland; for the methods, see Sata *et al.* (1989) and Taatjes *et al.* (1987a).

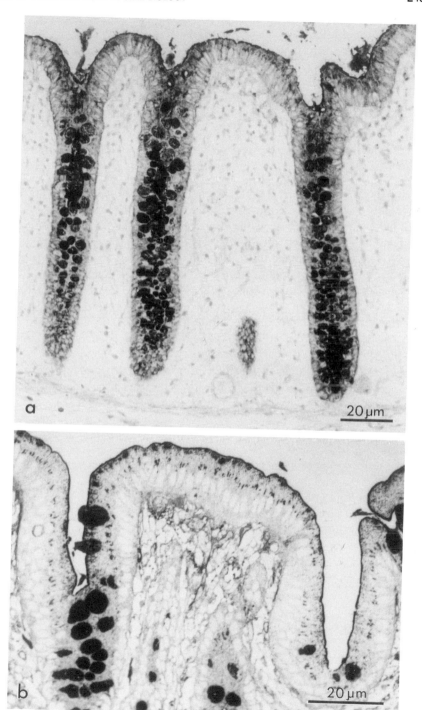

surface of living cells in real time (Geerts *et al.*, 1987). Using this method, De Brabander *et al.* (1991) have demonstrated that it is possible to observe the motion of individual specific cell surface components, thus providing a new tool to study the dynamic reorganization of the cell membrane during locomotion. For instance, 40-nm particles coupled to different types of poly-L-lysine were used in Nanovid microscopy to study the lateral diffusion and retrograde movements of negative cell surface components of single mobile PTK_2 cells and 3T3 fibroblasts. The particles, coupled to short poly-L-lysing molecules (4 kDa), displayed Brownian motion. Steric hindrance of this label could in principle restrict movement. However, their mobility was in the same range as the values obtained with fluorescent probes.

Recently, Schafer *et al.* (1991) were able to observe transcription by single molecules of RNA polymerase by video-enhanced differential interference contrast light microscopy. These authors were able to analyze the behavior of 40-nm gold particles attached to the ends of DNA molecules that were being transcribed by RNA polymerase immobilized on a glass surface.

Multiple staining procedures have been described in light microscopy. For example, the IGSS method (Holgate *et al.*, 1983) provides excellent color contrast with chromogens used in immunoenzymatic methods. Double and triple immunocytochemical staining has been reported at the light microscope level in histopathology using black (IGSS), red-brown (immunoperoxidase), and/or blue (immunoalkaline phosphatase) (Krenács *et al.*, 1989, 1990).

Different techniques in immunogold histochemistry have been proposed for both light and electron microscopy to study the presence and/or the precise localization of macromolecules in various type of cells and animal tissues at the cellular and ultrastructural level. Fluorescent colloidal gold was introduced by Horisberger and Vonlanthen (1979b). Roth *et al.* (1980) also demonstrated that gold particles labeled with FITC–protein A could be used as a second-step reagent for the visualization of cellular antigens in fluorescent and electron microscopy. The use of the same label and sections from the same block was proposed by Lucocq and Roth (1985). More recently, Cornelese-ten Velde and Prins (1990) have presented a technique for the precise localization of antigens in the same cells or tissue compartments by gold-immunolabeling at the light and electron microscopy level, using sequential semithin and ultrathin sections. The presence of gold particles or silver-enhanced gold particles could be successfully detected on ultrathin sections by reflection contrast microscopy.

B. Scanning Electron Microscopy

Scanning electron microscopy (SEM) cytochemistry can provide valuable information that is otherwise difficult to obtain by TEM when a large surface area of tissues or a number of cells needs to be examined. The gold method for SEM was introduced by Horisberger *et al.* (1975), taking advantage of the high secondary electron emission of gold particles. This property permits visualization of gold particles bound to a cell surface even when the specimen is not coated with metal (Fig. 5).

SEM, in the backscattered electron imaging (BEI) mode, was introduced by De Harven *et al.* (1984) to study cell surfaces labeled with colloidal gold; it takes advantage of the striking contrast generated by the particles when observed in this mode. Initially, SEM had a limited use in immunocytochemistry, owing to instrumentation and specific preparation technologies. Since labeling efficiency is increased with smaller markers, methods have been developed to detect small probes (1–10 nm). A new high-resolution field emission SEM has permitted the detection of 4- to 20-nm gold particles on cell surfaces by secondary electron emission in samples coated with carbon (Albrecht *et al.*, 1989). With field emission SEM, 5- to 15-nm gold particles can now be localized by BEI at 10–20 kV. Furthermore, with the development of an improved single-crystal (YAG–Ce type) detector for BEI (Autrata, 1989), gold particles as small as 1 nm can be detected (Müller and Hermann, 1990). The silver enhancement method has also increased the sensitivity. A number of applications in SEM have been reviewed by Horisberger (1981a) and Hodges *et al.* (1987). The gold method has been applied in SEM to a number of problems, such as cell surface receptor–ligand interactions, expression of cell surface lectin-binding sites, surface distribution of extracellular matrix components, and visualization of cytoskeletal elements (reviewed by Hodges *et al.*, 1987). Cationic gold complexes have been used to study the surface distribution of anionic sites (Fig. 5). The resolution of individual particles is improved with stereomicrographs, which facilitate the interpretation of the results (Fig. 5).

1. Single- and Multistep Labeling Procedures

In a number of cases, the direct (one-step) method gives satisfactory results (Fig. 5). However, it can be hampered by steric hindrance since cell surface binding sites may not be accessible to the larger gold particles. Therefore, one has to either use smaller particles and specialized instrumentation or indirect marking procedures. Two- or three-step procedures generally lead to denser marking when the successive probes interact with high affinities. The direct procedure can also be applied to study not only the lateral, but also the longitudinal location of surface binding sites. Indeed, binding sites close to the bilayer will interact only with the smaller probes (Horisberger, 1978).

2. Multiple Marking

Using gold particles of different sizes, multiple marking has been achieved by many investigators using SEM (Horisberger *et al.*, 1975; Horisberger and Rosset,

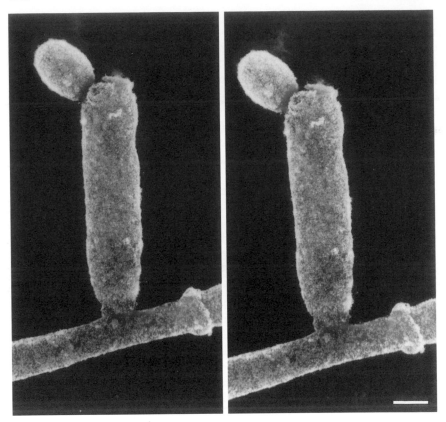

FIG. 5   Stereomicrographs of the mycelial form of the yeast *Candida albicans* marked for anionic sites by the direct method with chitosan conjugated to 44-nm gold particles. Marking is dense over the whole cell surface area. Tilt angle, 6°. Bar, 1 μm. From Horisberger (1989b; Fig. 1, p. 222).

1977a; Hoyer *et al.*, 1979; De Harven *et al.*, 1990). In some instances, silver enhancement has facilitated the observation of gold particles (De Harven *et al.*, 1990). Bot the secondary emission and the BEI modes have been used. However, the usefulness of the information obtained in multiple marking remains to be demonstrated.

C. Transmission Electron Microscopy

Gold markers are easily recognized on thin sections owing to their distinct shape and electron-opaque properties. Both direct and indirect methods are used in pre- and postembedding techniques. However, the considerable size of an antibody/secondary antibody–gold complex or of the antibody–protein A complex may restrict immunogold applications at the subcellular level. In order to improve the precision of localization, the size of the antibody–gold complex was reduced by labeling small gold particles (2–3 nm) with Fab fragments (reviewed by Baschong and Wrigley, 1990). Baschong and Wrigley also provided evidence on the applicability of the method for targeting single protein units in regular arrays such as in bacteriophage T4 polyheads.

1. Preembedding Methods

Preembedding detection is often the procedure of choice in situations where the amount of target molecule is small or it is sensitive to the process of tissue preparation, such as fixation, dehydration, and embedding.

Preembedding staining procedures primarily detect cell surface components. They are not ideally suited for labeling antigens expressed in the cytoplasm due to the poor penetration of commonly used particles (5–20 nm). However, 1-nm particles have better penetration and may prove more useful when detected by the silver intensification method. The technique has been successfully applied by Chan et al. (1990) to detect the catecholamine-synthesizing enzyme tyrosine hydroxylase, which shows little immunoreactivity in central axons following plastic embedding (Pickel et al., 1981). Preembedding multiple marking has also been described and may lead to a better understanding of the topography of cell membrane components (Horisberger and Vonlanthen, 1979a) (Fig. 6).

Beside the use of gold particles of widely different sizes, the colloidal gold/silver staining method can be employed in double-labeling procedures. In the preembedding routine of van den Pol (1985), immunogold–silver staining, used as the first marking procedure, is combined with an immunoenzyme method as the second procedure. Another approach is to localize antigenic sites on the cell surface with gold particles labeled with protein A, which are then silver-enhanced before embedding, whereas nuclear and cytoplasmic antigens are visualized again on thin sections after embedding (Zelechowska and Mandeville, 1989). In a third application, the first gold probe is enlarged with silver to a predetermined size. The second gold marker is then applied and not augmented (Bienz et al., 1986; Pettitt and Humphris, 1991).

2. Postembedding Methods

Postembedding cytochemistry with gold-labeled reagents in light and electron microscopy was reviewed by Roth (1986) for the detection of antigens and glycoconjugates with the protein A–gold and the lectin–gold techniques. Examples of postembedding lectin–gold labeling are shown in Fig. 7.

Comparison of the results obtained in TEM with peroxidase and gold as labels shows that the latter has advantages, in so far as it allows the underlying structures

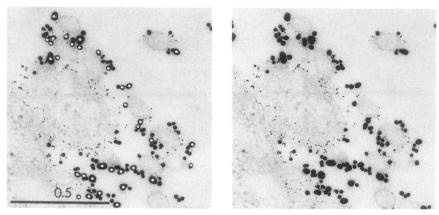

FIG. 6   Stereoscopic transmission electron micrograph of mouse embryo fibroblast marked with three lectins. The cells were successively incubated with 5-, 17-, and 26-nm gold particles labeled, respectively, with concanavalin A, *Ricinus communis* lectin, and wheat germ lectin (white dots). When the stereopair is examined, most of the particles are bound by spatially separated sites. It is best viewed with an optical viewer. However, most people with a little determination can achieve stereopsis by crossing the eyes. Bar, 0.5 μm. From Horisberger and Tacchini-Vonlanthen (1983a; Fig. 1, p. 252).

to be identified. In the peroxidase preparation, the underlying structures can be covered by the enzyme reaction product.

In a number of studies, the direct (one-step) method gives satisfactory results (Figs. 7 and 8). However, the highest density of marking is generally obtained by multistep procedures (Fig. 8).

Although all procedures, including tissue fixation, processing, embedding, and labeling, can influence the level of sensitivity of the colloidal gold method, embedding in resins is the major factor that affects sensitivity. In this respect, hydrophilic resins show much greater levels of labeling intensity than epoxy resins (Shida and Ohga, 1990). Most antigens do not tolerate osmium tetroxide as secondary fixative. In osmicated tissue. Bendayan and Zollinger (1983) have shown that sodium metaperiodate can restore protein antigenicity.

To increase access to scarcely distributed antigenic sites, procedures have been devised to remove the embedding material prior to immunogold cytochemistry (Johnson and Bettica, 1989; Baigent and Müller, 1990; Nickerson *et al.*, 1990). With such techniques, Nickerson *et al.* (1990) have studied the precise localization of specific cytoskeleton and nuclear matrix proteins throughout entire sections.

3. Multiple Labeling

Multiple labeling in TEM can be achieved using gold probes of different sizes. Double labeling was introduced in 1977 by Horisberger and co-workers (Horis-

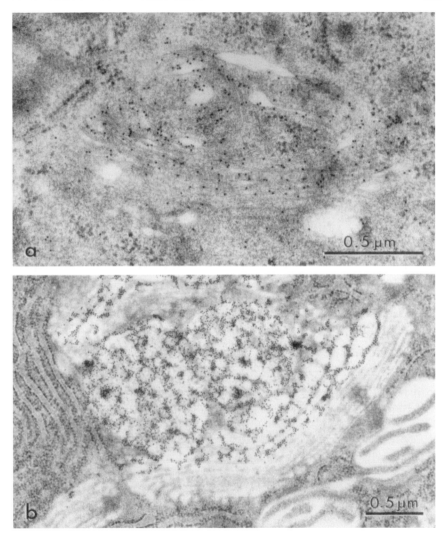

FIG. 7   Electron microscopic postembedding lectin–gold labeling. (a) Diffuse distribution of sialic acid residues in the Golgi apparatus cisternal stack of an absorptive enterocyte as visualized with the *Limax flavus* lectin in a Lowicryl K4M thin section of rat colon. (b) Sialic acid residues are present in *trans* cisternae of the Golgi apparatus cisternal stack of a goblet cell. Mucus droplets are intensely labeled. Courtesy of J. Roth, University of Zürich, Switzerland; for methods, see Taatjes and Roth (1990).

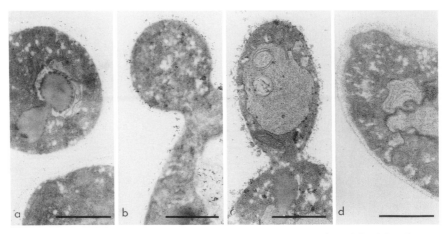

FIG. 8   In an example of the direct method of using gold markers, thin sections of *Candida utilis* were marked for mannan with concanavalin A (a) and anti-mannan antibodies (b) conjugated to 5-nm gold particles. The indirect method is illustrated by sections which were incubated with anti-mannan antiserum and then with goat anti-rabbit IgG (c) and protein A (d) conjugated to 5-nm particles. Mannan was found in the cell walls and in some vacuoles. Although the results were similar in all cases, the highest density of marking was achieved by the protein A–gold method. Bars, 1 μm. From Horisberger (1981b; Fig. 3, p. 93).

berger and Rosset, 1977a) and triple labeling was introduced in a preembedding procedure in 1979 (Horisberger and Vonlanthen, 1979a) (Fig. 6). To avoid spurious interpretation, stereoelectron micrographs are taken, since particles which seem to be closely associated may, in fact, be in different planes (Fig. 6). In preembedding procedures, varying the size of the probes will provide further information on the topography of cell membranes. For example, when erythrocyte membranes were first labeled with wheat germ agglutinin (WGA) conjugated to 18-nm gold particles and then reincubated with WGA conjugated to 5-nm particles, additional sites were labeled with the smaller probe, indicating that not all lectin-binding sites (glycophorin) were available to the larger probe due to steric hindrance (Horisberger and Rosset, 1977a). In a postembedding procedure, double labeling was first reported in 1977 (Horisberger and Vonlanthen, 1977).

Since then, a number of variations have been proposed for TEM, including applications of the gold–silver staining technique both in pre- and postembedding methods (Bienz *et al.*, 1986; Pettitt and Humphris, 1991). In the latter, the biotin–streptavidin system and the colloidal gold–silver staining methods were applied for the consecutive localization of two intracellular cytoplasmic binding sites in cells and tissues embedded in acrylic plastic.

Double immunogold labeling can be achieved according to the two-face methods of Bendayan (1982). First, the grids are floated on drops of the incubation solution with the section side facing downward. After washing and drying, the grids are turned over and floated on drops with the section side facing upward. During all the incubation steps, wetting the opposite side of the grid should be carefully avoided to prevent cross-contamination of the immunological reagents. Using 5- and 10-nm gold particles, two antigens can be detected unambiguously (Holm *et al.*, 1988; Stransky and Gay, 1991).

Rutter *et al.* (1988) have developed a procedure for obtaining high-resolution topographical information on replica about the spatial distribution of antigens at both sides of isolated plasma membranes. Cell surface antigens are tagged by the protein A–gold method. After splitting the cells attached to a cationized coverslip, the antigens on the cytoplasmic surface were visualized by the antibody bridge method, using immunoglobulins coupled to gold particles of different sizes. The method permits concomitant localization of antigens present at the inner and outer leaflets of the plasma membrane.

D. Cryotechniques

1. Cryo-ultramicrotomy

One method frequently used for the localization of macromolecules at the ultra-structural level is cryo-ultramicrotomy in combinating with immunogold labeling as developed by Tokuyasu (1973, 1986). In this procedure, biological samples are fixed in aqueous fixatives and frozen. Ultrathin section are prepared from the frozen material. Subsequently, the cyrosections are thawed and processed for immunogold labeling. As an example, the intracellular distribution of epidermal growth factor receptor in cryosections of cultured A431 cells is shown in Fig. 9. Several groups have shown that cryo-ultramicrotomy provides sections for post-embedding immunoelectron microscopy in which there is greater preservation of antigenic determinants than can be achieved using resin embedding. The technique often results in excellent preservation of membranes. Although gold immunolabeling of ultrathin cryosections is a highly sensitive method, not all tissues are easily amenable to cryosectioning and fine structure can be difficult to correlate with what is seen in specimens prepared by conventional techniques. Among other applications, the method has been used in cell biology for studying cytoskeleton (Geiger *et al.*, 1981), membranes (Griffiths *et al.*, 1982), and protein secretion (Slot and Geuze, 1984). It has also been applied to human biopsies (Burt *et al.*, 1990). The procedure has been critically reviewed several times (see Boonstra *et al.*, 1991, for the latest references). Beside fixation and accessibility of antigenic determinants, Boonstra *et al.* have also discussed a number of other parameters which may affect label efficiency. The practical aspects of cryo-ultramicrotomy and immunogold labeling with respect to specimen preparation and staining have been described by Leunissen and Verkleij (1989).

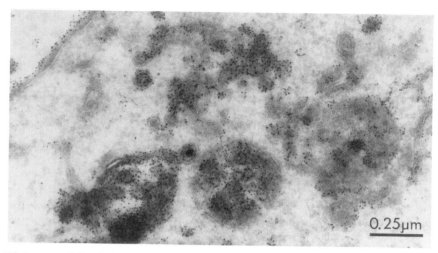

FIG. 9 Intracellular distribution of epidermal growth factor (EGF) receptor in cryosections of cultured A431 cells. Cryosections were successively incubated with a monoclonal antibody directed against EGF receptor and protein A conjugated to 10-nm gold particles. Intracellularly, gold particles were observed in large vesicles which probably arise from fusion of membrane folds with the plasma membrane (not shown). A higher number of particles are found on intracellular vesicular membrane structures and also in multivesicular bodies. These will be transferred ultimately to lysosomes. Courtesy of A. J. Verkleij, University of Utrecht, The Netherlands. First published in Boonstra et al. (1985; Fib. 4b, p. 214).

The successful combination of cryo-ultramicrotomy and immunogold labeling has not yet been fully exploited. Despite the fact that the biological material is fully hydrated during the incubations, there exists a very limited degree of penetrability of the gold particles. However, the use of the very small particles (approximately 1 nm) could result in an increase of the depth of immunolabeling and labeling efficiency (Leunissen and Verkleij, 1989). Cryo-ultramicrotomy gives information on the presence of antigenic sites or receptors on the cell surface, but it does not provide reliable information on the lateral distribution of these entities. In order to visualize the planar distribution of cell surface-located proteins, a number of methods have been developed, such as surface replica combined with cryotechniques (see the next sections).

2. Freeze-Substitution

In contrast to cryo-ultramicrotomy, freeze-substitution permits immunocytochemical labeling to be achieved without prior chemical fixation of the specimen (reviewed by Humbel and Schwarz, 1989). However, freeze-substitution media consist of cold acetone and fixatives such as glutaraldehyde, paraformaldehyde, osmium tetroxide, and acrolein. The substituted specimens are then heated stepwise and subsequently embedded in epoxides in the usual way. In 1982, Carlemalm *et al.* introduced low-temperature embedding and UV-light polymerization of acrylic acid esters at low temperatures. An example is given in Fig. 10, where acyltransferase was localized in the fungus *Penicillium chrysogenum* following cryofixation, freeze-substitution, and embedding at low temperature.

As chemical fixation is generally believed to decrease immunoreactivity of antigens, Usuda *et al.* (1990) have described a method for immunogold microscopy where tissues are processed by rapid freezing and freeze-substitution fixation without the use of any chemical fixatives. This method was applied to the localization of catalase in rat hepatocytes. When embedded in Lowicryl K4M at −20°C, the freeze-substituted tissues gave the most intense gold labeling density when compared to freeze-substituted tissues processed conventionally. A similar approach was described by Monaghan and Robertson (1990) using both solid

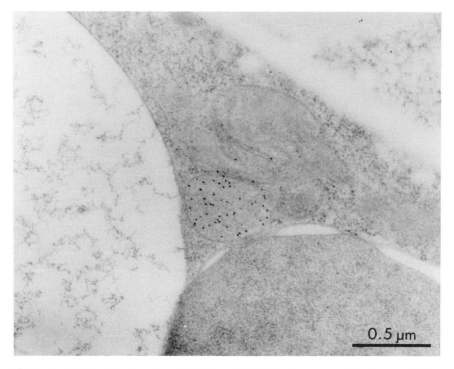

FIG. 10 Localization of acyltransferase in the fungus *Penicillium chrysogenum* following cryofixation and freeze-substitution. Specimens were prepared for immunoelectron microscopy by cryofixation, freeze-substitution in 0.5% uranyl acetate in methanol, and low-temperature embedding in HM20. After ultramicrotomy, sections were incubated with anti-acyltransferase antibodies followed by goat anti-rabbit antibodies conjugated to 10-nm gold particles. The organelle containing the acyltransferase is wedged by a vacuole, nucleus, and mitochondrion. Courtesy of A. J. Verkleij, University of Utrecht, The Netherlands; see Müller *et al.* (1991).

tissues (mouse small intestine and human kidney) and a human tumor cell line grown *in vitro*. Well-preserved ultrastructure was observed in the outer 10–15 μm of all samples. All these new procedures warrant further investigation to assess their potential better. Indeed, the type of specimen plays a more important role in freeze-substitution than in cryo-ultramicroscopy.

3. Freeze-Etching

Freeze-etching, the first approach for combining cytochemistry with freeze-fracture, was developed to relate the distribution of a cell surface group, antigen, or receptor to that of the membrane particles revealed by freeze-fracture of biological membranes (reviewed by Pinto da Silva, 1989). In theory, it is the simplest method for the visualization of plasma-associated proteins, provided that these proteins are labeled at their external domain. Freeze-etching immunogold cytochemistry is performed with specimens whose actual surface can be exposed by sublimation. Following immunolabeling and washing, the specimens (cells or isolated cell membranes) are rapidly frozen (without the use of cryoprotectants) and freeze-fractured. This is followed by a period of controlled sublimation and the cell surfaces are subsequently replicated. The method has relatively few applications in cytochemistry, since attractive alternatives are now available, such as fracture-label, a method developed by Pinto da Silva and Kan (1984) (see Section IV, D, 5).

4. Surface Replica

Surface replica techniques have become important for the investigation of biological structures in TEM, especially in combination with cryotechniques and the immunogold method (see Hohenberg, 1989, for a review). For surface replication, an electron-opaque metal is evaporated onto the biological material at an angle oblique to the average orientation of the biological surface. The heavy metal film is then reinforced by deposition of a supporting layer of carbon, usually evaporated perpendicular to the surface. After shadowing, the heavy-metal carbon film is cleaned by dissolving the biological matrix. When specimens are immunogold labeled, the gold particles bound to the surface remain in the replica subsequent to the dissolution of the biological matrix. In the label fracture technique (see below), which does not include replica cleaning by oxidation and hydrolysis in acids and alkalis, the replica is shown together with the labeled biological matrix. The gold particles should be as small as possible in order to decrease steric hindrance and promote high labeling density. However, unless the particles are silver-enhanced, a minimum size is required to still permit the identification of the particles in the evaporated heavy-metal carbon film. Under proper experimental conditions, the loss of particles during detachment from the support of the labeled and shadowed species is minimal. As shown by Hohenberg (1989), there is no direct loss of gold particles from the replica, but rather a dissolution of the particles when exposed to harsh acids. Figure 11 illustrates the combination of replica and freeze-fracture techniques with the use of ultrasmall gold particles (1 nm) and silver enhancement. In this striking example, the epidermal growth factor was mapped on A431 cells.

5. Fracture-Label

In this method, introduced by Pinto da Silva *et al.* in 1981, the components of plasma membranes or of intracellular membranes can be labeled *in situ*, after freeze-fracture (reviewed by Pinto da Silva, 1989). The method permits the labeling of cell surfaces and extracellular matrices, as well as components of the cytoplasm and nucleoplasm. For fracture-label, cells and tissues are chemically fixed, impregnated with 25–30% glycerol, and frozen. The biological material is then processed for two types of fracture label: thin-section fracture-label and critical-point drying fracture-label. In the latter, the material is replicated by platinum/carbon evaporation.

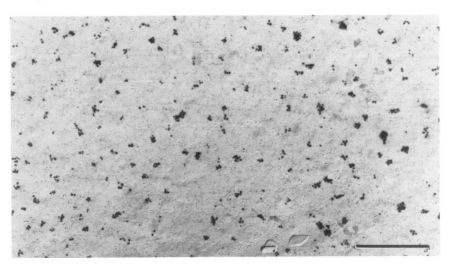

FIG. 11   Direct visualization of biotinylated epidermal growth factor (EGF) binding to the EGF receptor on A431 cells. Cells were successively incubated with biotinylated EGF, rabbit anti-biotin antibody, and goat anti-rabbit antibody conjugated to 1-nm gold particles. Following silver enhancement, the frozen specimens were fractured and shadowed. Replicas of exoplasmic fractures were examined in a Philips CM10 electron microscope. The use of ultrasmall gold particles, silver enhancement, freeze-fracture, and replica are strikingly combined in this example. Bar, 1 μm. Courtesy of A. ter Avest and P. J. Rijken, University of Utrecht, The Netherlands.

6. Label-Fracture

According to Pinto da Silva (1989), who introduced label-fracture in 1984 (Pinto da Silva and Kan, 1984), the objective of this method is to associate in one single, coincident image the distribution of a chemical group, receptor, or antigen on the surface of a membrane and the conventional freeze-fracture image of the same membrane. Diluted cell suspensions are treated as for freeze-etching, but, instead of etching, the labeled cells are only freeze-fractured and platinum/carbon replicated. In this technique, there is no digestion of the biological material by oxidation and hydrolysis. The replica is just washed with water. The label-fracture image is formed by the superimposition of two images: (1) the conventional platinum/carbon replica of the exoplasmic face, and (2) the distribution of the electron-opaque marker (colloidal gold) attached to sites at the outer surfaces. Stereo pairs should be used in most label-fracture studies as the two images are not apposed: they are separated by a thickness that corresponds to one membrane "half" (Pinto da Silva, 1989). In this procedure, isolated organelles as well as cell monolayers and epithelia can be used. Pavan et al. (1989a,b) used label-fracture to study the dynamics and relationships between intramembrane particles and the surface distribution and redistribution of surface antigens. Due to the hydrophobic nature of the replica, the outer monolayer of the plasma remains attached to the replica during the procedure and is also amenable to immunogold labeling (Boonstra et al., 1991).

7. Fracture-Flip

In the fracture-flip method, the exoplasmic halves of membranes are stabilized by, and remain attached to, carbon replicas of freeze-fractured specimens. Inversion ("flipping") of these carbon replicas exposes the actual outer surfaces of plasma membranes, which are then shadowed by platinum/carbon evaporation (Andersson Forsman and Pinto da Silva, 1988). Immunogold labeling can be combined with fracture-flip by labeling the specimens either before (Andersson Forsman and Pinto da Silva, 1988) or after freeze-fracture (Pimenta et al., 1989). With fracture-flip, it is possible to observe larger areas than with freeze-etching. The method recently provided the first extended, high-resolution stereo views of the lymphocyte cell surface during capping of two transmembrane proteins (Pavan et al., 1990).

E. Other Techniques

1. Silver Enhancement

A major advance for the use of immunogold techniques in light microscopy was achieved by the application of photochemical reactions (Danscher, 1981). It provided considerable amplification of the gold particle signal, resulting in enhanced sensitivity compared to immunoperoxidase (Holgate *et al.*, 1983; Springall *et al.*, 1984). Colloidal gold–silver staining involves the controlled precipitation of concentric metallic silver shells around the gold particle, thereby increasing the diameter and hence the visibility of the marker. The technique is employed for the demonstration of antigens either on intact enkaryotic cells (De Waele *et al.*, 1986), or on tissue sections (Holgate *et al.*, 1983; Springall *et al.*, 1984). Nonmicroscopical applications include electroimmunoblotting (Brada and Roth, 1984).

The silver enhancement method has also found applications in TEM, both in pre- and postembedding routines and in SEM. The various practical applications of the technique have been reviewed by Scopsi (1989) and Hacker (1989). Illustrations of the application of the method are given in Figs. 4 and 11.

2. *In Situ* Hybridization

Numerous methods have been described to visualize labeled gene probes at the light and electron microscope level. Hutchison *et al.* (1982) introduced the immunogold method for *in situ* hybridization studies. These authors hybridized biotinylated mouse satellite DNA and cRNA to whole-mount chromosomes using anti-biotin antibody and a secondary antibody–gold complex for detection. Binder *et al.* (1986) were the first to apply the protein A–gold method with silver enhancement for *in situ* hybridization on semithin and ultrathin sections, using biotinylated probes. Volkers *et al.* (1988) demonstrated the application of the immunogold/silver staining (IGSS) method for detecting biotinylated DNA probes in tissues. A similar technique was described by Jackson *et al.* (1989) for paraffin sections. They used a two-layer technique with a 5-nm immunogold reagent. The same group (Jackson *et al.*, 1990) extended their studies with 1-nm immunogold reagents (two-layer technique). Furthermore, biotinylated DNA probes were detected directly by a 1-nm labeled goat antibiotin antibody without loss of labeling intensity. This method may be preferable to the two-layer technique, which is more time-consuming.

The use of biotinylated cRNA probes to localize cellular mRNA makes it possible to identify subcellular sites of hybridization with streptavidin–gold, as suggested by Childs *et al.* (1989). However, Jackson *et al.* (1989) reported that the method did not work in their hands. This was probably due to the large molecular size of the streptavidin–gold complex. Steric hindrance may render the binding of this complex impossible when the biotin is positioned on only a short 11-atom spacer arm as used in their study. Nick translation of probes with a biotin nucleotide incorporating a longer spacer arm may lead to more successful applications (Jackson *et al.*, 1989).

For *in situ* hybridization studies, the advantage of the immunogold method combined with silver enhancement are the following: (1) low background, (2) stable reaction products, (3) rapid signal detection. Moreover, when compared to radioactive probes, the method offers (1) stability, (2) high resolving power,

(3) no hazard, (4) no disposal problems, (5) a signal-to-noise ratio equivalent to radiolabeling, (6) superior morphological preservation and spatial resolution (Binder *et al.*, 1986). In all cases, the method is suitable for microwave incubation to combine sensitivity and rapidity (Van den Brink *et al.*, 1990). The signal is insoluble in common laboratory dehydrating and clearing agents and can be enhanced by epipolarization (De Waele *et al.*, 1988; Jackson *et al.*, 1989).

F. Quantitative Methods

The quantitative assessment of gold labeling was introduced by Horisberger and Rosset (1977a). Several methods are now available, such as direct counting on micrographs, spectrophotometry, radioassay, and X-ray analysis. The number of gold particles per volume unit and per absorbance unit can easily be determined in the colloid (Horisberger, 1985) once the size distribution of particles has been measured and assuming that gold chloride is completely reduced by the reducing agent used in the preparation of the colloid. As a consequence, the number of particles bound onto a cell and the affinity constant (Scatchard plot) can be determined by spectrophotometric measurements (Horisberger, 1981a, 1985).

On thin sections, the quantitative assessment of gold labeling suffers from a number of experimental variables, such as fixation, embedding, and the use of cryosections. Variations are also tissue- and antigen-related. Immunolabeling procedures (one-step versus multistep) considerably influence the labeling density, which depends also on the number of ligand molecules carried per particle (Horisberger, 1985; Horisberger and Clerc, 1985).

Most of the quantitative methods available are relative. An approach to absolute quantification of antigens has been proposed by Slot *et al.* (1989) using cryosections of tissue embedded in 30% polyacrylamide. As a reference, a mixture of 10% gelatin with a known concentration of the antigen to be studied is similarly fixed and integrated into the polyacrylamide-embedded tissue blocks. In this system, a constant labeling efficiency is achieved and the absolute antigen concentration in cell structures can be measured *in situ*. However, the sensitivity of this technique is still low.

Agreement between quantitative immunocytochemistry on thin sections and immunoblotting was reported by Beier *et al.* (1988). For this demonstration, they studied the effect of a peroxisome-proliferating hypolipidemic drug on six different enzymes in rat liver peroxisomes. The drug induced selectively the β-oxidation enzymes, while the other peroxisomal enzymes were reduced. There was an excellent agreement between quantitative immunocytochemistry (protein A method) and immunoblotting from normal and drug-treated animals, provided special precautions were taken with uricase. This enzyme was localized exclusively in the electron-dense region of the peroxisomes, whereas all other enzymes investigated spared the core region and were localized in the surrounding matrix.

Because of the small size of the core region, its sparing was not evident in all micrographs, owing to the fact that immunogold labeling occurs only on the surface of sections (Bendayan *et al.*, 1987). If the core lies deep within the section, the labeling of the surrounding matrix, extending to the surface, can give the impression of partial labeling of the core.

V. Applications in Cell Biology

A. Virus and Microbial Cells

Beesley (1988) has reviewed the methodology of colloidal gold and some of its applications in microbiological immunocytochemistry, mainly with respect to localization of viral, bacterial, and fungal antigens. In this section, due to the vast amount of literature available, only recent examples are discussed to demonstrate the usefulness of the colloidal gold method in studying host relationships with viruses and microbial cells.

1. Virus

Immunogold labeling, especially when combined with replica techniques, has proved to be very useful for investigating dynamic processes in cell–virus interactions (Mannweiler *et al.*, 1989). Such investigations can not only increase our knowledge of the pathogenesis of viral diseases, but also provide insight into functions of cellular structures and the regulation of cellular pathways. In this respect, enveloped viruses have been extensively used because synthesis, post-translational processing, and transport of viral proteins are indistinguishable from processes for normal cellular membrane-bound glycoproteins.

For the detection of virus-specific antigens at the plasma membrane and for studying their spatial arrangement, fracture-label and surface replica techniques using immunogold markers are well suited (Torrisi *et al.*, 1990). Using paramyxoviruses as a model, Mannweiler *et al.* (1989) have shown how these techniques permit one to follow the involvement of the plasma membrane and cytoskeleton in the process of virus assembly.

Many researchers have proposed that cytoskeletal structures participate in the replication cycle of a variety of RNA and DNA viruses (for reference, see Mannweiler *et al.*, 1989). The replica technique combined with immunogold labeling permitted the demonstration of the association of simian virus polypeptides with actin filaments (Kasamatsu *et al.*, 1983) and the visualization of the presence of bluetongue virus structures on vimentin filaments (Eaton *et al.*, 1987).

For an ultrastructural analysis of the interaction of paramyxoviruses with the cytoskeleton, Bohn *et al.* (1986) developed a method for preparing platinum carbon pseudoreplicas of cytoskeletons obtained from measles virus-infected cells grown on coverslips. The pseudoreplicas showed a high degree of three-dimensional preservation in TEM. The viral components were found associated with the outer part of the cytoskeleton network. However, the visualization of the arrangement in the interior of complex structures could only be obtained by examining ultrathin sections of embedded cytoskeleton. In this case, actin filaments were shown to extend into the paramyxovirus structures.

The immunogold method has permitted the study of the localization of viral antigens in infected cells such as the 65-kDa DNA-binding protein of herpes simplex virus type 1 (Goodrich *et al.*, 1989), the influenza virus matrix protein and nucleoprotein (Patterson *et al.*, 1988), the adult T cell leukemia-associated antigens (Tanaka *et al.*, 1984), and the bluetongue virus antigen (Nunamaker *et al.*, 1990). In the latter example, immunogold labeling studies demonstrated that the antigen could be detected within developing oocytes of the fly *Culicoides variipennis*, thus demonstrating that viral antigens can penetrate the ovarian sheath of this biting fly, probably through micropinocytosis during yolk formation.

Attention has been recently directed at the budding process and maturation of human immunodeficiency virus (HIV) using immunogold techniques. While the budding process and the fine structures of HIV have been extensively studied by many researchers, the relationship between the area of localized budding and intracellular structures is still largely unknown. Katsumoto *et al.* (1990) have employed pre- and postembedding immunogold procedures to differentiate events occurring outside and inside the cell membrane and demonstrated that the budding process was restricted to a localized area at the membrane adjacent to the Golgi apparatus.

While out of the scope of this review, the different aspects of the methodology of immunogold labeling combined with negative staining as applied to a number of plant, animal, and human viruses and viral antigens have been reviewed by Kjeldsberg (1989).

2. Bacteria

Immunogold–silver staining was shown to be useful for studying bacterial antigen variation and the uptake of bacteria by eukaryotic cells using light microscopy (van Putten *et al.*, 1990). In this application, based on the impermeability of the eukaryotic plasma membrane to antibodies and the gold-labeled conjugate, extracellular bacteria were selectively immunosilver-stained while intracellular bacteria and eukaryotic cells were counterstained by crystal violet. These techniques were applied to *Neisseria gonorrhoeae*, which has a high frequency of antigen variation and a capacity to enter human epithelial cells.

Immunogold electron microscopy was used to study the invasion-related antigens of *Campylobacter jejuni* in Hep-2 cells (Konkel *et al.*, 1990) and the entry and multiplication of *Listeria monocytogenes* in the human colonic carcinoma cell line Caco-2. Direct evidence was obtained that the intracellular bacteria were enveloped with a thick layer of F-actin (Mounier *et al.*, 1990).

The intracellular growth of pathogenic mycobacteria has been linked to the presence of a capsule which surrounds the phagocytosed bacteria and prevents the diffusion of lysosomal enzymes in infected macrophages. Under special fixation and embedding conditions, the capsule could be preserved and was shown by immunogold cytochemistry to contain mycobacterial antigens (*Mycobacterium intracellulare*), thus confirming its mycobacterial origin (Rastogi and Hellio, 1990).

3. Yeast

The composition and the molecular aspects of yeast cell wall synthesis and morphogenesis are relatively well known (Cabib *et al.*, 1982). In budding yeast such as *Saccharomyces cerevisiae*, carbohydrate components form the greatest part of the cell wall and consist of β-glucans, mannan as a proteoglycan, and chitin. Their proportions may vary in different yeasts. A model has been proposed for the organization of polysaccharides in the cell wall of budding yeast (Cabib *et al.*, 1982). Using the immunogold technique (Figs. 5 and 8), both in TEM and SEM, Horisberger and colleagues have confirmed some features of this model in budding yeasts such as *S. cerevisiae* (Horisberger and Vonlanthen, 1977; Horisberger and Clerc, 1987), *Saccharomyces rouxii* (Horisberger *et al.*, 1985), and *Candida utilis* (Horisberger and Vonlanthen, 1977). At present, the following picture emerges: mannoproteins and β-glucan interweave to a considerable degree in the lateral wall, the septum, and the bud scars. The mannoprotein, which is believed to act as a "cement," and glucan are not uniformly distributed in the cell wall. They are more concentrated at the periphery of the wall and near the outer layer of the plasmalemma. Chitin is largely confined to the forming septum and the bud scars (Figs. 5 and 8).

The immunogold method also provided a very useful tool to study the cell wall architecture of *Candida albicans*. This important human pathogen exists in the form of yeast cells (blastoconidia) (Fig. 12a), germ tubes, and mycelium (Fig. 12b). The hyphal form is believed to be the most invasive by most investigators. The cell wall composition and architecture of *C. albicans* have been reviewed by Shepherd (1987). Blastoconidia cell walls are mainly composed of β-glucans and mannoproteins and of a small amount of chitin principally located in the bud scars, the primary and secondary septa, and near the plasmalemma. The outer layer of the cell surface is composed of mannoproteins overlapping β-glucans and chitin. The cell wall mannoproteins play an important role in the host–parasite relationship. They are the major cellular antigens and mediate adhesion of the microorganism to host cell surfaces (Torosantucci *et al.*, 1990). Some studies have demonstrated remarkable changes in the surface expression of mannoproteins during fungal growth and morphogenesis (Poulain *et al.*, 1985; Horisberger and Clerc,

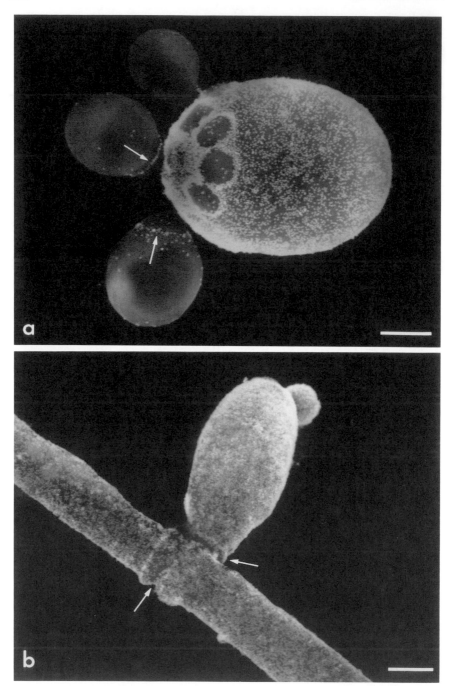

1988a). Anionic sites, probably associated with the phosphorylated mannoproteins, were detected by a cationic gold marker at a remarkable high density on the
hyphae (Horisberger and Clerc, 1988a) (Figs. 5 and 12). They may be involved
in the adhesion process, as the mycelial form of *C. albicans* has the greatest
virulence.

The immunogold method was also used to probe by TEM and SEM the cell wall
architecture of *Schizosaccharomyces pombe*, a yeast dividing by fission. The cell
walls of *S. pombe* contain branched and linear glucans and galactomannan but no
chitin. On the cell surface, in exact opposition to α-galactomannan, β-glucan was
detected only on walls generated by fission, but not on those growing by extension,
indicating that β-glucan is directly implicated in the fission process (Horisberger
and Rouvet-Vauthey, 1985).

B. Plant Cells

Because plant cells are impermeable to antibodies, the postembedding approach
must be used. It has been used to investigate the molecules localized include seed
storage proteins (Horisberger *et al.*, 1986; Krishnan *et al.*, 1991; zur Nieden *et al.*,
1982), lectin (Horisberger and Vonlanthen, 1980) (Fig. 13), protease inhibitors
(Horisberger and Tacchini-Vonlanthen, 1983b; Rasmussen *et al.*, 1990), and various enzymes, such as lipoxygenases 1 and 2 (Vernooy-Gerritsen *et al.*, 1984). The
immunogold localization of intra- and extracellular proteins and polysaccharides
of plant cells has been reviewed by Craig *et al.* (1987).

Although much information is available on the structural features of plant cell
wall polysaccharides and on the ultrastructure of plant cell walls, relatively little
information is available on the localization at the ultrastructural level of individual
cell wall components. The immunogold labeling of plant cell wall components has
been achieved using polyclonal and monoclonal antibodies against purified xyloglucans and pectic polymers. Another promising approach is enzyme–gold cytochemistry (for references, see Vian *et al.*, 1991). Vian *et al.* have localized
xyloglucans in seed cell walls using a novel *endo*-(1→4)β-glucanase and a β-D-
galactosidase conjugated to colloidal gold. When gold complexes were prepared
from the active enzymes they retained enzyme activity and gave extremely weak
section labeling or no labeling at all. Surprisingly, the complexes prepared from
heat-deactivated enzymes gave strong, specific labeling of xyloglucans in ultrathin

---

FIG. 12   Localization of anionic sites on *Candida albicans*. (a) Scanning electron micrograph of
blastoconidia cells marked with chitosan conjugated to 47-nm gold particles. The mother cell is densely
marked except for the bud scars. The emerging buds are free of marking with the exception of an
occasional ring (arrows). (b) Scanning electron micrograph of *C. albicans* hyphae. Hyphae, blasto-
conidium, and emerging bud are densely marked except for the septal region (arrows). Bars, 1 μm. From
Horisberger and Clerc (1988a; Figs. 2a and 4d, pp. 447 and 450).

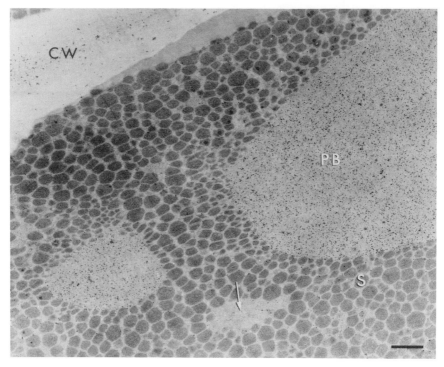

FIG. 13   Localization of soybean lectin (SBL) on a thin section of *Glycine max* (soybean) var. Altona by the direct method using anti-SBL antibodies conjugated to 12-nm gold particles. SBL was found in most protein bodies (PB) but not in all (arrow). Spherosomes (S) were free of SBL. The cell walls (CW) and the space between the spherosomes were weakly stained nonspecifically. Bar, 1 μm. From Horisberger and Vonlanthen (1980; Fib. 1b, p. 183).

sections. The authors concluded that the enzyme–gold complexes which retain high activity against the substrate to be localized may be unsuitable as cytochemical probes because they may cause *in situ* substrate modification. These observations warrant further studies to determine whether the phenomenon is general.

The most thoroughly characterized secretory tissue in plants is the barley aleurone. However, little is known about the process in aleurone cells that segregates proteins targeted to vacuoles from those in the pathway for secretion. Holwerda *et al.* (1990) have identified a protein in aleurone cells that could be used to study targeting mechanisms. This protein, aleurain, originally described from its cDNA as a thiol protease, was characterized as a glycoprotein and shown to be targeted to a distinct vacuolar compartment in aleurone cells using the immunogold method and silver enhancement.

The immunogold method was also applied to study the cellular routes of protein body formation in cereals. Krishnan *et al.* (1988, 1990) have provided evidence for the condensing role of the Golgi apparatus in prolamin trafficking in cereal caryopses. Initially, these authors used 10-nm particles labeled with protein A. IN a more recent study (Krishnan *et al.*, 1991), the same authors have confirmed that the labeling efficiency of gold markers on thin sections decreases as the size of the particle increases (Horisberger, 1985; Geuze *et al.*, 1981).

In an entirely different application, colloidal gold of prescribed sizes (1.6 to 18 nm in diameter) was used as an apoplastic tracer for the determination of pore size in woody plant tissue (Schaffer and Wisniewski, 1990). All colloidal gold particles in the range 1.5–18 nm were able to pass through intervessel pit membranes, whereas particles greater than 4 nm were excluded from xylem parenchyma pit membranes of willow (*Salix babylonica*). This method warrants further investigation since the gold particles were not stabilized by macromolecules and thus could flocculate in the tissue. The same approach was used by Reddy and Locke (1990) to study the penetration of gold particles (6–27 nm in size) through insect basal laminae (larvae of *Calpodes ethlius*).

C. Animal Cells

In this subsection, the topics have been selected for their particular emphasis on dynamic events and to illustrate not only the scope and the various aspects of the colloidal gold method but also to discuss potential pitfalls.

1. Vertebrate Photoreceptor Cells

Rod and cone photoreceptor cells serve as the primary sites of visual excitation in the retina. They are highly differentiated, elongated cells and contain a specialized photoreceptor organelle (outer segment) which is joined by a thin cilium to the mitochondria-rich inner segment of the cell body. The outer segment consists of an assembly of hundreds of stacked disks surrounded by a separate plasma membrane in the rod cell and a continuous infolding of the plasma membrane in the cone cell (Molday and Laird, 1989).

Immunogold–dextran-labeling techniques were almost exclusively used by the group of Molday to study the organization of rhodopsin and several minor proteins in the membranes of rod and cone cells (reviewed by Molday and Laird, 1989). Both pre- and postembedding labeling studies indicated a dense distribution of rhodopsin along the extracellular surface of the rod outer segment plasma membrane. Only scattered labeling, however, was found along the surface of the inner segment plasma membrane using monoclonal antibodies against different epitopes. The degree of labeling varied for different monoclonal antibodies against different epitopes and was probably due to differential accessibility of these epitopes (Hicks and Molday, 1986). In the same study, the mapping of the distribution of rhodopsin within rod and cone cells of frog and bovine retina were compared. It was shown that significant homology exists between bovine and frog rhodopsin, but more limited homology is apparent between rhodopsin and cone opsin. Immunogold–dextran-labeling techniques were also used to map the distribution of minor proteins within the rod outer segments.

Phagocytosis of immunogold–dextran-labeled rod outer segments was studied in cultured retired pigment epithelial cells and analyzed both by TEM and SEM (secondary and backscattered electron imaging). Phagocytosed gold-labeled rod outer segments were unambiguously identified and localized using the backscatter mode (Molday and Laird, 1989).

The outer segment of the vertebrate rod photoreceptor cell contains cation-selective channels which are directly gated by cGMP. These channels are localized in the rod outer segment plasma membrane. Molday *et al.* (1990) have shown by immunogold–dextran labeling that this channel is associated with a 240-kDa protein exhibiting immunochemical cross-reactivity with spectrin. According to these authors, the 240-kDa protein does not appear to be directly involved in the channel activity, but may be part of a cytoskeletal system that serves to maintain the organization of the channel complex with the rod outer segment plasma membrane.

2. Cytoskeleton

The cytoskeleton in higher eukaryote cells is composed of three major classes of protein network: microtubules composed of tubulin polymers, actin filaments, and intermediate filaments whose proteins can be classified into four types. Together, these systems are believed to influence the structural and functional properties of the cell (such as motility, intracellular transport, cell shape, and polarity) and to play important roles in the differentiation and morphogenesis of particular cell lineages. Knowledge of the interconnections between the major filament systems of the cytoskeleton as well as the elucidation of cytoskeletal dynamics are necessary for understanding the cytomatrix organization and the mechanisms by which cytoskeleton contributes to cell differentiation and morphogenesis.

Cytoskeletal filaments were first displayed by immunofluorescent techniques, but major advances were achieved more recently by the wide use of gold conjugates. De Mey *et al.* (1981b) introduced colloidal gold as a marker for cytoskeletal proteins in cultured cells. As pointed out by Langanger and De Mey (1989), the labeling of microtubules is relatively easy compared to that of microfilament-associated proteins. Their localization presents a particular problem. They are either present in fine and delicate networks in close association with plasmalemma, and are very susceptible to detergent extraction, or occur in densely packed stress fibers, which are more resistant structures, but less accessible. The localization of intermediate filament-associated proteins depends very much on the reactivity of the primary antibodies. It has also been reported that nonspecific binding of gold probes to cell components, such as extracellular matrix and intermediate filaments, may cause disturbing background problems (Birrell *et al.*, 1987).

Studies of detergent-permeabilized cells and *in vitro* organelle motility assays have provided much useful information about components involved in organelle motility. However, information is less precise on the control mechanisms that determine location and movements of intracellular organelles in nonneural cells under *in vivo* conditions. Therefore, there is a current need for studying spatial relationships and molecular interactions between the cytoskeleton and the organelles by high-resolution immunocytochemistry, microinjections of labeled proteins, and related methods within the environment of nonneural cells (Saetersdal *et al.*, 1990).

Detergent-permeabilization followed by rapid-freezing, rapid-drying, and metal shadowing has been used to obtain high-resolution images of the cytoskeleton. In combination with immunogold cytochemistry, this approach has provided sharply defined views of the membrane skeleton of the resting platelet and has demonstrated its substructure and how it connects with cytoplasmic active filaments (Hartwig and De Sisto, 1991).

Detergents are usually required to obtain a degree of permeabilization adequate for penetration of the probe into intact cells. This, however, can cause a severe loss of lipids and cytoplasmic constituents. The penetration of 1-nm particles was shown to be superior to that of 5-nm particles for the intracellular labeling of tubulin in while mounts of PtK_2 cells in the absence of permeabilization agents. Some labeling of microtubules could be seen, but was restricted only to tubules immediately below the cell surface (van de Plas and Leunissen, 1989).

Preembedding procedures have been described by Langanger and De Mey (1989). Despite many drawbacks, they are still widely used in immunogold cytochemistry on detergent-permeabilized cells. One limitation for structures such as the cytoskeleton or nuclear matrix is that their form is apparent only when observed in three dimensions.

A post embedding cryoprocedure has been described by Katsuma *et al.* (1988). More recently, Nickerson *et al.* (1990) proposed an immunogold postembedding procedure for staining cytoskeletal and nuclear matrix proteins in resinless electron microscopic sections. The method permitted the localization of specific proteins throughout the entire section in stereoscopic micrographs. However, it required very careful handling of delicate sections through the staining and the washing steps.

In cardiac myocytes, immunogold cytochemistry has allowed the identification of microtubule constitutive proteins. It has emerged that desmin filaments are particularly concentrated in the intermyofibrillar space of the Z disk area, where microtubules are few, and at the intercalated disk areas, which are devoid of microtubules. Desmin label never appears to reach the nuclear membrane (Rappaport and Samuel, 1988).

The importance of microtubule-to-organelle translocation has been demonstrated in a number of reports (for references, see Saetersdal *et al.*, 1990). The association between β-tubulin and mitochondria in cardiac myocytes isolated from the rat has been examined by immunofluorescence and high-resolution immunogold electron microscopy (preembedding method and cryo-ultramicrotomy). The complexity of the microtubule network was shown to vary considerably among myocytes, thus reflecting microtubule dynamic instability (Saetersdal *et al.*, 1990).

3. Prosomes

Prosomes ("program-O-somes") are a novel type of ribonucleoprotein (RNP) particles associated with messenger RNA in eukaryotic cells. They were first observed by Scherrer and co-workers in 1969 (reviewed by Scherrer, 1990). These RNP particles are made up of a combination of 20–25 protein components and contain one or more small RNA molecules which have the ability to hybridize to mRNA. Prosomes associate *in vitro* with mRNA and inhibit cell-free protein synthesis by inducing a mRNA structure unable to interact with ribosomes. Their function(s) is still unknown.

In addition to microfilaments and microtubules, vertebrate cells contain a third cytoskeletal component, the intermediate-sized filaments (IF). Prosomes were shown to be present on the cytokeratin networks in PtK and HeLa cells but not in the vimentin network (or the actin-type filaments) also present in such transformed tissue culture cells (Grossi de Sa *et al.*, 1988a). The correspondence of the two types of network was demonstrated by Grossi de Sa *et al.* in PtK cells by double-label immunofluorescence and techniques using anti-cytokeratin antibodies and the anti-p27K and -p29K prosomal monoclonal antibodies. These observations were confirmed by double-label immunogold experiments on HeLa cells. A particularly intriguing observation was that the prosomal network bridges cells at specific points (Scherrer, 1990).

Immunofluorescence, as well as electron microscopic techniques using prosomal monoclonal antibodies and peroxidase- or gold-labeled secondary antibodies, have shown that prosomal antigens also exist in the nucleus (Grossi de Sa *et al.*, 1988b). It seems that particular types of prosomal antigens, and thus prosomes of specific protein composition, may preferentially occupy different compartments in the same cell (Scherrer, 1990).

The discovery that the prosomes (subcomplexes of the messenger RNP particles) are linked to the IF networks of the cytoskeleton might sustain the proposition that the IF are involved in the mechanism of gene expression. They might have a function in controlling the transport, distribution, and regulation of activity of specific mRNAs in the cell (Scherrer, 1990).

4. Receptors

Among different mechanisms, animal cells make a large use of specific receptors to target extracellular macromolecules to intracellular destinations. Macromolecules such as proteins, hormones, toxins, and viruses are then internalized (receptor-mediated endocytosis) by a selective process against a concentration gradient and packaged in vesicles. Immunocytochemical techniques have been used to a large extent for analyzing such intracellular processes as receptor recycling or transcytosis and for investigating cells *in situ* (Mommaas-Kienhuis and Vermeer, 1989). Much of the information available on receptor-mediated endocytosis is derived from studies on the metabolism of low-density lipoproteins (LDL) in cultured fibroblasts.

In the case of low-molecular-weight ligands, bioactive complexes are only obtained by using a bridge protein such as horseradish peroxidase between the ligands and the colloidal gold particles. Among others using this approach, Ackerman and Wolken (1981) studied the distribution of the insulin receptor, and Beppu (1989), with bovine serum albumin as a bridge protein, localized the progesterone and estradiol receptors in cells of the uterus and the liver (they were absent in the colon). The steroid hormone receptors were mainly localized in the cell nuclei of the target organs but the density of labeling was low.

In an elegant study, Vázquez-Nin *et al.* (1991) investigated the localization and quantitative changes of estradiol receptor (ER) by the immunogold method. Indirect labeling was performed on ultrathin sections using as secondary reagents either a protein A–gold complex or an antibody–gold complex. The primary antibodies were a polyclonal antibody directed against an aminoacid sequence representing the DNA-binding site of ER, a monoclonal antibody against hmRNP core protein, and an anti-DNA antibody. From double-labeling experiments of thin sections of epithelial cells, muscle cells, and fibroblasts of rat uterus using 10- and 30-nm gold particles, these authors concluded that (1) ER is mainly nuclear but is also present in the cytoplasm, (2) ER binds to nuclear particles containing newly synthesized RNA, and (3) the binding to RNPs does not block the DNA-binding domain of the ER. Labeling was intense on RNP fibrils with the anti-ER antibody.

a. The Low-Density Lipoprotein Receptor   In earlier studies, the visualization of the interaction of LDL with cultured fibroblasts was achieved using a direct technique (ferritin conjugated to LDL) and the indirect immunoperoxidase method. Handley *et al.* (1981a,b) introduced a procedure for conjugating LDL to colloidal gold where 20-nm particles are decorated by 7–9 LDL molecules. A large mount of evidence points to the fact that conjugation to colloidal gold does not alter the biological activity of LDL (Mommaas-Kienhuis and Vermeer, 1989).

Due to the size of the LDL–gold complexes (~60 nm) for the small pinocytic vesicles (vesicle neck ~30 nm), the technique is not suitable for investigating transcytosis in vascular (capillary) endothelial cells (Mommaas-Kienhuis and Vermeer, 1989). In that case, the intracellular detection of both LDL and LDL receptors can be achieved on thin sections by the protein A–gold method or by the use of secondary antibodies conjugated to colloidal gold (Mommaas-Kienhuis and Vermeer, 1989).

Most of the cell types studied (cultured human fibroblasts, cultured human umbilical endothelial cells, human monocyte-derived macrophages, isolated rat liver endothelial cells) bind lipoprotein–gold complexes in coated pits, i.e., indentations on the plasma membrane with a bristle coat on the cytoplasmic site (see Mommaas-Kienhuis and Vermeer, 1989, for references). However, in other cells, such as isolated rat liver Kupffer cells and various cultured epithelial tumor cells, binding of LDL–gold particles was predominantly seen in uncoated regions of the plasma membrane. From these studies, it was concluded that the coated pit is not a general feature in receptor-mediated endocytosis.

Robenek et al. (1982) were the first to introduce replication techniques for studying the topographical distribution of LDL receptors by the gold method. The advantage of the replication technique is that it reveals large areas of cell surface which can be examined at high resolution by TEM and viewed in three dimensions. By surface replica, Robenek and Hesz (1983) could visualize the spatial arrangement of LDL receptors in cultured mouse peritoneal macrophages. The lipoprotein–gold complexes were preferentially bound in the plasma membrane region around the nucleus. Using the same technique, the authors were also able to visualize the dynamics of LDL receptors in cultured human skin fibroblasts and suggested that the site of coated-pit formation coincides with the site of receptor insertion in the plasma membrane. After the completion of binding in pits (coated or uncoated), these pits pinch off to form coated or uncoated vesicles, which are located close to the plasma membrane. At the early stage of endocytosis, the gold complexes are still attached to the endosomal membrane; later, they are found in the matrix of the endosome. In all cell types investigated, free pools of accumulated gold, apparently no longer attached to lipoprotein, were found in lysosomes after prolonged exposure at 37°C. LDL–gold conjugates were also used to demonstrate that estrogen treatment of animals enhanced LDL–receptor expression in parenchymal and endothelial cells in the lining of the sinuosidal cavity (Handley et al., 1981b, 1983).

The lipoprotein Lp[a] is composed of an LDL-like lipoprotein to which the glycoprotein apo[a] is attached through disulfide bridging to apoβ-100. Several studies have addressed the question as to whether Lp[a] can be catabolized through the LDL–receptor pathway (see Armstrong et al., 1990, for references). Hesz et al. (1985) have observed that Lp[a] and LDL–gold particles were concentrated in discrete clusters on the cell surface of human skin fibroblasts and demonstrated the presence of both LDL and Lp[a] in coated pits. These authors found no qualitative differences in the specificity or in the binding of Lp[a] to the cell surface compared with LDL. Using highly purified Lp[a] fractionated into two species with different affinities for lysine-Sepharose and surface replica techniques, Armstrong et al. (1990) observed that Lp[a]–gold particles of both species bound in discrete clusters to the cell surface of human skin fibroblasts. However, contrary to Hesz et al. (1985), they observed that the extent of labeling with LDL–gold was far in excess of that seen with Lp[a]–gold. Taken together with earlier findings, Armstrong et al. (1990) concluded that the LDL receptor pathway does not play a major role in the metabolism of Lp[a].

b. The Epidermal Growth Factor Receptor   Polypeptide growth factors have been recognized as important determinants in the regulation of cellular proliferation and differentiation. An important aspect for the understanding of the regulatory role of polypeptide growth factors concerns the identification and localization of growth factors and their receptors at the ultrastructural level (Boonstra *et al.*, 1989).

The receptor for epidermal growth factor (EGF) is the receptor most studied at the ultrastructural level by the colloidal gold method. A number of techniques have been proposed to understand the mechanism of action of this receptor better. Cryo-ultramicrotomy has been used to localize intracellular EGF receptors (Fig. 9). Freeze-etching and label feature permitted the analysis of the lateral distribution of EGF receptors present on cell surfaces. Surface replication, dry cleavage, and the lysis-squirting method provided techniques to study, mostly in the epidermoid carcinoma A431 cell, the interaction of cell surface-located EGF receptors with the underlying cytoskeleton (reviewed by Boonstra *et al.*, 1989). By surface replica, EGF receptors were found clustered to a significant extent (Fig. 11). A high density of gold label was observed in the cell surface area containing microvili.

In cryosections of A431 cells, a dense labeling is observed on membranes of the Golgi complex and endoplasmic reticulum. Furthermore, a dense labeling is also found in lysosomes, multivesicular bodies, and a network of vesicular and tubular membranes (Boonstra *et al.*, 1985). It should be noted that, in cryosections of A431 human carcinoma cells, cell surface labeling appears in areas where gelatin (i.e., the embedding material) is retracted from the cell surface. This artifact is attributed to the size of the gold probes, which are limiting in their penetration, and to differences in the matrix of various organelles (Boonstra *et al.*, 1989, 19910.

Observation of cells treated with Triton-X-100 immediately after immunogold labeling suggested a structural association between cell surface-located EGF receptors and the underlying cytoskeleton. Isolated cytoskeletons were shown to contain EGF receptors. Some of these results were confirmed by the dry cleavage method, which allows a view from the inside of the cells toward the outside, and by the lysis-squirting method, which permits immunolabeling of both the cytoplasmic domain of the EGF receptor and the membrane-associated cytoskeleton (Boonstra *et al.*, 1989).

A number of experiments were performed to study the effect of EGF on the lateral distribution of EGF receptors, which is thought to play an important role in the biological response to EGF. From pulse–chase experiments, Miller *et al.* (1986) demonstrated that EGF caused a redistribution of EGF receptors from the plasma membrane to the peripheral endosome compartment and then to the pericentriolar compartment and lysosomes.

The presence of EGF receptors in cell organelles belonging to the endocytotic pathway of untreated A431 cells and human fibroblasts indicates a continuous internalization of EGF receptors, without requiring binding of EGF. These results indicate that EGF does not induce endocytosis of EGF receptors, but enhances its rate.

c. IgE Receptor on Mast Cells   Despite the fact that rat basophilic leukemia (RBL-1) cells possess receptors with high affinity for mouse monoclonal IgE (Mendoza and Metzger, 1976), they do not bind IgE–gold complexes (Clerc *et al.*, 1988). This is not surprising according to the model proposed by Holowka *et al.* (1985), where the IgE molecule must bend in order to bind to its receptor through the F_c portion. Such bending would be mostly prevented when IgE is adsorbed on a solid surface since the molecule loses flexibility. Mapping of cell membrane IgE receptors was therefore achieved at the TEM level by incubating RBL-1 cells sensitized with IgE successively with anti-IgE antibodies and a protein A–gold marker at 4°C (Clerc *et al.*, 1988). For the first time, these authors demonstrated at the TEM level that surface clusters develop when mast cells sensitized with IgE are warmed at 37°C (the formation of surface caps is the first step which normally initiates a cascade of responses leading to degranulation) (Fig. 14). This method should also be useful to study the fate of IgE receptor complexes of RBL-1 cell mutants which degranulate.

Stump *et al.* (1989) developed methods to map IgE receptors on RBL-2H3 cells at high-resolution SEM. Mapping of the IgE receptor required the development of a fixative that would preserve cell surface architecture and not impair the binding of any of the reagents. Sensitized cells were successively incubated with secondary antibodies and protein A–gold and examined with detectors for both secondary and backscattered electrons (Fig. 15a,b).

d. Fibrinogen Receptor   In a series of investigations, Albrecht *et al.* (1989) used fibrinogen coupled to colloidal gold to evaluate the organization of fibrinogen receptors (glycoprotein IIb/IIIa) after surface activation of human platelets. Scanning electron microscopy was used in conjunction with stereo-pair, high-voltage TEM as well as video-enhanced light microscopy to correlate changes in shape and surface structure with changes in the internal structure of platelets. Using fibrinogen–colloidal gold and colloidal gold–anti-IIb/IIIa (an antibody specific for the fibrinogen receptor site), this group has shown that active GP IIb/IIIa appears to be present at all stages of shape change. However, the binding of fibrinogen to its receptor could be observed only after the cells had transformed into dentritic platelets. During conversion from dentritic to fully spread forms, the fibrinogen–gold probe moves from the pseudopods and the body of platelets, where it is relatively uniformly distributed, to centers concentrated directly over the subjacent cytoskeletal inner filamentous zone. These results were confirmed and extended by White *et al.* (for references, see White and Escolar, 1990). These authors have given evidences that the central movement of fibrinogen–gold particles was driven by the *particles* rather than by the *coupled fibrinogen*, even though fibrinogen was essential for binding gold particles to the GPIIb/IIIa receptors. These findings

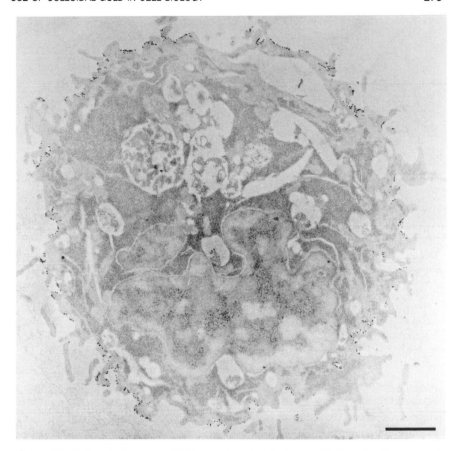

FIG. 14   Distribution of F_c receptor of IgE on rat basophilic leukemia cells. RBL-1 cells were sensitized with mouse IgE antibody monoclonal anti-bovine milk β-lactoglobulin and successively incubated at 4°C with anti-IgE antibodies and 12-nm protein A–gold. The cells were warmed at 37°C and developed surface clusters as seen by TEM following embedding. Bar, 1 μm. From Clerc *et al.* (1988; Fig. 6, p. 348).

underline the fact that labeling studies with unfixed cells must be interpreted with caution.

D. Cellular Events

1. Endocytosis

During receptor-mediated endocytosis of many plasma proteins, a ligand binds to a specific cell surface receptor, and the receptor–ligand complex is taken into cells

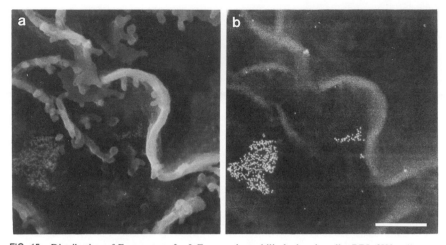

FIG. 15   Distribution of F_c receptors for IgE on rat basophilic leukemia cells. RBL-2H3 cells were primed with IgE and incubated for 15 min at 37°C with anti-IgE, fixed, and labeled with 15-nm protein A–gold. The samples were silver-enhanced before processing for SEM secondary (a) and backscattered (b) electron imaging. Most of the gold particles are found in clusters always confined to the interlamellar membrane region. Bar, 1 μm. Courtesy of J. M. Oliver, University of New Mexico, Albuquerque. First published in Stump *et al.* (1989; Fig. 6g,h, p. 138). Copyright © 1989 Alan R. Liss, Inc.; reprinted by permission of Wiley-Liss, a division of John Wiley and Sons, Inc.

in clathrin-coated pits and vesicles (Pastan and Willingham, 1985). The phenomenon of receptor-mediated endocytosis was first identified in studying the LDL receptor system. Subsequently, it was shown that the same type of receptor-mediated endocytosis applies to a variety of other nutrients and regulatory molecules (Brown and Golstein, 1985).

Most ligands dissociate from their receptor in the low pH environment of prelysosomal vesicles (Geuze *et al.*, 1983, 1984; Klausner *et al.*, 1983) and are then transported to lysosomes for degradation. Generally, the receptor is not degraded and recycles to the cell surface, as do the receptors for asialoglycoprotein (Schwartz *et al.*, 1982), LDL (Jaekle *et al.*, 1989), α-macroglobulin (McVey Ward *et al.*, 1989), and insulin (Marshall, 1985; Trischitta *et al.*, 1989). However, the fate of transferrin is different. After endocytosis of iron–transferrin complexes, transferrin is not degraded but is released from the cells by exocytosis, whereas the iron originally bound to the transferrin remains in the cell.

Due to their opacity to electrons, colloidal gold particles coated with specific proteins have been extremely useful in following the uptake and intracellular transport of various ligands by receptor-mediated endocytosis. The use of colloidal gold in double-label endocytosis experiments has also demonstrated the advantages of the method (Willingham, 1989). In most studies on endocytosis with colloidal gold, probes have been applied to living cells in order to elucidate the intracellular routes followed by ligands that are normally directed to lysosomes, such as asialoglycoproteins, α_2-macroglobulin, and LDL. Since in these studies gold probes were transported to lysosomes in a manner consistent with the behavior of the native ligands (as determined by other methods), it was assumed that attachment to gold particles did not significantly change ligand uptake and processing.

Using cells of the EJ human bladder line growing in monolayer, McIntosh et al. (1990) studied the intracellular movement and cycling of ricin by labeling the toxin *after* its binding to the cell surface. The toxin was visualized with gold-labeled secondary antibodies. Using the same cell line, endocytosis and exocytosis of radiolabeled ricin were studied to ascertain whether the ricin gold complex was handled by the cell in the same manner as native ricin. The ricin/anti-ricin antibody–gold complex and the toxin itself were bound, capped, and internalized via coated pits. This was unlike the reported mode of entry of ferritin-labeled ricin (Nicolson et al., 1975) and was not found to be the dominating endocytic uptake route of a ricin–horseradish peroxidase conjugate (Gonatas et al., 1980). Following treatment with ricin or the ricin–gold complex, the cells liberated ricin on their external surface both free, attached to, and associated with shed vesicles, apparently from the multivesicular bodies. Parallel studies with radiolabeled ricin supported the apparent cycling of ricin through the cells as observed in the morphological studies. In the study of McIntosh et al. (1990), rabbit anti-ricin IgG was used to label the toxin molecule *after* its binding to the cell surface. The toxin/antitoxin complex was visualized with a gold-labeled second antibody.

However, endocytosis and internalization of ligands could be perturbed by using ligand–gold complexes due to (1) the size of the particle, (2) the multivalency of ligand complexes, and (3) the increased affinity constant of the complexes. Reggio et al. (1983) have shown that large particles may perturb endocytosis. While a gold particle up to 5 nm in diameter may complex only one ligand molecule (Horisberger, 1978), larger particles can accommodate many ligand molecules. As a consequence, the valency of these complexes is greater than the ligands alone. Therefore, the increased affinity of multivalent gold complexes may affect endocytosis and internalization of ligands. Gold conjugates of lectins and protein A indeed have affinity contants several orders of magnitude higher than their native molecules (Horisberger and Rosset, 1977b; Horisberger and Tacchini-Vonlanthen, 1983a; Horisberger and Clerc, 1985).

It has been reported that the correct routing of ligand in receptor-mediated endocytosis may depend on the monovalency of the ligand. For example, van Deurs et al. (1986, 1988) have described studies on the sorting of internalized ricin to the Golgi complex and found differences in the cellular handling of monovalent and polyvalent conjugates. In another example, it has been reported that insulin–gold conjugates can behave as a multivalent ligand and thus be misdirected after internalization (Smith et al., 1988). However, by reducing the number of insulin molecules adsorbed per 10-nm particles to 5–7, these authors obtained a monovalent-behaving gold complex. The organization and distribution of insulin receptors occupied by this complex were virtually identical to previous observations using monomeric ferritin–insulin. The normal intracellular pathway of transferrin appears to be alterated by its associated with a colloidal gold particle (Neutra *et al.*, 1985; Griffiths *et al.*, 1988). However, despite the possible misdirection of a fraction of colloidal gold-labeled transferrin to a lysosomal compartment, recent studies with gold-labeled transferrin have revealed the stages of the endocytic pathway and demonstrated the presence of two subpopulation sof multivesicular endosomes that are morphologically indistinguishable (Woods *et al.*, 1989). In one subpopulation, considered to correspond kinetically to early endosomes, reside recycling receptors such as transferrin. The other subpopulation represents a later endosomal compartment en route to lysosomes.

In an elegant study, Smith and Hunt (1990) showed that hemopexin joins transferrin as representative members of a distinct class of receptor-mediated endocytic transport systems. Hemopexin and transferrin were colocalized intracellularly using, on the one hand, radioiodinated hemopexin and horseradish peroxidase conjugates of transferrin and, on the other hand, gold particles of different sizes labeled by both proteins, respectively. The gold particles were found together in coated pits, coated vesicles, endosomes, and multivesicular bodies. Taken together, these results indicate that hemopexin and transferrin act by a similar receptor-mediated mechanism in which the transport protein recycles after endocytosis from the cell to undergo further rounds of intracellular transport. From this study, the use of colloidal gold conjugates would appear justified, despite their multivalency.

Endocytosis can be studied by the gold method essentially according to three procedures: (1) in many studies, the ligand is conjugated to gold particles and traced by examining thin sections directly at the TEM level; (2) the ligand is bound to its cell surface receptor and then labeled with a gold probe. Following endocytosis of the gold complex, the tracer is observed again directly on thin sections (McIntosh *et al.*, 1990); (3) the ligand is endocytosed and visualized on thin sections by protein A–gold complexes (Stang *et al.*, 1990). For reasons outlined above, each of the procedures has its particular advantages and disadvantages. The last procedure has the least artifacts but is also the least sensitive.

2. Liver Cells

The sinusoidal endothelial cells in the liver have receptors that mediate the endocytic uptake of several blood components, such as mannose-terminated glycoproteins, glycosaminoglycans, collagen, insulin, acetyl-LDL, and formaldehyde-treated serum albumin (reviewed by De Leeuw *et al.*, 1989). Horisberger *et al.* (1978) have demonstrated that the binding of some lectin–gold conjugates to hepatocytes becomes insignificant above a certain size. Similar results were found with asialoceruloplasmin binding to its hepatic cell surface receptor (Horisberger and Vonlanthen, 1978). It was hypothesized that the receptor (galactose-binding protein) was located between narrowly spaced glycoprotein brushes.

Kempka and Kolb-Bachofen (1988) studied endocytosis of mannan-coated gold particles and found that both endothelial cells and Kupffer cells are involved in clearing these particles from the bloodstream. As ligands, these authors used gold particles of 5, 17, and 35 nm. They demonstrated that, irrespective of size, these particles can pass through the sinusoidal cells and be endocytosed by hepatocytes. In double-labeling experiments with particles of different sizes, glycoproteins with mannosyl or galactosyl residues were found to be present in the same coated pit of endothelial cells, while, on macrophages, the clustered binding occurred at different plasma membrane areas.

To overcome the problem of multivalency, which may affect the processing of gold-labeled ligands, Stang et al. (1990) used ovalbumin to trace the endocytic pathway. Since ovalbumin does not normally occur in the rat, the labeling pattern is not influenced by proteins being produced by any of the liver cells or circulating in the plasma. This allows exact timing of the endocytic events. Endocytosis of ovalbumin is most probably due to binding to the mannose receptor (Magnusson and Berg, 1989). The fate of ovalbumin was determined by labeling cryosections with secondary antibodies and protein A–gold particles (Stang et al., 1990). The cryotechnique was used because it has been shown to be superior both in specificity and intensity compred to other techniques (Griffiths and Hoppeler, 1986; Hemming et al., 1983).

The results indicated that it is mainly the sinusoidal endothelial liver cells that are responsible for the removal of ovalbumin from the bloodstream. Stang et al. (1990) showed that endocytosis most likely occurs through coated pits, but labeling was also observed along uncoated parts of the cell membrane. Therefore, the presence of receptor was not restricted to coated pits, as observed by Kempka and Kolb-Bachofen (1988) with mannan–gold conjugates. Some of the discrepancies have been attributed by Stang et al. (1990) to nonspecific endocytosis of particulate matter, which occurs in Kupffer cells.

3. Internalization

Labeling of intracellular compartments by various gold-conjugated ligands is a commonly used method, especially in ultrastructural studies of receptor-mediated internalization. For instance, this approach has been used to study pulmonary surfactant (Kalina and Socher, 1990), transferrin and asialo-orosomucoid (Neutra et al., 1985), insulin (Smith et al., 1988), and ricin (van Deurs et al., 1986).

In some studies, only a small percentage of the organelles in a cell compartment is labeled and the density of gold labeling is low. As pointed out by Hammel and Kalina (1991), these results, which could reflect a true cellular phenomenon, could also be due to the probability of random sectioning of gold-labeled and unlabeled regions. Using a stereological approach for studying the internalization of gold-labeled native surfactant into lamellar bodies of cultured pulmonary type II cells, Hammel and Kalina (1991) demonstrated that the percentage of labeled organelles can indeed be explained as a result of the probability of random sectioning through the labeled areas.

4. Routing of Secretory Proteins

Exocrine, endocrine, and neuronal cells all show the ability to store and concentrate specific secretory products in membrane-limited granules whose release is regulated by external secretagogues. These secretory cells can also export other products, including plasma membrane components, that are not stored or externally regulated, but are transported rapidly to the cell surface in small vesicles. The terms "regulated" and "constitutive" have been applied to the former and latter pathways, respectively, referring to the two modes of release (reviewed by Burgess and Kelly, 1987). A few model cultured cell systems have been used to examine the mechanisms underlying entry of secretory proteins into these pathways (Burgess and Kelly, 1987).

Transgenic mice offer a promising alternative to *in vitro* systems. Trahair *et al.* (1989) have made much use of gold immunocytochemistry. They showed that transgenic mice containing a gastrointestinal tract-specific promoter linked to the gene encoding human growth hormone provide a valuable model system for examining the sorting and intracellular transport of secretory proteins in diverse, highly polarized, intestinal epithelial cells.

VI. Conclusion

The aim of this review has been to illustrate the scope and variety of applications of the colloidal gold method, including some of the pitfalls. The examples have been chosen to demonstrate (1) quantitative aspects in the preparation of gold conjugates, (2) preparative techniques of specimens where the method has been widely used, (3) various gold systems to map specific ligands, and (4) applications at all levels of microscopy, with a particular emphasis on dynamic events in cellular biology.

Due to the vast amount of literature published, the review has concentrated on the most recent findings. Even under this premise, the selection is somewhat arbitrary and may reflect the bias of the author since many other promising avenues are being actively explored in gold immunocytochemistry, including applications in pathological diagnoses, but could not be examined within the scope of this review.

Owing to the wide range of molecules that can form stable conjugates with colloidal gold, the diversity of probes is enormous and the number of applications constantly increasing at all level of microscopy. It is doubtful that particulate probes that are superior to colloidal gold will be developed in the near future. However, newer applications of colloidal gold will depend on a better understanding of the physicochemistry underlying the preparation of probes. It is imperative to obtain more information on the nature of the interaction of macromolecules with gold particles. Although colloidal probes have been successfully prepared with ligands of relatively high molecular weight, it has occasionally been reported that functional, specific probes cannot be obtained with low-molecular-weight ligands (Elliott and Dennison, 1990). New opportunities may occur when further information becomes available.

The trend has been to reduce the size of gold particles to increase penetration and limit steric hindrance. More studies should be made on the reduction of the size of macromolecules to be conjugated, as the precision of localization depends not only on the size of the gold particles but also on the size of the whole complex for high-resolution mapping. Conjointly, more studies should be directed at improving the sensitivity of direct (one-step) mapping techniques. Multistep procedures can only increase the size of the whole complex, with a consequent loss of mapping resolution.

Different approaches have been proposed to solve the problem of antigenic reduction due to masking by resins in sections (reviewed by Stirling, 1990). They involve removal of the embedding media or resins, immunolabeling, and, in some cases, reembedding. These techniques are still in their infancy and warrant further studies because they are still complex and tedious. In cryotechniques, attempts are being made to avoid chemical fixatives (aldehydes and osmium) at all stages.

With respect to instrumentation, the topographic imaging capabilities of the new generation of scanning electron microscopes, which possess both field-emission guns and low-aberration lenses will further promote the use of smaller gold probes. Indeed, spatial resolution may be limited only by the size of the immunogold complex and the technical aspects of specimen preparation (Erlandsen et al., 1990a,b). In the future, this approach may rival replica techniques.

The study of the reorganization of cell surface material on living mobile cells calls for an accurate dynamic visualization of specific components on the cell surface. Such studies can now be undertaken in light microscopy, since it is possible to view nanometer-sized particles. With differential interference contrast optics, the image of a single moving colloidal gold particle can be analyzed. In one of the most fascinating developments, the technique has permitted observation of transcription by single molecules of RNA polymerase (Schafer et al., 1991). This opens new possibilities for studying single molecules of processive enzymes.

Cryo-ultramicrotomy combined with immunogold labeling will continue to provide important data in fundamental cell biology. In addition, it will be used in developmental biology, because it allows the precise localization of particular proteins in developing embryos. Finally, according to Boonstra et al. (1991), it can be foreseen that the gold method will be used routinely as a screening test, for example, to detect transformed cells in tissue biopsies. The method still faces serious difficulties and needs to be much improved. For example, the method cannot be applied to the study of lung and fat tissues due to the fact that it is impossible at present to obtain ultrathin cryosections of these tissues.

Gold labeling, particularly immunogold labeling, offers a very powerful technique not only for mapping statically a variety of ligands but especially for following the route of transport and determining the ultimate location of molecules in cells. The precise analysis of intracellular processes such as receptor recycling and transcytosis depends greatly on reliable immunoelectron microscopy methods. In this respect, the gold method has provided information that is otherwise impossible to obtain by other approaches.

Although gold mapping seems to have evolved into a series of standardized procedures, there is no single method that can be recommended and applied in general. Each new situation must be examined for itself on the basis of past experience. Taken alone, this statement would already justify this review.

Acknowledgments

The author thanks Dr. A. J. Verkleij, A. ter Aves, and P. J. Rijken (University of Utrecht), Dr. J. Roth (University of Zürich), and Dr. J. M. Oliver (University of New Mexico) for providing photographic material from their own work. The author is also very thankful to Ms. M. F. Clerc for the art and editorial work and to Ms. C. Mordasini for typing the manuscript.

References

Ackerman, G. A., and Wolken, K. W. (1981). *J. Histochem. Cytochem.* **29**, 1137–1149.
Åkerström, B., Brodin, T., Reis, K., and Björck, L. (1985). *J. Immunol.* **135**, 2589–2592.
Albrecht, R. M., Goodman, S. L., and Simmons, S. R. (1989). *Am. J. Anat.* **185**, 149–164.
Andersson Forsman, C., and Pinto da Silva, P. (1988). *J. Cell Sci.* **90**, 531–541.
Armstrong, V. W., Harrach, B., Robenek, H., Helmhold, M., Walli, A. K., and Seidel, D. (1990). *J. Lipid Res.* **31**, 429–441.
Autrata, R. (1989). *Scanning Microsc.* **3**, 739–763.
Baigent, C. L., and Müller, G. (1990). *J. Microsc.* **158**, 73–80.
Baschong, W., and Wrigley, N. G. (1990). *J. Electron Microsc. Tech.* **14**, 313–323.
Bauer, H., Horisberger, M., Bush, D. A., and Sigarlakie, E. (1972). *Arch. Mikrobiol.* **85**, 202–208.
Beesley, J. E. (1988). *Scanning Microsc.* **2**, 1055–1068.
Behnke, O., Ammitzbøll, T., Jessen, H., Klokker, M., Nilausen, K., Tranum-Jensen, J., and Olsson, L. (1986). *Eur. J. Cell Biol.* **41**, 326–338.
Beier, K., Völkl, A., Hashimoto, T., and Fahimi, H. D. (1988). *Eur. J. Cell Biol.* **46**, 383–393.
Bendayan, M. (1981). *J. Histochem. Cytochem.* **29**, 531–541.
Bendayan, M. (1982). *J. Histochem. Cytochem.* **30**,81–85.
Bendayan, M. (1985). *In* "Techniques in Immunocytochemistry" (G. R. Bullock and P. Petrusz, eds.), Vol. 3, pp. 179–201. Academic Press, London.
Bendayan, M. (1987). *J. Electron Microsc. Tech.* **6**, 7–13.
Bendayan, M. (1989). *In* "Colloidal Gold: Principles, Methods and Applications" (M. A. Hayat, ed.), Vol. 1, pp. 33–94. Academic Press, San Diego.

Bendayan, M., and Garzon, S. (1988). *J. Histochem. Cytochem.* **36**, 597–607.
Bendayan, M., and Zollinger, M. (1983). *J. Histochem. Cytochem.* **31**, 101–109.
Bendayan, M., Nanci, A., and Kan, F. W. K. (1987). *J. Histochem. Cytochem.* **35**, 983–996.
Benhamou, N. (1989). *In* "Colloidal Gold: Principles, Methods and Applications" (M. A. Hayat, ed.), Vol. 1, pp. 95–143. Academic Press, San Diego.
Beppu, K. I. (1989). *J. Electron Microsc.* **38**, 430–440.
Bertolatus, J. A. (1990). *J. Histochem. Cytochem.* **38**, 377–384.
Bienz, K., Eggar, D., and Pasamontes, L. (1986). *J. Histochem. Cytochem.* **34**, 1337–1342.
Binder, M., Tourmente, S., Roth, J., Renaud, M., and Gehring, W. J. (1986). *J. Cell Biol.* **102**, 1646–1653.
Birrell, G. B., Habliston, D. L., Nadakavukaren, K. K., and Griffith, O. H. (1985). *Proc. Natl. Acad. Sci. U.S.A.* **82**, 109–113.
Birrell, G. B., Hedberg, K. K., and Griffith, O. H. (1987). *J. Histochem. Cytochem.* **35**, 843–853.
Bohn, W., Rutter, G., Hohenberg, H., Mannweiler, K., and Nobis, P. (1986). *Virology* **149**, 91–106.
Boonstra, J., van Maurick, P., Defize, L. K. H., de Laat, S. W., Leunissen, L. M., and Verkleij, A. J. (1985). *Eur. J. Cell Biol.* **36**, 209–216.
Boonstra, J., van Belzen, N., van Bergen en Henegouwen, P. M. P., Hage, W. J., van Maurik, P., Wiegant, F. A. C., and Verkleij, A. J. (1989). *In* "Immunogold Labeling in Cell Biology" (A. J. Verkleij and J. L. M. Leunissen, eds.), pp. 259–276. CRC Press, Boca Raton, Florida.
Boonstra, J., van Bergen en Henegouwen, P. M. P., van Belzen, N., Rijken, P. J., and Verkleij, A. J. (1991). *J. Microsc. (Oxford)* **161**, 135–147.
Brada, D., and Roth, J. (1984). *Anal. Biochem.* **142**, 79–83.
Brown, M. S., and Goldstein, J. L. (1985). *Curr. Top. Cell. Regul.* **26**, 3–15.
Burgess, T. L., and Kelly, R. B. (1987). *Annu. Rev. Cell Biol.* **3**, 243–295.
Burt, A. D., Griffiths, M. R., Schuppan, D., Voss, B., and Macsween, R. N. M. (1990). *Histopathology* **16**, 53–58.
Cabib, E., Roberts, R., and Bowers, B. (1982). *Annu. Rev. Biochem.* **51**, 763–793.
Carlemalm, E., Garavito, R. M., and Villiger, W. (1982). *J. Microsc. (Oxford)* **126**, 123–143.
Chan, J., Aoki, C., and Pickel, V. M. (1990). *J. Neurosci. Methods* **33**, 113–127.
Chardin, H., Londono, I., and Goldberg, M. (1990). *Histochem. J.* **22**, 588–594.
Childs, G. N., Yamauchi, K., and Unabia, G. (1989). *Am. J. Anat.* **185**, 223–235.
Clerc, M. F., Granato, D. A., and Horisberger, M. (1988). *Histochemistry* **89**, 343–349.
Cornelese-ten Velde, I., and Prins, F. A. (1990). *Histochemistry* **94**, 61–71.
Coulombe, P. A., Kan, F. W. K., and Bendayan, M. (1988). *Eur. J. Cell Biol.* **46**, 564–576.
Craig, S., Moore, P. J., and Dunahay, T. G. (1987). *Scanning Microsc.* **1**, 1431–1437.
Danscher, G. (1981). *Histochemistry* **71**, 81–88.
De Brabander, M., Nuydens, R., Ishihara, A., Holifield, B., Jacobson, K., and Geerts, H. (1991). *J. Cell Biol.* **112**, 111–124.
Debray, H., Decout, D., Strecker, G., Spik, G., and Montreuil, J. (1981). *Eur. J. Biochem.* **117**, 41–55.
De Harven, E., Leung, R., and Christensen, H. (1984). *J. Cell Biol.* **99**, 53–57.
De Harven, E., Soligo, D., and Christensen, H. (1990). *Histochem. J.* **22**, 18–23.
De Leeuw, A. M., Praaning-van Dalen, D. P., Brouwer, A., and Knook, D. L. (1989). *In* "Cells of the Hepatic Sinusoid" (E. Wisse, D. L. Knook, and K. Decker, eds.), Vol. 2, pp. 94–98. Kupffer Cell Found., Rijswijk, The Netherlands.
De Mey, J. (1983). *In* "Immunocytochemistry: Practical Applications in Pathology and Biology" (J. M. Polak and S. van Noorden, eds.), pp. 82–112. J. G. Wright, Bristol.
De Mey, J., Moeremans, M., De Waele, M., Geuens, G., and De Brabander, M. (1981a). *Protides Biol. Fluids* **29**, 943–947.
De Mey, J., Moeremans, M., Geuens, G., Nuydens, R., and De Brabander, M. (1981b). *Cell Biol. Int. Rep.* **5**, 889–899.

De Valck, V., Renmans, W., Segers, E., Leunissen, J., and De Waele, M. (1991). *Histochemistry* **95**, 483–490.

De Waele, M., De Mey, J., Moeremans, M., De Brabander, M., and van Camp, B. (1983). *J. Histochem. Cytochem.* **31**, 376–381.

De Waele, M., De Mey, J., Renmans, W., Labeur, C., Reynaert, P., and van Camp, B. (1986). *J. Microsc. (Oxford)* **143**, 151–160.

De Waele, M., Renmans, W., Segers, E., Jochmans, K., and van Camp, B. (1988). *J. Histochem. Cytochem.* **36**, 679–683.

Eaton, B. T., Hyatt, A. D., and White, J. R. (1987). *Virology* **157**, 107–116.

Elliott, E., and Dennison, C. (1990). *Anal. Biochem.* **186**, 53–59.

Erlandsen, S. L., Bemrick, W. J., Schupp, D. E., Shields, J. M., Jarroll, E. L., Sauch, J. F., and Pawley, J. B. (1990a). *J. Histochem. Cytochem.* **38**, 625–632.

Erlandsen, S. L., Frethem, C., and Autrata, R. (1990b). *J. Histochem. Cytochem.* **38**, 1779–1780.

Faraday, M. (1857). *Philos. Trans. R. Soc. London* **147**, 145–181.

Faulk, W. P., and Taylor, G. M. (1971). *Immunochemistry* **8**, 1081–1083.

Ferrier, L. K., Richardson, T., and Olson, N. F. (1972). *Enzymologia* **42**, 273–283.

Frens, G. (1973). *Nature (London), Phys. Sci.* **241**, 20–22.

Frisch, E. B., and Phillips, T. E. (1990). *J. Electron Microsc. Tech.* **16**, 25–36.

Geerts, H., de Brabander, M., Nuydens, R., Geuens, S., Moeremans, J., De Mey, J., and Hollenbeck, P. (1987). *Biophys. J.* **52**, 775–782.

Geiger, B., Dutton, A. H., Tokuyasu, K. T., and Singer, S. J. (1981). *J. Cell Biol.* **91**, 614–628.

Geoghegan, W. D., and Ackerman, G. A. (1977). *J. Histochem. Cytochem.* **25**, 1187–1200.

Geoghegan, W. D., Scillian, J. J., and Ackerman, G. A. (1978). *Immunol. Commun.* **7**, 1–12.

Geuze, H. J., Slot, J. W., van der Ley, P. A., Scheffer, R. C. T., and Griffith, J. M. (1981). *J. Cell Biol.* **89**, 653–665.

Geuze, H. J., Slot, J. W., Strons, G. J. A. M., Lodish, H. F., and Schwartz, A. L. (1983). *Cell (Cambridge, Mass.)* **32**, 277–287.

Geuze, H. J., Slot, J. W., Strons, G. J. A. M., Peppard, J., von Figura, K., Hasilik, A., and Schwartz, A. L. (1984). *Cell (Cambridge, Mass.)* **37**, 195–204.

Ghitescu, L., and Bendayan, M. (1990). *J. Histochem. Cytochem.* **38**, 1523–1530.

Gonatas, J., Stieber, A., Olsnes, S., and Gonatas, N. K. (1980). *J. Cell Biol.* **87**, 579–588.

Goodrich, L. D., Rixon, F. J., and Parris, D. S. (1989). *J. Virol.* **63**, 137–147.

Griffiths, G., and Hoppeler, H. (1986). *J. Histochem. Cytochem.* **34**, 1389–1398.

Griffiths, G., Brands, R., Burke, B., Louvard, D., and Warren, G. (1982). *J. Cell Biol.* **95**, 781–792.

Griffiths, G., Hoflack, B., Simons, K., Mellman, I., and Kornfeld, S. (1988). *Cell (Cambridge, Mass.)* **52**, 329–341.

Grossi de Sa, M. F., Martins de Sa, C., Harper, F., Coux, O., Akhayat, O., Gounon, P., Pal, J. K., Florentin, Y., and Scherrer, K. (1988a). *J. Cell Sci.* **89**, 151–165.

Grossi de Sa, M. F., Martins de Sa, C., Harper, F., Olink-Coux, M., Huesca, M., and Scherrer, K. (1988b). *J. Cell Biol.* **107**, 1517–1530.

Gu, J., De Mey, J., Moeremans, M., and Polyk, J. (1981). *Regul. Pept.* **1**, 365–374.

Hacker, G. W. (1989). *In* "Colloidal Gold: Principles, Methods and Applications" (M. A. Hayat, ed.), Vol. 1, pp. 297–321. Academic Press, San Diego.

Hammel, I., and Kalina, M. (1991). *J. Histochem. Cytochem.* **39**, 131–133.

Handley, D. A. (1989). *In* "Colloidal Gold: Principles, Methods and Applications" (M. A. Hayat, ed.), Vol. 1, pp. 13–32. Academic Press, San Diego.

Handley, D. A., Chien, S. (1987). *Eur. J. Cell Biol.* **43**, 163–174.

Handley, D. A., Arbeeny, C. M., Eder, H. A., and Chien, S. (1981a). *J. Cell Biol.* **90**, 778–787.

Handley, D. A., Arbeeny, C. M., Witte, L. D., and Chien, S. (1981b). *Proc. Natl. Acad. Sci. U.S.A.* **78**, 368–371.

Handley, D. A., Arbeeny, C. M., and Chien, S. (1983). *Eur. J. Cell Biol.* **30**, 266–271.
Hartwig, J. H., and De Sisto, M. (1991). *J. Cell Biol.* **112**, 407–425.
Hemming, F. J., Mesguich, P., Morel, G., and Dubois, P. M. (1983). *J. Microsc. (Oxford)* **131**, 25–34.
Hesz, A., Robenek, H., Ingolie, E., Roscher, A., Kremplev, F., Sandhofer, F., and Kostner, G. M. (1985). *Eur. J. Cell Biol.* **37**, 229–233.
Hicks, D., and Molday, R. S. (1986). *Exp. Eye Res.* **42**, 55–71.
Hodges, G. M., Southgate, J., and Toulson, E. C. (1987). *Scanning Microsc.* **1**, 301–318.
Hohenberg, H. (1989). *In* "Immunogold Labeling in Cell Biology" (A. J. Verkleij and J. L. M. Leunissen, eds.), pp. 157–177. CRC Press, Boca Raton, Florida.
Holgate, C. S., Jackson, P., Cowen, P. N., and Bird, C. C. (1983). *J. Histochem. Cytochem.* **31**, 938–944.
Holm, R., Nesland, J. M., Attramadal, A., and Johannessen, J. V. (1988). *Ultrastruct. Pathol.* **12**, 279–290.
Holowka, D., Conrad, D. H., and Baird, B. (1985). *Biochemistry* **24**, 6260–6267.
Holwerda, B. C., Galvin, N. J., Baranski, T. J., and Rogers, J. C. (1990). *Plant Cell* **2**, 1091–1106.
Horisberger, M. (1978). *Experientia* **34**, 721–722.
Horisberger, M. (1979). *Biol. Cell.* **36**, 253–258.
Horisberger, M. (1981a). *Scanning Electron Microsc.* **2**, 9–31.
Horisberger, M. (1981b). *Gold Bull.* **14**, 90–94.
Horisberger, M. (1985). *In* "Techniques in Immunocytochemistry" (G. R. Bullock and P. Petrusz, eds.), Vol. 3, pp. 155–178. Academic Press, London.
Horisberger, M. (1989a). *In* "Immunogold Labeling in Cell Biology" (A. J. Verkleij and J. L. M. Leunissen, eds.), pp. 49–60. CRC Press, Boca Raton, Florida.
Horisberger, M. (1989b). *In* "Colloidal Gold: Principles, Methods and Applications (M. A. Hayat, ed.), Vol. 1, pp. 217–227. Academic Press, San Diego.
Horisberger, M. (1990). *Prog. Colloid Polym. Sci.* **81**, 156–160.
Horisberger, M., and Clerc, M. F. (1985). *Histochemistry* **82**, 219–223.
Horisberger, M., and Clerc, M. F. (1987). *Eur. J. Cell Biol.* **45**, 62–71.
Horisberger, M., and Clerc, M. F. (1988a). *Eur. J. Cell Biol.* **46**, 444–452.
Horisberger, M., and Clerc, M. F. (1988b). *Histochemistry* **90**, 165–175.
Horisberger, M., and Rosset, J. (1977a). *J. Histochem. Cytochem.* **25**, 295–305.
Horisberger, M., and Rosset, J. (1977b). *Scanning Electron Microsc.* **2**, 75–82.
Horisberger, M., and Rouvet-Vauthey, M. (1985). *Experientia* **41**, 748–750.
Horisberger, M., and Tacchini-Vonlanthen, M. (1983a). *In* "Lectins" (T. C. Bøg-Hansen and G. A. Spengler, eds.), Vol. 3, pp. 189–197. de Gruyter, Berlin and New York.
Horisberger, M., and Tacchini-Vonlanthen, M. (1983b). *Histochemistry* **77**, 313–321.
Horisberger, M., and Vauthey, M. (1984) *Histochemistry* **82**, 13–18.
Horisberger, M., and Vonlanthen, M. (1977). *Arch. Microbiol.* **115**, 1–7.
Horisberger, M., and Vonlanthen, M. (1978). *J. Histochem. Cytochem.* **26**, 960–966.
Horisberger, M., and Vonlanthen, M. (1979a). *J. Microsc. (Oxford)* **115**, 97–102.
Horisberger, M., and Vonlanthen, M. (1979b). *Histochemistry* **64**, 115–118.
Horisberger, M., and Vonlanthen, M. (1980). *Histochemistry* **65**, 181–186.
Horisberger, M., Rosset, J., and Bauer, H. (1975). *Experientia* **31**, 1147–1149.
Horisberger, M., Rosset, J., and Vonlanthen, M. (1978). *Experientia* **34**, 274–276.
Horisberger, M., Rouvet-Vauthey, M., Richli, U., and Farr, D. R. (1985). *Eur. J. Cell Biol.* **37**, 70–77.
Horisberger, M., Clerc, M. F., and Pahud, J. J. (1986). *Histochemistry* **85**, 291–294.
Hoyer, L. C., Lee, J. C., and Bucana, C. (1979). *Scanning Electron Microsc.* **3**, 629–636.
Hsu, Y. H. (1984). *Anal. Biochem.* **142**, 221–225.
Humbel, B. M., and Schwarz, H. (1989). *In* "Immunogold Labeling in Cell Biology" (A. J. Verkleij and J. L. M. Leunissen, eds.), pp. 115–134. CRC Press, Boca Raton, Florida.
Hunt, L. B. (1981). *Endeavour* **5**, 61–67.

Hutchinson, N. J., Langer-Safer, P. R., Ward, D. C., and Hamkalo, B. A. (1982). *J. Cell Biol.* **95,** 609–618.

Jackson, P., Lewis, F. A., and Wells, M. (1989). *Histochem. J.* **21,** 425–428.

Jackson, P., Dockey, D. A., Lewis, F. A., and Wells, M. (1990). *J. Clin. Pathol.* **43,** 810–812.

Jaekle, S., Brady, S. E., and Havel, R. J. (1989). *Proc. Natl. Acad. Sci. U.S.A.* **86,** 1880–1884.

Johnson, A. B., and Bettica, A. (1989). *Am. J. Anat.* **185,** 335–341.

Kalina, M., and Socher, R. (1990). *J. Histochem. Cytochem.* **38,** 483–492.

Kasamatsu, H., Lin, W., Edens, J., and Revel, J. P. (1983). *Proc. Natl. Acad. Sci. U.S.A.* **80,** 4339–4343.

Katsuma, Y., Marceau, N., Ohta, M., and French, S. W. (1988). *Hepatology (Baltimore)* **8,** 559–568.

Katsumoto, T., Asanaka, M., Kageyama, S., Kurimura, T., Nakajima, K., Nofo, A., Tanaka, H., and Sato, R. (1990). *J. Electron Microsc.* **39,** 33–38.

Kausche, G. A., and Ruska, H. (1939). *Kolloid-Z.* **89,** 21–26.

Kausche, G. A., Pfankuch, E., and Ruska, H. (1939). *Naturwissenchaften* **27,** 292–299.

Kempka, G., and Kolb-Bachofen, V. (1988). *Exp. Cell Res.* **176,** 38–48.

Kjeldsberg, E. (1989). *In* "Colloidal Gold: Principles, Methods and Applications" (M. A. Hayat, ed.), Vol. 1, pp. 433–449. Academic Press, San Diego.

Klausner, R. D., Van Renswonde, J., Ashwell, G., Kempf, C., Schechter, A. N., Dean, A., and Bridges, K. R. (1983). *J. Biol. Chem.* **258,** 4715–4724.

Konkel, M. E., Babakhani, F., and Joens, L. A. (1990). *J. Infect. Dis.* **162,** 888–895.

Kramarcy, N. R., and Sealock, R. (1991). *J. Histochem. Cytochem.* **39,** 37–39.

Krenács, T., Lászik, Z., and Dobó, E. (1989). *Acta Histochem.* **85,** 79–85.

Krenács, T., Krenács, L., Bozóky, B., and Iványi, B. (1990). *Histochem. J.* **22,** 530–536.

Krishnan, H. B., White, J. A., and Pueppke, S. G. (1988). *Protoplasma* **144,** 25–33.

Krishnan, H. B., White, J. A., and Pueppke, S. G. (1990). *Cereal Chem.* **67,** 360–366.

Krishnan, H. B., White, J. A., and Pueppke, S. G. (1991). *Cereal Chem.* **68,** 108–111.

Langanger, G., and De Mey, J. (1989). *In* "Immunogold Labeling in Cell Biology" (A. J. Verkleij and J. L. M. Leunissen, eds.), pp. 335–351. CRC Press, Boca Raton, Florida.

Langone, J. J. (1982). *Adv. Immunol.* **32,** 157–252.

Leunissen, J. L. M., and De Mey, J. R. (1989). *In* "Immunogold Labeling in Cell Biology" (A. J. Verkleij and J. L. M. Leunissen, eds.), pp. 3–16. CRC Press, Boca Raton, Florida.

Leunissen, J. L. M., and Verkleij, A. J. (1989). *In* "Immunogold Labeling in Cell Biology" (A. J. Verkleij and J. L. M. Leunissen, eds.), pp. 95–114. CRC Press, Boca Raton, Florida.

Lin, W. L. (1990). *Brain Res. Bull.* **24,** 533–536.

Lucocq, J. M., and Roth, J. (1985). *In* "Techniques in Immunocytochemistry" (G. R. Bullock and P. Petrusz, eds.), Vol. 3, pp. 203–236. Academic Press, London.

Magnusson, S., and Berg, T. (1989). *Biochem. J.* **257,** 651–656.

Mahdihassan, S. (1984). *Am. J. Chin. Med.* **12,** 32–42.

Mannweiler, K., Bohn, W., and Rutter, G. (1989). *In* "Immunogold Labeling in Cell Biology" (A. J. Verkleij and J. L. M. Leunissen, eds.), pp. 317–334. CRC Press, Boca Raton, Florida.

Marshall, S. (1985). *J. Biol. Chem.* **260,** 13517–13523.

McIntosh, D., Timar, J., and Davies, A. J. S. (1990). *Eur. J. Cell Biol.* **52,** 77–86.

McVey Ward, D., Ajioka, B., and Kaplan, J. (1989). *J. Biol. Chem.* **264,** 8164–8170.

Mendoza, G. R., and Metzger, H. (1976), *J. Immunol.* **117,** 1573–1578.

Miller, K., Beardmore, J., Kanety, H., Schlessinger, J., and Hopkins, C. R. (1986). *J. Cell Biol.* **102,** 500–509.

Moeremans, M., Daneels, G., van Dijck, A., Langanger, G., and De Mey, J. (1984). *J. Immunol. Methods* **74,** 353–360.

Moeremans, M., Daneels, G., De Reymaeker, M., De Wever, E., and De Mey, J. (1989). *In* "Immunogold Labeling in Cell Biology" (A. J. Verkleij and J. L. M. Leunissen, eds.), pp. 17–27. CRC Press, Boca Raton, Florida.

Molday, L. L., Cook, M. L., Kaupp, U. B., and Molday, R. S. (1990). *J. Biol. Chem.* **265**, 18690–18695.

Molday, R. S., and Laird, D. W. (1989). *In* "Immunogold Labeling in Cell Biology" (A. J. Verkleij and J. L. M. Leunissen, eds.), pp. 29–48. CRC Press, Boca Raton, Florida.

Mommaas-Kienhuis, A. M., and Vermeer, B. J. (1989). *In* "Immunogold Labeling in Cell Biology" (A. J. Verkleij and J. L. M. Leunissen, eds.), pp. 247–257. CRC Press, Boca Raton, Florida.

Monaghan, P., and Robertson, D. (1990). *J. Microsc. (Oxford)* **158**, 355–363.

Mounier, J., Ryter, A., Coquis-Rondon, M., and Sansonetti, P. J. (1990). *Infect. Immun.* **58**, 1048–1058.

Mühlpfordt, H. (1982). *Experentia* **38**, 1127–1128.

Müller, M., and Hermann, R. (1990). *In* "Electron Microscopy" (L. D. Peachy and D. B. Williams, eds.), Vol. 3, pp. 4–7. San Francisco Press, San Francisco.

Müller, W. H., van der Krift, T. P., Krouver, A. J. J., Wösten, H. A. B., van der Voort, L. H. M., Smaal, E. B., and Verkleij, A. J. (1991). *EMBO J.* **10**, 489.

Nagatani, T., and Saito, S. (1986). *In* "Electron Microscopy" (T. Imura, S. Maruse, and T. Suzuki, eds.), Vol. 3, pp. 2101–2104. Kyuwa Book, Tokyo.

Neutra, M. R., Clechanover, A., Owen, L. S., and Lodish, H. F. (1985). *J. Histochem. Cytochem.* **33**, 1134–1144.

Nickerson, J. A., Krockmalnic, G., He, D., and Penman, S. (1990). *Proc. Natl. Acad. Sci. U.S.A.* **87**, 2259–2263.

Nicolson, G. L., Lacorbière, M., and Hunter, T. R. (1975). *Cancer Res.* **35**, 144–155.

Nunamaker, R. A., Sieburth, P. J., Dean, V. C., Wigington, J. G., Nunamaker, C. E., and Mecham, J. O. (1990). *Comp. Biochem. Physiol. A* **96A**, 19–31.

Onetti-Muda, A., Crescenzi, A., Faraggiana, T., and Marinozzi, V. (1990). *Basic Appl. Histochem.* **34**, 189–197.

Pastan, I., and Willingham, M. C. (1985). *In* "Endocytosis" (I. Pastan and M. C. Willingham, eds.), pp. 1–44. Plenum, New York.

Patterson, S., Gross, J., and Oxford, J. S. (1988). *J. Gen. Virol.* **69**, 1859–1872.

Pauli, W. (1949). *Helv. Chim. Acta* **32**, 795–810.

Pavan, A., Mancini, P., Cirone, M., Frati, L., Torrisi, M. R., and Pinto da Silva, P. (1989a). *J. Histochem. Cytochem.* **37**, 1489–1496.

Pavan, A., Mancini, P., Frati, L., Torrisi, M. R., and Pinto da Silva, P. (1989b). *Biochim. Biophys. Acta* **978**, 158–168.

Pavan, A., Mancini, P., Lucania, G., Frati, L., Torrisi, M. R., and Pinto da Silva, P. (1990). *J. Cell Sci.* **96**, 151–157.

Pawley, J., and Albrecht, R. (1988). *Scanning* **10**, 184–189.

Pettitt, J. M., and Humphris, D. C. (1991). *Histochem. J.* **23**, 29–37.

Pickel, V. M., Beckley, S. C., Joh, T. H., and Reis, D. J. (1981). *Brain Res.* **225**, 373–385.

Pimenta, P. F. P., da Silva, R. P., Sacks, D. L., and Pinto da Silva, P. (1989). *Eur. J. Cell Biol.* **48**, 180–190.

Pinto da Silva, P. (1989). *In* "Immunogold Labeling in Cell Biology" (A. J. Verkleij and J. L. M. Leunissen, eds.), pp. 179–197. CRC Press, Boca Raton, Florida.

Pinto da Silva, P., and Kan, F. W. K. (1984). *J. Cell Biol.* **99**, 1156–1161.

Poulain, D., Hopwood, V., and Vernes, A. (1985). *CRC Crit. Rev. Microbiol.* **12**, 223–270.

Rappaport, L., and Samuel, J. L. (1988). *Int. Rev. Cytol.* **113**, 101–143.

Rasmussen, U., Munck, L., and Ullrich, S. E. (1990). *Planta* **180**, 272–277.

Rastogi, N., and Hellio, R. (1990). *FEMS Microbiol. Lett.* **70**, 161–166.

Reddy, J. T., and Locke, M. (1990). *J. Insect. Physiol.* **36**, 397–407.

Reggio, H., Webster, P., and Louvard, D. (1983). *In* "Methods in Enzymology" (S. Fleischer and B. Fleischer, eds.), Vol. 98, pp. 379–395, Academic Press, New York.

Robenek, H., and Hesz, A. (1983). *Eur. J. Cell Biol.* **31**, 275–282.

Robenek, H., Rassat, J., Hesz, A., and Grunwald, J. (1982). *Eur. J. Cell Biol.* **27**, 242–250.

Romano, E. L., and Romano, M. (1977). *Immunochemistry* **14**, 711–715.

Roth, J. (1982). *In* "Techniques in Immunocytochemistry" (G. R. Bullock and P. Petrusz, eds.), Vol. 1, pp. 107–132. Academic Press, London.

Roth, J. (1983a). *In* "Techniques in Immunocytochemistry" (G. R. Bullock and P. Petrusz, eds.), Vol. 2, pp. 215–284. Academic Press, London.

Roth, J. (1983b). *J. Histochem. Cytochem.* **31**, 547–552.

Roth, J. (1986). *J. Microsc. (Oxford)* **143**, 125–137.

Roth, J. (1987). *Scanning Microsc.* **1**, 695–704.

Roth, J., and Wagner, M. (1977). *J. Histochem. Cytochem.* **25**, 1181–1184.

Roth, J., Bendayan, M., and Orci, L. (1978). *J. Histochem. Cytochem.* **26**, 1074–1081.

Roth, J., Bendayan, M., and Orci, L. (1980). *J. Histochem. Cytochem.* **28**, 55–57.

Roth, J., Taatjes, D. J., and Warhol, M. J. (1989). *Histochemistry* **92**, 47–56.

Rutter, G., Bohn, W., Hohenberg, H., and Mannweiler, K. (1988). *J. Histochem. Cytochem.* **36**, 1015–1021.

Saetersdal, T., Greve, G., and Dalen, H. (1990). *Histochemistry* **95**, 1–10.

Sata, T., Lackie, P. M., Taatjes, D. J., Peumans, W., and Roth, J. (1989). *J. Histochem. Cytochem.* **37**, 1577–1588.

Schafer, D. A., Gelles, J., Sheetz, M. P., and Laudick, R. (1991). *Nature (London)* **352**, 444–448.

Schaffer, K., and Wisniewski, M. (1990). *J. Electron Microsc. Techn.* **15**, 318–319.

Scherrer, K. (1990). *Mol. Biol. Rep.* **14**, 1–9.

Schwartz, A. L., Fridovich, S. E., and Lodish, H. F. (1982). *J. Biol. Chem.* **257**, 4230–4237.

Scopsi, L. (1989). *In* "Colloidal Gold: Principles, Methods and Applications" (M. A. Hayat, ed.), Vol. 1, pp. 251–295. Academic Press, San Diego.

Sheetz, M. P., Turney, S., Qian, H., and Elson, E. L. (1989). *Nature (London)* **340**, 284–288.

Shepherd, M. G. (1987). *CRC Crit. Rev. Microbiol.* **15**, 7–25.

Shida, H., and Ohga, R. (1990). *J. Histochem. Cytochem.* **38**, 1687–1691.

Skutelsky, E., and Roth, J. (1986). *J. Histochem. Cytochem.* **34**, 693–696.

Slot, J. W., and Geuze, H. J. (1984). *In* "Immunolabeling for Electron Microscopy" (J. M. Polak and I. M. Varndell, eds.), pp. 129–142. Elsevier, Amsterdam.

Slot, J. W., and Geuze, H. J. (1985). *Eur. J. Cell Biol.* **38**, 87–93.

Slot, J. W., Posthuma, G., Chang, L. Y., Crapo, J. D., and Geuze, H. J. (1989). *In* "Immunogold Labeling in Cell Biology" (A. J. Verkleij and J. L. M. Leunissen, eds.), pp. 135–155. CRC Press, Boca Raton, Florida.

Smith, A., and Hunt, R. C. (1990). *Eur. J. Cell Biol.* **53**, 234–245.

Smith, R. M., Goldberg, R. I., and Jarett, L. (1988). *J. Histochem. Cytochem.* **36**, 359–365.

Springall, D. R., Hacker, G. W., Grimelius, L., and Polak, J. M. (1984). *Histochemistry* **81**, 603–608.

Stang, E., Kindberg, G. M., Berg, T., and Roos, N. (1990). *Eur. J. Cell Biol.* **52**, 67–76.

Stirling, J. W. (1990). *J. Histochem. Cytochem.* **38**, 145–157.

Stransky, G., and Gay, S. (1991). *J. Histochem. Cytochem.* **39**, 185–191.

Stump, R. F., Pfeiffer, J. R., Schneebeck, M. C., Seagrave, J. C., and Oliver, J. M. (1989). *Am. J. Anat.* **185**, 128–141.

Taatjes, D. J., and Roth, J. C. (1990). *Eur. J. Cell Biol.* **53**, 255–266.

Taatjes, D. J., Schaub, U., and Roth, J. (1987a). *Histochem. J.* **19**, 235–245.

Taatjes, D. J., Chen, T. H., Åkerström, B., Björck, L., Carlemalm, E., and Roth, J. (1987b). *Eur. J. Cell Biol.* **45**, 151–159.

Tanaka, H., Haga, S., Takatsuki, K., and Yamaguchi, K. (1984). *Cancer Res.* **441**, 3493–3504.

Teradaira, R., Kolb-Bachofen, V., Schlepper-Schäfer, J., and Kolb, H. (1983). *Biochim. Biophys. Acta* **759**, 306–310.

Tokuyasu, K. T. (1973). *J. Cell Biol.* **57**, 551–565.

Tokuyasu, K. T. (1986), *J. Microsc. (Oxford)* **143**, 139–149.

Torosantucci, A., Boccanera, M., Casalinuovo, I., Pellegrini, G., and Cassone, A. (1990). *J. Gen. Microbiol.* **136,** 1421–1428.

Torrisi, M. R., Pavan, A., Lotti, L. V., Migliaccio, G., Pascale, M. C., Covelli, E., Leone, A., and Bonatti, S. (1990). *J. Histochem. Cytochem.* **38,** 1421–1426.

Trahair, J. F., Neutra, M. R., and Gordon, J. I. (1989). *J. Cell Biol.* **109,** 3231–3242.

Trischitta, V., Wong, K. Y., Brunetti, A., Scalisi, R., Vigneri, R., and Goldfine, I. D. (1989). *J. Biol. Chem.* **264,** 5041–5046.

Usuda, N., Ma, H., Hanai, T., Yokota, S., Hashimoto, T., and Nagata, T. (1990). *J. Histochem. Cytochem.* **38,** 617–623.

van Bergen en Henegouwen, P. M. P., and Leunissen, J. L. M. (1986). *Histochemistry* **85,** 81–87.

van den Brink, W. J., Zijlmans, H. J., Kok, L. P., Bolhuis, P., Volkers, H. H., Boon, M. E., and Houthoff, H. J. (1990). *Histochem. J.* **22,** 327–334.

van den Pol, A. N. (1985). *Science* **228,** 332–335.

van de Plas, P. F. E. M., and Leunissen, J. L. M. (1989). AuroFile No. 2. Janssen Life Sciences Products.

van Deurs, B., Tønnessen, T. I., Petersen, O. W., Sandvig, K., and Olsnes, S. (1986). *J. Cell Biol.* **102,** 37–47.

van Deurs, B., Sandvig, K., Petersen, O. W., Olsnes, S., Simons, K., and Griffiths, G. (1988). *J. Cell Biol.* **106,** 253–267.

Van Putten, J. P. M., Hopman, C. T. P., and Weel, J. F. L. (1990). *J. Med. Microbiol.* **33,** 35–41.

Vázquez-Nin, G. H., Echeverría, O. M., Fakan, S., Traish, A. M., Wotiz, H. H., and Martin, T. E. (1991). *Exp. Cell Res.* **192,** 396–404.

Vernooy-Gerritsen, M., Leunissen, J. L. M., Veldink, G. A., and Vliegenthart, J. F. G. (1984). *Plant Physiol.* **76,** 1070–1079.

Vian, B., Nairn, J., and Reid, J. S. G. (1991). *Histochem. J.* **23,** 116–124.

Volkers, H. H., Van den Brink, W. J., Bolhuis, P., Leunissen, J., and Houthoff, H. J. (1988). *J. Histochem. Cytochem.* **36,** 888.

Vorbrodt, A. W. (1987). *J. Histochem. Cytochem.* **35,** 1261–1266.

Weiser, H. B. (1933). *Inorg. Colloid Chem.* **1,** 21–57.

White, J. G., and Escolar, G. (1990). *Arteriosclerosis* **10,** 738–744.

Willingham, M. C. (1989). *Am. J. Anat.* **185,** 109–127.

Wolber, R. A., Beals, T. F., Lioyd, R. V., and Haassab, H. F. (1988). *Lab. Invest.* **59,** 144–151.

Woods, J. W., Goodhouse, J., and Farquhar, M. G. (1989). *Eur. J. Cell Biol.* **50,** 132–143.

Zelechowska, M. G., and Mandeville, R. (1989). *Anticancer Res.* **9,** 53–58.

zur Nieden, U., Neumann, D., Manteuffel, R., and Weber, E. (1982). *Eur. J. Cell Biol.* **26,** 228–233.

Tracheary Element Formation as a Model System of Cell Differentiation

Hiroo Fukuda

Biological Institute, Faculty of Science, Tohoku University, Aramaki, Aoba-ku, 980 Sendai, Japan

I. Introduction

Differentiation in higher plants is plastic. Differentiated cells can redifferentiate *in vitro* to other types of cells and eventually form tissues, organs, and even whole plants. Tracheary element (TE) differentiation is an excellent example of re-differentiation occurring at the cellular level in higher plants (Torrey *et al.*, 1971; Roberts, 1976; Shininger, 1979; Barnett, 1979, 1981; Phillips, 1980; Fukuda and Komamine, 1985; Aloni, 1987; Roberts *et al.*, 1988; Fukuda, 1989b; Sugiyama and Komamine, 1990). The TEs are cells that have localized thickenings of secondary cell wall with a reticulate, spiral, annular, or pitted pattern, which is their most characteristic morphological feature. TEs are the distinctive constituents of xylem and are derived *in situ* from cells of the procambium of roots and shoots in primary xylem and from cells produced by the vascular cambium in secondary xylem. In *in vitro* culture, the differentiation of parenchyma cells into TEs can be induced with relative ease by phytohormones. This easy induction of differentiation, and detection of differentiated cells by their morphological features, offer a great advantage for the study of cell differentiation. In addition, the establishment of a system in which single cells differentiate directly into TEs without intervening cell division (Fig. 1) (Fukuda and Komamine, 1980a) and recent detailed studies on biochemical markers of the differentiation (Bolwell and Northcote, 1981; Fukuda and Komamine, 1982, 1983; Dalessandro *et al.*, 1986; Fukuda, 1987; Kobayashi *et al.*, 1987; Cassab and Varner, 1987; Church and Galston, 1988c; Thelen and Northcote, 1989; Keller *et al.*, 1989a; Roberts and Haigler, 1989) have made TE differentiation the best model system for studying cell differentiation in higher plants. This article reviews recent progress on TE differentiation, with particular emphasis on the control mechanism of differentiation at cellular and molecular levels.

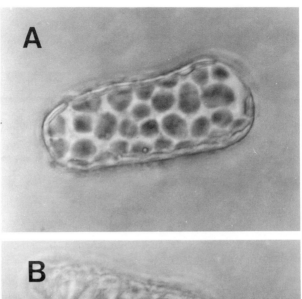

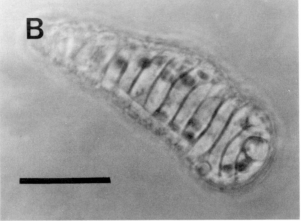

FIG. 1   (A) A single isolated *Zinnia* mesophyll cell. (B) A tracheary element (TE) formed at 58 hr of culture without intervening cell division. Bar, 25 μm. From Fukuda and Kobayashi (1989).

II. Experimental Systems

A. *In Vivo* Conditions

The primary xylem is formed from tissues derived from the apical meristems of the root and shoot, called procambium (Torrey *et al.*, 1971). Linear rows of TEs, which are differentiated continuously from cells produced by the procambium of both root and shoot, form a continuous axial conducting system. The secondary xylem is derived from cells produced by the vascular cambium, which is a lateral meristem found in gymnosperms and dicotyledons. There are numerous reports on TE formation *in situ* (Torrey *et al.*, 1971; Roberts, 1976; Shininger, 1979; Barnett, 1979, 1981; Aloni, 1987; Roberts *et al.*, 1988). Among them, roots of the water fern *Azolla* provide a combination of advantages for a cytological study of TE differentiation *in situ* (Gunning *et al.*, 1978a–c): The roots are small enough to make comprehensive examination by electron microscopy feasible, and in the root, every file of cells can be identified, that is, the cell lineage is clear (Fig. 2). Gunning and colleagues (Hardham and Gunning, 1978, 1979, 1980; Gunning *et al.*, 1978a–c) have demonstrated the patterns of cell differentiation and the organization of microtubules in differentiating cells in *Azolla*.

Xylem is formed by the redifferentiation of parenchyma cells as well as by programmed differentiation of procambium or cambium. This occurs where lateral roots interconnect with the vascular system of the main axis *in situ* (Aloni, 1987). Vöchting (1892), Simon (1908), and Freundlich (1909) demonstrated that severing vascular bundles induces differentiation of parenchyma cells into TEs around the wound, resulting in the reconstitution of xylem strands which join the original strands. Young internodes of *Coleus* have been used for a study of the wound-induced redifferentiation of TEs (Simon, 1908; Sinnott and Bloch, 1944, 1945; Jacobs, 1952, 1954; Aloni and Jacobs, 1977). Jacobs (1952, 1954) first showed that a natural auxin, indoleacetic acid (IAA), produced by expanding leaves in *Coleus*, played a crucial role in the regeneration of xylem strands around a wound. Sachs (1969) also showed, with elegant surgical techniques, that basipetally transported IAA was a limiting factor for the formation of xylem strands in pea roots. Through a series of studies, Sachs (1969, 1981, 1986) proposed the hypothesis that canalization of auxin flux determines the orderly pattern of vascular differentiation in intact plants. In other words, auxin moves initially from expanding leaves by diffusion, the auxin flux induces cell differentiation, and differentiated cells promote further auxin transport, leading to canalization of further auxin flow along a narrow file of cells.

Wound-induced redifferentiation of TEs occurs in many dicotyledonous species, including *Coleus* (Simon, 1908), pea (Robbertse and McCully, 1979), and tobacco (Sussex *et al.*, 1972). In contrast, monocotyledons showed only a weak regeneration activity for xylem differentiation (Simon, 1908), and the regeneration was limited to the very young tissues (Aloni and Plotkin, 1985).

B. Callus Cultures

TE formation in intact plants was reproduced *in vitro* by Wetmore and Sorokin (1955). This was accomplished by grafting a bud into *Syringa* callus tissue grown on an agar medium. The bud could be replaced by an agar block containing IAA and sucrose (Wetmore and Rier, 1963), and this suggested that IAA was a limiting

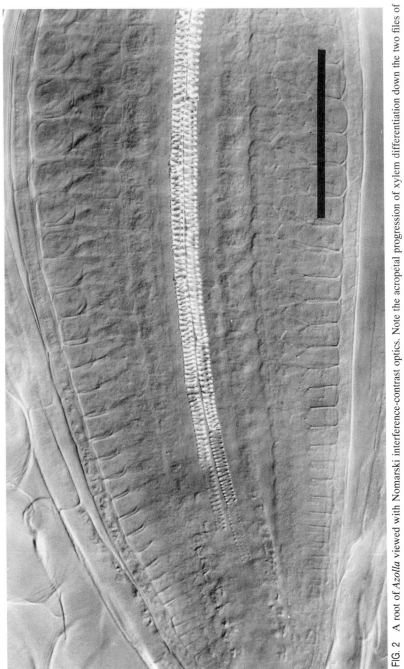

FIG. 2 A root of *Azolla* viewed with Nomarski interference-contrast optics. Note the acropetal progression of xylem differentiation down the two files of xylem during root growth. Bar, 50 μm. Courtesy of A. R. Hardham.

factor of TE differentiation in callus tissues as well as in intact plants. Jeffs and Northcote (1967) showed that both IAA and sucrose move along diffusion gradients in *Phaseolus* callus and that TE differentiation was induced at the region where the concentration of IAA and sucrose became optimum for the initiation of differentiation.

Induction of TE formation has been reported for callus tissues grown on agar from a wide range of plant species, including *Acer pseudoplatanus* (Wright and Northcote, 1973), *Allium sativum* (Havránek and Movák, 1973), *Eucalyptus* (Sussex and Clutter, 1968), *Fraxinus* (Doley and Leyton, 1970), *Glycine max* (Fosket and Torrey, 1969), *Nicotiana tabacum* (Bornman and Ellis, 1971), *Parthenocissus* (Rier and Beslow, 1967), *Phaseolus vulgaris* (Jeffs and Northcote, 1967), and *Vinca rosea* (Datta *et al.*, 1979). These callus cultures have some advantages for a study of TE differentiation: (1) A fairly uniform physiological and cytological state of callus can be made by devising culture conditions. (2) Some types of callus are composed of fairly homogeneous cells. (3) Large amounts of callus can be used as experimental materials.

However, there is an inevitable disadvantage in using callus cultures. TEs are formed as nests or nodules in discrete limited areas in callus tissues, probably because of the gradient of concentration of inducers, such as auxin or sugars, supplied exogenously within the callus (Wetmore and Rier, 1963; Jeffs and Northcote, 1967). This results in asynchrony and a low frequency of TE differentiation, making the analysis of differentiation at cellular level difficult.

The use of suspension cultures consisting of small clusters of cells in a defined liquid medium may surmount the disadvantages of callus cultures on agar. Many attempts have been made to establish suspension cultures suitable for the study of TE differentiation using *Asparagus* (Albinger and Beiderbeck, 1983), bean (Bevan and Northcote, 1979a), *Centaurea* (Torrey, 1975), peanut (Verma and van Huystee, 1970), *Pelargonium* (Reuther and Werckmeister, 1973), and tobacco (Kuboi and Yamada, 1978). The suspension cells of *Centaurea* (Torrey, 1975) differentiated as single elements or small colonies, whereas TE formation occurred internally as nests or nodules in many other suspension cultures. However, the frequency and synchrony of differentiation in *Centaurea* suspension cells are insufficient for a detailed analysis of TE differentiation.

Another difficulty with callus systems is that cultures often lose their potential for differentiation with time. This is true not only for TE differentiation, but also other types of differentiation, including embryogenesis. In callus (Haddon and Northcote, 1975, 1976b) or suspension culture (Bevan and Northcote, 1979a,b) of *Phaseolus vulgaris*, the ability to form vascular nodules containing TEs, when transferred to an induction medium, declined during 5–10 subcultures on a maintenance medium that contained 2,4-dichlorophenoxy acetic acid (2,4-D) as the only plant hormone. This decline was accompanied by a decrease in induced levels of phenylalanine ammonia-lyase (PAL) and *O*-methyltransferase, which were related to the synthesis of lignin (Haddon and Northcote, 1976b; Bevan and

Northcote, 1979b). Cells grown in the presence of coconut milk and 2,4-D retained the ability to induce PAL and TEs. A detailed comparison of the two cultures suggested that the loss of the potency for differentiation was not due to selective growth of noninducible cells, but to reversible changes in hormonal requirements of the cells necessary for induction (Bevan and Northcote, 1979b).

C. Primary Cultures

Parenchyma cells in plants can redifferentiate to TEs when excised from intact plants and cultured under adequate conditions. Fosket and Roberts (1964) demonstrated induction of TEs in 2-mm-thick slices of *Coleus* stems cultured on a semisolid medium supplemented with 2% sucrose and 0.05 mg/liter of IAA. Since then, several workers have used tissues excised from different organs of various species of plants in redifferentiation experiments. In particular, parenchyma tissues of Jerusalem artichoke tubers (Dalessandro, 1973; Minocha and Halperin, 1974; Phillips and Dodds, 1977), cortical tissues of pea roots (Phillips and Torrey, 1973), parenchyma tissues of lettuce pith (Dalessandro and Roberts, 1971), and phloem parenchyma tissues of carrot roots (Mizuno *et al.*, 1971) are well analyzed materials. Excised parenchyma tissues are valuable for the study of cell differentiation, because these cells have never been committed to TE differentiation. In particular, hormonal induction of TE formation in excised tissues has been studied extensively, for example, auxin with tuber slices of Jerusalem artichoke (Dalessandro, 1973; Minocha and Halperin, 1974; Phillips and Dodds, 1977), cytokinin with phloem tissues of carrots (Mizuno and Komamine, 1978), and ethylene with pith explants of lettuce (Roberts and Miller, 1982). Some excised tissues are composed of fairly homogeneous cells. For example, cells in excised tuber slices of Jerusalem artichoke are all at the G1 phase of the cell cycle (Dodds and Phillips, 1977). This homogeneity gave rise to the synchronous differentiation of TEs (Phillips and Dodds, 1977). Excised tissues are under the influence of the physiological state of the original organs. The maturity of the tubers affected the ability to differentiate TEs in explants of Jerusalem artichoke tubers (Phillips, 1981a). TE differentiation in explants, as in callus, is influenced by the gradient of concentration of exogenously supplied inducers within tissues and by cell-to-cell interaction.

D. Single-Cell Cultures

A single-cell culture is the simplest system in which single cells differentiate to TEs directly without intervening cell division. Torrey (1975) showed, from plating experiments, that single cells derived from *Centaurea* cell suspensions differentiated directly into single isolated TEs. Unfortunately, this occurred asynchronously and at low frequency. In the same year, Kohlenbach and Schmidt (1975) reported that mechanically isolated mesophyll cells of *Zinnia elegans* could differentiate into TEs directly. Improvement of the *Zinnia* system by Fukuda and Komamine (1980a) led to the establishment of an efficient experimental system with single cells (Fig. 1).

In this system, 30–60% of isolated cells differentiated to TEs synchronously between 60 and 80 hr of culture in the presence of two phytohormones: 0.1 mg/liter of 1-naphthaleneacetic acid (NAA) and 1 mg/liter of benzyladenine (BA) (Fig. 3) (Fukuda and Komamine, 1980a, 1985; Fukuda, 1989b; Sugiyama and Komamine, 1990). In addition, the isolated cells are very homogeneous, because they are composed mainly of palisade cells, with a small number of spongy cells. All cells are at the 2C level of DNA, that is, at the G1 phase (Fukuda and Komamine, 1981a). More than 60% of TEs are formed directly without cell division (Fukuda and Komamine, 1980b). These characteristics enabled us to study redifferentiation of parenchyma cells into TEs at the cellular level without the effects of cell-to-cell interaction. More recent studies of TE differentiation at the cellular level have been carried out mainly with the *Zinnia* system (Dodds, 1980; Fukuda and Komamine, 1982; Burgess and Linstead, 1984; Falconer and Seagull, 1985a; Haigler and Brown, 1986; Fukuda, 1987; Sugiyama *et al.*, 1986; Kobayashi *et al.*, 1987; Church and Galston, 1988a; Thelen and Northcote, 1989; Roberts and Haigler, 1989; Lin and Northcote, 1990).

There are reports on modifications of the *Zinnia* system. The use of a homo-blender instead of a pestle and mortar for the isolation of cells resulted in a much higher yield of viable cells (Fukuda and Komamine, 1982). Phosphate limitation or γ-irradiation was found to suppress cell division without interfering with TE

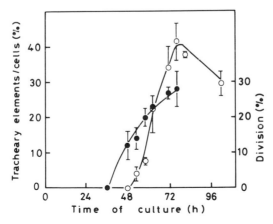

FIG. 3   Time course of TE formation (○) and cell division (●) in isolated mesophyll cells of *Zinnia elegans*. From Fukuda (1987).

differentiation in *Zinnia* cell culture (Sugiyama *et al.*, 1986). In contrast, Iwasaki *et al.* (1986, 1988) established a *Zinnia* cell culture in which only cell division occurred, without any TE differentiation, in order to study the induction of the first cell division cycle. More recently, Church and Galston (1989) showed that peeling the epidermis of a *Zinnia* leaf disk induced TE formation from mesophyll cells *in situ*. Interestingly, many epidermal cells in the peeled disks also differentiated into TEs (Fig. 4).

Protoplasts prepared with wall-degrading enzymes, namely, cellulase and pectinase, are single isolated cells in the strict sense. Kohlenbach and Schöpke (1981) showed that mesophyll protoplasts isolated from *Zinnia* can be induced to differentiate into TEs. Protoplasts isolated from cultured haploid cells of *Brassica napus* were also found to differentiate into TEs (Kohlenbach *et al.*, 1982). The protoplasts of the two plant species transformed to TEs without cell division in the presence of auxin and cytokinin. Using molecular biological techniques such as gene transfer, such protoplasts have a potential to become a good system for studying differentiation in the future when the frequency and synchrony of differentiation increase to a level sufficient for analysis.

III. Induction of Differentiation

A. Wounds

Wound stress must be an important factor for the induction of TE differentiation, since severing vascular bundles in roots or stems induced the redifferentiation of parenchyma cells into TEs around the wound. Mechanical wounding induces a variety of rapid responses in plants, including the generation and transmission of action potentials (Pickard, 1973), changes in ion transport (Gronewald and Hanson, 1980), membrane lipid breakdown (Theologis and Laties, 1981), polysome formation (Davies and Schuster, 1981), and enhanced production of ethylene (Saltveit and Dilley, 1978). Studies have shown that wounding activates the transcription of a set of genes related to defense mechanisms. The following wound-inducible gene products have been identified: (1) PAL (Liang *et al.*, 1989b), 3-deoxy-D-*arabino*-heptulosonate 7-phosphate synthase (DAHP) (Dyer *et al.*, 1989), and chalcone synthase (Ryder *et al.*, 1987), which play a role in biosynthesis of secondary metabolites; (2) 1-aminocyclopropane-1-carboxylic acid synthase, which is a key enzyme of ethylene synthesis (Nakajima *et al.*, 1988); (3) β-1,3-glucanase and chitinase (Shinshi *et al.*, 1987); and (4) protease inhibitors (Graham *et al.*, 1986; Sánchez-Serrano *et al.*, 1986). In addition, wounding induced the genes of wall proteins such as a hydroxyproline-rich glycoprotein (Lawton and Lamb, 1987) and a glycine-rich protein (Condit and Meagher, 1987). The mechanism of wound-induced gene expression has been pursued with gene transfer techniques (Chen and Varner, 1985; Stanford *et al.*, 1989; Keil *et al.*, 1990), indicating the importance of the 5' upstream sequence of the genes. The transferred DNA (T-DNA) genes of *Agrobacterium* tumor-inducing plasmid are actively expressed in transformed plant cells. The nopaline synthase gene in T-DNA was reported to be induced in plant cells by wounding (An *et al.*, 1990). Deletion analysis of the nopaline synthase promoter indicated that a 10-base element is essential for the wound and auxin response (An *et al.*, 1990). However, a common DNA motif responsible for the expression of all wound-induced genes has not yet been found.

The isolation of mesophyll cells from *Zinnia* leaves causes the rapid expression of a set of genes regardless of the presence or absence of phytohormones (H. Fukuda, unpublished data). The inhibition of synthesis of RNA and protein at this stage blocked TE differentiation (Fukuda and Komamine, 1983). These results suggested that the products of wound-inducible genes play a key role in the induction of TE differentiation.

Wounding affects first the wounded cells, then cells around the wound, and finally cells far from the wound site. For example, a protease inhibitor accumulates both in the damaged leaf and in other unwounded leaves in tomato (Ryan, 1978). The signal for the accumulation was transmitted from the wounded leaf to unwounded leaves in tomato plants as early as 20 min after wounding (Nelson *et al.*, 1983). Because TE formation is induced in unwounded parenchyma cells around the wound by severing vascular bundles, the signal substance(s) should be transmitted from the wound site to the parenchyma cells. The signal may be endogenous phytohormones, including IAA (Roberts, 1988). The wounding disturbs the normal movement of hormones and causes the release of endogenous hormones. The leakage of hormones around the wound initiates the redifferentiation of parenchyma cells. Church and Galston (1989) showed that only a few TEs were formed from mesophyll cells near the wounded site when leaf disks of *Zinnia* were cultured in a liquid medium containing both auxin and cytokinin, whereas large numbers of TEs were formed not only in the mesophyll but also in the remaining epidermis when leaf disks whose upper or lower epidermis was peeled off were cultured in the same medium (Fig. 4). This peeling effect can explain the high frequency of TE formation in isolated mesophyll cells of *Zinnia* (Fukuda and Komamine, 1980a). Ryan's group demonstrated that cell wall fragments, especially polygalacturonides, are released *in situ* during wounding and function as inducers of the expression of protease inhibitor gene (Bishop *et al.*, 1984). The polygalacturonide was found to enhance *in vitro* phosphorylation of proteins in plasma membrane of tomato, suggesting that protein kinase may play an important role in the signaling mechanism for events occurring in response to wounding (Farmer *et al.*, 1989). TE differentiation in *Zinnia* might involve such a signaling process induced by wounding. Severe mechanical wounding might produce sufficient amounts of signal substances, such as cell wall fragments, to induce the differentiation. The signal substances might induce a secondary messenger via a

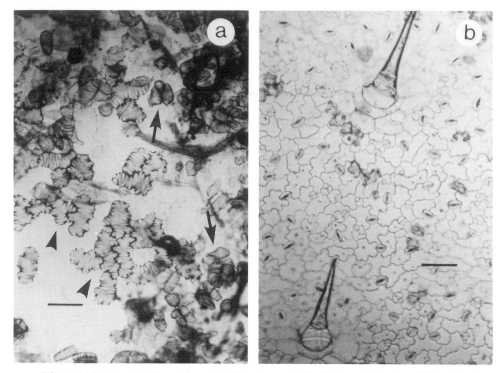

FIG. 4   (a) Patches of epidermal (arrowheads) and mesophyll cells (arrows) that have differentiated into TEs. A *Zinnia* leaf disk (lower epidermis removed) was cultured for 7 days in medium containing 5 μM naphthaleneacetic acid + 5 μM benzyladenine. (b) Peel of lower epidermis from fresh *Zinnia* leaf. From Church and Galston (1989).

signal-transducing system on the plasma membrane, and then the messenger might act on the initiation of the differentiation.

B. Phytohormones

1. Auxin and Cytokinin

Auxin is a plant hormone that plays a wide range of roles in plant development, such as cell elongation, cell division, and cell differentiation. An auxin require-ment for TE differentiation is well confirmed in almost all cells and tissues that have been tested so far (Roberts, 1976; Shininger, 1979; Phillips, 1980; Savidge and Wareing, 1981; Fukuda and Komamine, 1985; Little and Savidge, 1987; Roberts *et al.*, 1988). The pioneering experiments demonstrating its requirement were performed by Jacobs (1952, 1954), who found a crucial role for IAA in TE formation in *Coleus* stem wounds. Later, Wetmore and Sorokin (1955) showed that IAA could also induce TE differentiation in *Syringa* callus tissue. In *in vitro* culture, the exogenous supply of auxin is absolutely necessary for the induction of differentiation. An exception is cells containing high levels of endogenous IAA, such as habituated cultured cells (Phillips, 1980) and plant tumor cells (Basile *et al.*, 1973). In these cells, TEs were formed even in the absence of an exogenous supply of auxin. Treatment with an anti-auxin or an inhibitor of auxin transport inhibited TE differentiation in *Zinnia* cells (Burgess and Linstead, 1984; Church and Galston, 1988b). In intact plants, the polar transport of IAA from young leaves toward the roots triggers organized TE differentiation, resulting in the formation of a xylem strand along the flow of IAA (Aloni, 1987). IAA produced as a consequence of the autolysis of TEs may be involved in an initiation of further differentiation in adjacent cells (Sheldrake and Northcote, 1968a,b).

Cytokinin promotes TE differentiation in a variety of plant species (Roberts, 1976; Shininger, 1979; Phillips, 1980; Fukuda and Komamine, 1985; Roberts *et al.*, 1988). Cytokinin acts only in combination with auxin, and its requirement varies in different tissues. While an exogenous supply of cytokinin is not necessary for the induction of TE differentiation in *Syringa* callus (Wetmore and Sorokin, 1955), *Zinnia* mesophyll cells have an absolute requirement for cytokinin for the induction (Fukuda and Komamine, 1980a). Mizuno *et al.* (1971) and Mizuno and Komamine (1978), using cultured phloem slices of carrot, found that the difference of cytokinin requirement for TE differentiation in different cultivars reflected endogenous levels of cytokinin. Cultivars such as Kuroda-gosun, in which TE differentiation was induced solely by auxin, possessed a high endogenous level of cytokinin in the form of zeatin ribonucleoside, while cultivars such as Hokkaido-gosun, in which TE differentiation required an exogenous supply of both auxin and cytokinin, did not possess measurable endogenous cytokinin. Consequently, cytokinin may be a general inducer of differentiation along with auxin, regardless of whether the source of cytokinin is exogenous or endogenous.

Investigation of the time sequence of hormonal effects on the induction of TE differentiation in *Zinnia* mesophyll cells revealed that the process of differentiation started at the time when both auxin and cytokinin were supplied (Fukuda and Komamine, 1985). These experiments also suggested the presence of an early process, which can occur without the hormones, during a 12-hr period following the onset of culture. This period is coincident with the period when the wound response occurs, as described in Section III,A, suggesting that wounding is the first factor responsible for the induction, and, subsequently, auxin and cytokinin together play an essential role in the induction. Both hormones are necessary for the sequence of differentiation as well as the induction, but the requirements of the two hormones for the sequence seem to be different (Minocha, 1984; Fukuda and Komamine, 1985; Tucker *et al.*, 1986; Phillips, 1987; Church and Galston, 1988a). The presence of auxin was required continuously until the late stage of differentiation, while cytokinin was necessary only for a brief period in the early stage of differentiation in *Zinnia* cells (Fukuda and Komamine, 1985; Church and Galston, 1988a). It may be argued that the early process of differentiation is regulated by both auxin and cytokinin, but the succeeding progression of differentiation is controlled mainly by auxin. This suggestion is, however, still tentative, because of a lack of information on the endogenous level of hormones and the difficulty of completely removing the hormones by washing in the pulse experiments. A similar result was obtained in lettuce pith explants (Tucker *et al.*, 1986).

What is the molecular mechanism of phytohormonal induction of TE differentiation? During the past several years, there has been a noticeable increase in the number of studies on the molecular mechanism of auxin action. Many genes have been isolated that are expressed rapidly in response to auxin treatment. These genes are suggested to be related to cell division (van der Zaal *et al.*, 1987; Takahashi *et al.*, 1989) or cell elongation (Theologis *et al.*, 1985; Hagen *et al.*, 1988; Ainley *et al.*, 1988; Alliotte *et al.*, 1989; McClure *et al.*, 1989), but the functions of their products are still unknown. Among them, the products of a gene group called SAURs are candidates for the early messenger of auxin. They code polypeptides of 9–10.5 kDa and are activated as early as 2.5 min after auxin treatment to soybean hypocotyls (McClure and Guilfoyle, 1987; McClure *et al.*, 1989). A tissue print technique revealed that they were expressed much more in the lower side than in the upper side of horizontal soybean hypocotyls, coinciding with the hypothesis of geotropism that a higher concentration of auxin in the lower side causes prominent growth in the lower side (McClure and Guilfoyle, 1989). Also, studies on auxin receptors have been progressing. A membrane-bound auxin-binding protein and its gene have been isolated from *Zea mays* (Shimomura *et al.*, 1986; Inohara *et al.*, 1989; Hesse *et al.*, 1989). A similar protein was suggested to exist in the plasma membrane of tobacco (Barbier-Brygoo *et al.*, 1989). Binding of an antibody to the protein prevented the auxin function, suggesting that this is a general auxin receptor (Barbier-Brygoo *et al.*, 1989). Also, several soluble auxin-binding proteins have been isolated and some of them promoted transcriptional activity of all genes or some specific genes in the presence of auxin (Kikuchi *et al.*, 1989). Unfortunately, the possible relationship between these auxin-related proteins and TE differentiation has not been elucidated. We expect that these proteins will be properly characterized and the transductional mechanism of auxin will be elucidated in the near future, enabling us to understand the function of auxin in TE differentiation. At the same time, we should analyze genes that may be expressed in response to auxin and/or cytokinin during TE differentiation. This comprehensive analysis may lead to new findings of auxin-inducing genes specific for TE differentiation.

2. Gibberellin, Ethylene, and Brassinolide

The exogenous supply of other plant hormones, such as gibberellin, abscisic acid, ethylene, and brassinosteroid, does not induce TE differentiation *in vitro*. However, endogenous contribution of these substances to induction or promotion of the differentiation is likely to occur.

Minocha and Halperin (1974) indicated marked inhibition of TE differentiation in explants of Jerusalem artichoke by exogenously supplied gibberellin. The suppression or noneffectiveness of exogenous gibberellin has been shown with other plant materials (Haddon and Northcote, 1976b; Fukuda and Komamine, 1985). In contrast, it has been reported for lettuce and Jerusalem artichoke that exogenous gibberellin promotes the differentiation induced by auxin and cyto-kinin (Dalessandro, 1973; Phillips and Dodds, 1977; Pearce et al., 1987). The conflicting findings may be due to the endogenous level of gibberellin (Fukuda and Komamine, 1985). Measurement of the endogenous level of gibberellin and its exogenous supply at low concentration indicated that the timing and concentra-tion of exogenous application may be critical for the promotion of differentiation, supporting the hypothesis that gibberellin has a positive role in TE differentiation (Pearce et al., 1987).

Ethylene may also be a hidden inducer or promoter of TE differentiation (Roberts and Miller, 1982). Although the exogenous application of ethylene is not necessary for the induction of TE differentiation, ethylene inhibitors such as $AgNO_3$, $Co(NO_3)_2$, and aminoethoxyvinylglycine inhibited TE differentiation in lettuce pith explants (Miller and Roberts, 1984; Miller et al., 1984, 1985), soybean callus (Miller and Roberts, 1982), Jerusalem artichoke explants (Koritsas, 1988), and Zinnia cells (Y. Shoji, M. Sugiyama, and A. Komamine, unpublished data). Methionine, S-adenosylmethionine, and 1-aminocyclopropane-1-carboxylic acid, which are ethylene precursors, promoted the differentiation and overcame the suppression of the ethylene inhibitors (Miller and Roberts, 1984; Koritsas, 1988).

Ethylene production is one of the earliest reactions caused by wounding (Boller and Kende, 1980). As discussed in Section III,A, wounding causes the formation of wound vessel members. A key part of the induction by wounding may be the biosynthesis of ethylene, which may, in turn, induce the differentiation process. Koritsas (1988) discussed the possibility that ethylene stimulates protein phos-phorylation, which regulates TE differentiation. However, the idea is premature and the molecular mechanism of the induction of differentiation by ethylene is still unknown.

Recently, Iwasaki and Shibaoka (1991) found that an exogenous supply of brassinolide accelerated the appearance of TEs in Zinnia cells. Uniconazole, which is known to reduce the content of an endogenous brassinosteroid (Yokota et al., 1991), inhibited TE differentiation without affecting cell division in Zinnia cells (Iwasaki and Shibaoka, 1991). An exogenous supply of brassinolide at surprisingly low concentrations (above 0.2 nM) reversed the inhibition. These results suggest that endogenous brassinosteroids function as a key factor in the process of differentiation into TEs from Zinnia mesophyll cells. An urgent priority is to examine whether brassinosteroids play a key role in TE differentiation in other cells and tissues.

C. Calcium and Calmodulin

The function of calcium as a second messenger in animal cells has been established. Recent studies have indicated that the developmental process of plant cells is also regulated by calcium (Hepler and Wayne, 1985). Roberts and Baba (1987) reported that chlorpromazine and trifluoperazine, which are calmodulin antagonists, inhibited TE differentiation without suppressing callus formation in lettuce pith explants at concentration of 50 and 100 μM, respectively. In contrast, chlorpromazine sulfoxide, a nonantagonistic chlorpromazine analog, did not block differentiation. In *Zinnia* cells, calmodulin antagonists were found to be effective only on the induction or the very early process of TE differentiation (Roberts and Haigler, 1990). Roberts and Haigler speculated that calmodulin may be involved in determination of differentiation caused by cytokinin or in calcium uptake occurring in the early process. Calmodulin may also play a role in secondary wall formation, because the immunolocalized distribution of calmodulin in some TEs in pea and *Zinnia* appeared to be limited between thickenings (Dauwalder *et al.*, 1986; Fukuda and Kobayashi, 1989). This idea does not conflict with the observation that pulse-treatment with trifluoperazine was effective just before secondary wall formation (H. Kobayashi, unpublished observation).

Calcium depression is known to cause inhibition of TE differentiation (Hewitt, 1983; Roberts and Haigler, 1990). Histochemical analysis with chlorotetracycline, a fluorescent chelate probe for membrane-associated calcium, revealed a greater amount of calcium in differentiation cells (the calcium is thought to localize mainly in the endoplasmic reticulum of differentiating TEs) than in nondifferentiating cells (Roberts and Haigler, 1989). Calcium channel blockers such as La^{3+}, nifedipine, and $(-)202$-791 compound strongly inhibited TE differentiation in *Zinnia* cells and also caused the formation of thinner wall bands in differentiating TEs (Roberts and Haigler, 1990). These results suggested that calcium uptake across the plasma membrane and sequestration in intracellular compartments are required for the formation of the secondary wall.

IV. Early Stages of Differentiation

Cells induced to differentiate to TEs enter the differentiating process. During the ontogenetic stages of TE differentiation, cell enlargement occurs after the induction of differentiation in primary meristems and is followed by secondary wall formation and autolysis. In excised tissues or cells which have not been predetermined to TE differentiation, the differentiating process after the induction is thought to be divided into early and late processes. The late process involves various events specific for TEs, such as secondary wall formation and autolysis, and it occurs irreversibly. The early process is poorly characterized, because of limited information on events which occur in this process. The *Zinnia* system, in which isolated single mesophyll cells differentiate directly into TEs, that is, predetermined cells differentiate to new types of cells, has great advantages for an analysis of the early process. Studies using the *Zinnia* system have indicated that the early process is a complex process including dedifferentiation-related events that are required for TE differentiation (Fukuda, 1989b; Sugiyama and Komamine, 1990).

A. Relationship between Differentiation, Cell Cycle, and DNA Synthesis

In multicellular organisms, cell division and cytodifferentiation are like two wheels of a cart, which should be regulated interdependently. Cell division yields a large number of cells, which differentiate into various mature forms with different tasks *in situ*. In *in vitro* systems, cell division very often precedes TE cytodifferentiation (Clutter, 1980; Torrey and Fosket, 1970; Ronchi and Gregorini, 1970; Dalessandro and Roberts, 1971; Phillips and Torrey, 1973; Comer, 1978; Malawer and Phillips, 1979; Phillips, 1981b; Hardham and McCully, 1982a). Fosket (1968, 1970) found that inhibitors of DNA synthesis, such as fluorodeoxyuridine and mitomycin C, and colchicine, which inhibits cell division, prevented TE differentiation in cultured *Coleus* stem segments in which cell division preceded the appearance of TEs. This suggested that cell division is essential for the induction or sequence of TE differentiation. Since then, the attention of many workers has been drawn to the relationship between cell division and TE differentiation (reviewed by Torrey *et al.*, 1971; Roberts, 1976; Shininger, 1978, 1979; Phillips, 1980; Dodds, 1981a,b; Fukuda and Komamine, 1985; Gahan, 1988; Fukuda, 1989b; Sugiyama and Komamine, 1990).

While there are reports consistent with Fosket's suggestion, many workers have reported the occurrence of TE differentiation without intervening cell division in colchicine- or caffeine-treated or γ-irradiated cells (Hammersley and McCully, 1980; Fukuda and Komamine, 1980b; Hardham and McCully, 1982b; Phillips, 1981a,b; Sugiyama *et al.*, 1986). Chlorosulfuron is known to arrest cell cycle progression in the G1 and G2 phases, without affecting the S and M phases (Rost, 1984), by inhibiting the biosynthesis of isoleucine and valine (Ray, 1984). Chlorosulfuron inhibited cell division without having any inhibitory effect on TE differentiation in cell suspension cultures of *Solanum carolinense* (Reynolds, 1986). Fukuda and Komamine (1980b) provided direct evidence for TE differentiation without cell division by serial observation of the process of differentiation from single cells isolated from *Zinnia* leaves. These results indicated that cell division preceding TE differentiation is not necessary for differentiation. Shininger (1975, 1978, 1979), based on his results that inhibitors of DNA synthesis prevented cells of pea roots from differentiation, presented a hypothesis that DNA replication preceding TE differentiation, but not mitosis, is essential and functions in primary signals, inducing differentiation at the molecular level. Many cells isolated from the *Zinnia* mesophyll, however, have been found to differentiate to TEs directly from the first G1 phase of the cell cycle without DNA replication in the S phase (Fukuda and Komamine, 1981a). Other types of TEs were also formed from *Zinnia* mesophyll cells. The relationship between TE differentiation and the cell cycle is shown in Fig. 5 (Fukuda and Komamine, 1981a). It is clear that neither DNA replication in the S phase nor mitosis is required for the transdifferentiation from parenchyma cells to TEs. Sugiyama *et al.* (1986) reported that low doses of γ-irradiation preferentially blocked cell division through the inhibition of DNA replication without affecting TE differentiation, supporting the hypothesis that TE differentiation is independent of DNA replication.

Based on the results obtained from *Zinnia* and Jerusalem artichoke cells, Dodds (1981a,b) presented a new hypothesis, that critical events for the induction of TE differentiation occur in the "early" G1 phase, and then cells in the G1 phase before the "early" stage differentiate directly. Cells in any phase after the "early" stage go along the cell cycle to reach the "early" G1 phase and receive a signal for differentiation. However, this hypothesis cannot explain the observation that double TEs (b in Fig. 5) that differentiate after one round of the cell cycle appear simultaneously with single TEs (c in Fig. 5) that are initiated directly in the first G1 phase (Fukuda and Komamine, 1985). The results rather support the hypothesis of Fukuda and Komamine (1981a, 1985) that the initiation of differentiation occurs in the first G1 phase, but not during the S phase or mitosis, and that the early sequence of cytodifferentiation proceeds regardless of the cell cycle; that is, cytodifferentiation is compatible with the cell cycle. However, it is

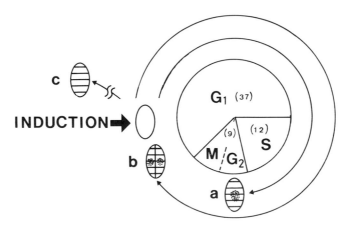

FIG. 5   Diagram of the relationship between TE differentiation and the cell cycle in a culture of mesophyll cells isolated from *Zinnia elegans*. From Fukuda and Komamine (1981b).

unknown whether the induction is limited to the first G1 phase in any other plant cell.

Following the early sequence, there could be a late sequence during which differentiating cells withdraw from the cell cycle and are engaged in cytodifferentiation. This process is thought to be irreversible. Thus, cytodifferentiation *in vitro* could be divided into two stages: one is a "prerestricted stage" in which the sequence of differentiation can proceed independently of and concurrently with the cell cycle, and is probably reversible; the other is a "restricted stage" in which differentiating cells engage only in cytodifferentiation, and is irreversible (Fukuda, 1989b). Early and late processes referred to in this review could correspond fairly to prerestricted and restricted stages, respectively.

Although, as described above, replication of DNA in the S phase is not required for TE differentiation in *Zinnia* cells, all inhibitors of DNA synthesis tested (fluorodeoxyuridine, mitomycin C, arabinosylcytosine, fluorouracil, hydroxyurea, and aphidicolin) blocked TE differentiation (Fukuda and Komamine, 1981b; Sugiyama and Komamine, 1990; Sugiyama *et al.*, 1990). Because these chemicals act at different sites in the biosynthesis of DNA, it is unlikely that the inhibition is the result of a side effect. The results could also be interpreted as an indication that some minor amount of DNA synthesis, but not DNA replication in the S phase, is essential for TE differentiation. In the early stage of differentiation between 4 and 36 hr of culture, minor DNA synthesis seems to be required for the sequence of differentiation (Sugiyama *et al.*, 1990). An autoradiographic study demonstrated a low level of incorporation of [³H]thymidine over the nuclei of cells out of the S phase at the early stages in culture (Sugiyama and Komamine, 1987b). This DNA synthesis during the early stage was found to be repair-type DNA synthesis by a double-labeling method with bromodeoxyuridine and [³H]thymidine (M. Sugiyama, personal communication). This was also supported by the fact that inhibitors of ADP-ribosyltransferase, associated with DNA excision repair, prevented TE differentiation in *Zinnia* cells without having any significant effects on cell division (Sugiyama and Komamine, 1987a). Similar results were reported for tuber explants of Jerusalem artichoke (Hawkins and Phillips, 1983) and pea root segments (Phillips and Hawkins, 1985). This repair-type DNA synthesis might be involved in the step of transdifferentiation from parenchyma cells to TEs, that is, the step at which gene expression in the parenchyma cells is reprogrammed toward the formation of a new type of cell (Fukuda, 1989b; Sugiyama and Komamine, 1990).

Italian scientists have indicated that there is amplification of ribosomal cistrons and methylation of genomic DNA during root development of *Allium cepa* (Avanzi *et al.*, 1973; Durante *et al.*, 1977, 1990). They suspected this phenomenon to be involved in the differentiation process of the metaxylem cell line in onion roots. Although there was no direct evidence, amplification of some specific genes or activation of such genes by methylation/demethylation might have occurred during some early phase of TE differentiation.

B. Cytological Changes

In the early process of differentiation there is little morphological change, but intracellular organization changes dynamically. There are many studies using electron microscopy that have observed cytological changes during TE differentiation. To describe these in detail is beyond the scope of this review (see Cronshaw, 1965; Wardrop, 1965; O'Brien, 1974, 1981; Hepler, 1981; Fukuda and Komamine, 1985). In brief, during the early process, proliferation of various organelles such as mitochondria, ribosomes, Golgi apparatus, the endoplasmic reticulum, and Golgi vesicles occurs (Srivastava and Singh, 1972; Fukuda and Komamine, 1985). In the late process, some of these organelles localize at specific sites in differentiating cells and are involved in the synthesis and transport of wall materials (Fukuda and Komamine, 1985). The early process may include a role of preparing materials for differentiation-specific events which occur in the late process.

Actin filaments may play a role in the early process of differentiation to TEs from *Zinnia* mesophyll cells. It is known that, during the process in which the isolated *Zinnia* cells were losing photosynthetic function, chloroplasts became located farther from the plasma membrane. This may be involved in the reorientation of actin filaments from reticulate arrays over chloroplasts, probably for anchoring chloroplasts to the plasma membrane (Fig. 6a,b), to a three-dimensional network over the whole length of the cell (Fig. 6c) (Kobayashi *et al.*, 1987). Thick filaments in cells cultured for 48 hr (Fig. 6c) are thought to function in cytoplasmic streaming.

Freshly isolated *Zinnia* mesophyll cells form a random sparse network of microtubules (Fig. 7B) (Fukuda and Kobayashi, 1989). The number of microtubules increases rapidly during the early process, and, in association with the increase, arrays of microtubules change from a random to an ordered pattern, especially to a pattern parallel to the long axis of the cell (Fig. 7B) (Fukuda and Kobayashi, 1989). An increase in the number of microtubules in the early process was also reported for *Azolla* roots (Hardham and Gunning, 1979). This increase was due to the synthesis of tubulin, which is described in detail in Section V,A,1 of this review.

C. Gene Expression

Cytochemical analysis of esterase activity, using naphthol AS-D acetate as a substrate, revealed that high esterase activity was localized in a region of future xylem or phloem formation in an intact root tip (Rana and Gahan, 1982) and the cortex of wounded root of pea (Rana and Gahan, 1983). This esterase was found to be a carboxylesterase (Gahan *et al.*, 1983). Determination of the induction of esterase activity and of secondary wall formation was considered to occur within 10 and 20 hr after wounding, respectively, and each determination preceded cell

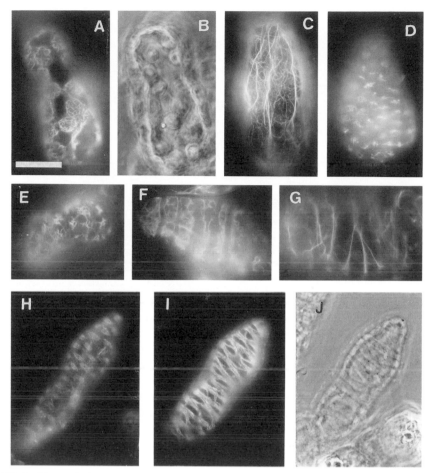

FIG. 6   Changes in the organization of actin filaments during TE differentiation from *Zinnia* cells. (A) Actin filaments in a freshly isolated cell; (B) phase-contrast image of the cell shown in A; (C) actin filaments in a cell at 48 hr; (D, E, F) actin filaments in a cell at 60 hr; (G) actin filamnents in a cell at 75 hr; (H) actin filaments in a cell at 66 hr; (I) microtubules in the cell shown in H; (J) phase-contrast image of the cell shown in H. Bar, 20 μm. From Kobayashi *et al.* (1987, 1988).

division (Rana and Gahan, 1983). Esterase may be used as a marker of TE differentiation at an early process, although it is unknown as yet whether it occurs in other plants.

Synthesis of both RNA and protein increased markedly during the early process of differentiation into TEs from *Zinnia* mesophyll cells (Fukuda and Komamine, 1983). Fukuda and Komamine (1983) analyzed qualitative changes in cytoplasmic protein by two-dimensional electrophoresis and found that the protein pattern on

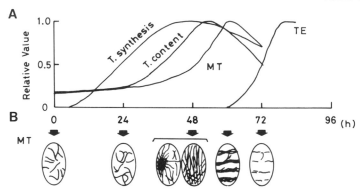

FIG. 7 Sequential changes in microtubule organization during the process of TE differentiation from *Zinnia* mesophyll cells. (A) Changes in the number of microtubules (MT), tubulin content (T. content), and tubulin synthesis (T. synthesis). (B) Changes in microtubule organization. From Fukuda and Kobayashi (1989).

the electrophoretograms was the same with the TE-inductive culture and the control culture. This implied that "switch on" or "off" of main proteins may not occur specifically for differentiation in the early process. There is no report showing clearly the expression of differentiation-specific genes in the early process. However, synthesis of some polypeptides seems to be more active in TE-inductive cultures than in the controls. Tubulin, which is a major component of microtubules, was found to increase in the early process of differentiation in *Zinnia* cells (Fig. 7A) (Fukuda, 1987). The synthesis of tubulin started as early as within 4–8 hr of culture, and continued with a gradual increase until 48 hr. An analysis with a cDNA probe indicated that tubulin synthesis may be regulated at the transcriptional level (K. Takase and H. Fukuda, unpublished data). Although the increases in protein and mRNA of tubulin occurred in both TE-inductive cultures and the controls, the increasing level was higher in TE-inductive cultures (H. Fukuda, unpublished data). This might be due to a differentiation-specific isotype of tubulin genes, as is discussed in Section V,A,1. Activity of tubulin degradation was low during this period, and the low degradation activity was also responsible for the rapid increase in tubulin content in the early process (Fukuda, 1989a).

V. Late Stages of Differentiation

The late stages involve dramatic events specific for TEs: the formation of patterned secondary cell walls and autolysis. These events have attracted the attention of many people who have studied TE differentiation. Consequently, much more information on these events than on events that occur in the early stages has been accumulated.

A. Synthesis of Secondary Cell Wall

TEs have, inside their primary wall, a localized secondary wall with a reticulate, spiral, annular, or pitted pattern, which is their most characteristic morphological feature. The formation of the secondary cell wall is one of the best analyzed events occurring during TE differentiation. The secondary cell wall is composed of cellulose microfibrils arranged parallel to one another and to the bands of secondary wall, and of encrusting substances that contain lignin, hemicellulose, pectin, and protein, which add strength and rigidity to the wall (Torrey et al., 1971). These substances are synthesized and deposited in cooperation during secondary wall formation.

1. Involvement of Cytoskeleton

The secondary wall pattern of TEs is due to the organization of cellulose microfibrils. Many microscopical studies have shown that cortical microtubules become grouped exclusively under the ridge of the secondary wall and are oriented parallel to the bands of the wall in developing TEs (Hepler and Newcomb, 1964; Wooding and Northcote, 1964; Hepler and Fosket, 1971; Hepler, 1981; Burgess and Linstead, 1984). Disruption of microtubules by colchicine caused a highly aberrant deposition of cellulose without the inhibition of cellulose synthesis itself in many kinds of plant cells (Pickett-Heaps, 1967; Hepler and Fosket, 1971; Hardham and Gunning, 1980; Fukuda and Komamine, 1980b). These results have led to the suggestion that microtubules control the pattern of the secondary wall by predicting the location and orientation of cellulose microfibrils in the wall (see review by Gunning and Hardham, 1982).

Falconer and Seagull (1985a,b), using an immunofluorescence technique with an anti-tubulin antibody, indicated a dynamic change in microtubule arrays during TE differentiation in cultured Zinnia cells. Microtubule arrays changed from random or longitudinal to transverse just before secondary wall formation, and aggregated laterally to form bundles over which secondary wall bands would be formed (Fig. 7B). As the wall bands thickened, the microtubules disappeared. As to the mechanism of the transition of microtubule organization, Falconer and Seagull (1985b, 1986, 1988) presented the hypothesis that the transition occurs due to a gradual shift of cortical microtubules from random or longitudinal to transverse, which is associated with cell elongation but not directly with differentiation, and subsequently, microtubules form lateral aggregates. However, the following results suggest that the transition is an active process that occurs in close association with differentiation, probably depending on polymerization and depolymerization of microtubules (Fukuda and Kobayashi, 1989): Cell elongation prior to secondary wall formation only occurs to a negligible extent in *Zinnia* cells (H. Fukuda, unpublished data). Treatment with taxol, which suppresses depolymerization of microtubules, fixed cortical microtubules to longitudinal orientation in differentiating *Zinnia* cells and then caused the formation of longitudinal bands of secondary wall (Falconer and Seagull, 1985b).

Actin filaments are another essential components of the cytoskeleton. Recent studies with actin-specific fluorescent phallotoxin derivatives or anti-actin antibodies have shown the existence of actin filaments in a wide variety of cells from higher plants (Parthasarathy *et al.*, 1985; Kakimoto and Shibaoka, 1987; Seagull *et al.*, 1987). Kobayashi *et al.* (1987) demonstrated dynamic changes in the organization of actin filaments during the cytodifferentiation to TEs from isolated *Zinnia* mesophyll cells, and suggested an involvement of actin filaments in pattern formation of secondary wall (Fig. 6). Double staining of microtubules and actin filaments in differentiating cells showed that, at the time when deposition of the wall began, reticulate bundles of microtubules and aggregates of actin filaments emerged simultaneously adjacent to the plasma membrane, and the aggregates existed exclusively between microtubules (Fig. 6) (Kobayashi *et al.*, 1988). Subsequently, when microtubule bundles became oriented transversely to the long axis of the cell, actin filaments also ran transversely as bands between the microtubule bundles. Disruption of actin filaments by treatment with cytochalasin B produced TEs with longitudinal bands of secondary wall along which microtubule bundles were aligned, while TEs with transverse wall bands were formed in untreated culture (Kobayashi *et al.*, 1988). These results suggest a coordinated mechanism in which actin filaments are involved in the reorganization of microtubules from longitudinal to transverse which, in turn, regulate the spatial disposition of secondary wall (Fukuda and Kobayashi, (1989).

A rapid increase in the number of cortical microtubules occurred prior to reorganization of the microtubules during TE differentiation in cultured *Zinnia* cells (Fig. 7A) (Fukuda, 1987). A measurement of the amount of tubulin, using a sensitive immunoblotting method (Fukuda and Iwata, 1986), showed that the increase in tubulin content was responsible for the increase in microtubule number (Fukuda, 1987). This increase in tubulin level was found to be caused by *de novo* tubulin synthesis, which started as early as 4 to 8 hr after the onset of culture, and preceded the increase in tubulin level with a similar time course (Fig. 7) (Fukuda, 1987). Tubulin transcripts were also elevated during the differentiation process (K. Takase and H. Fukuda, unpublished data). This indicates that the expression of the tubulin genes is possibly regulated at a transcriptional level during TE differentiation. Tubulin genes are known to form a multigene family in higher plants as well as animals (Oppenheimer *et al.*, 1988; Ludwig *et al.*, 1988). Two hypotheses have been presented on the divergency of a family of tubulin genes in animals (Cleveland, 1987). First, the individual tubulin genes may encode functionally divergent polypeptides which can confer some unique property to the final microtubule polymers. Second, multiple genes encode polypeptides which are functionally equivalent, but have evolved to possess different regulatory sequences for activation of transcription during alternative programs of differentiation. Hoyle and Raff (1990) demonstrated that two β-tubulin genes which were expressed in different tissues in *Drosophila*, were not functionally equivalent. In contrast, highly divergent β-tubulins which are expressed in different developmental stages of *Aspergillus nidulans* are found to be functionally interchangeable (May, 1989). Therefore, in some instances, both hypotheses could be correct. The process of differentiation of *Zinnia* mesophyll cells into TEs involves the expression of at least three different transcripts of tubulin (T. Yoshimura and H. Fukuda, unpublished data). Further study of these tubulin genes may throw a new light on the unique function of tubulin polypeptides in microtubule organization and/or controlling the mechanism of their expression during TE differentiation.

Both the rate of degradation and the synthesis of tubulin have been suggested to be coupled with microtubule organization during differentiation in cultured *Zinnia* cells (Fukuda, 1989a). Tubulin protein was degraded most rapidly in the 8-hr period between 46 and 54 hr of culture, during which time the orientation of microtubules changed dramatically. The degradation was specific for tubulin (Fukuda, 1989a). The mechanism of degradation of tubulin is unknown, but it may involve degradation systems that are specific for individual species of proteins, such as the ubiquitin system (Hershko and Ciechanover, 1982).

2. Polysaccharides

Calcofluor white, which stains wall polysaccharides such as cellulose, can reveal secondary wall deposition in differentiating *Zinnia* TEs about 10 hr earlier than with a phase-contrast microscope (Falconer and Seagull, 1985a; Ingold *et al.*, 1988). The cellulose content of the cell increases during the formation of the secondary cell wall in *Zinnia* cells (Ingold *et al.*, 1988). Cellulose is thought to be synthesized on the exterior surface of the cell by a plasma-membrane-bound enzyme complex, cellulose synthase, which is observed as a rosette structure in electron micrographs of freeze-fractured cells of higher plants. These rosettes are located in the plasma membrane over regions of secondary wall thickenings, and are absent between thickenings in TEs of *Zinnia elegans* (Haigler and Brown, 1986) and *Lepidium sativum* (Herth, 1985). The rosettes are also present in the Golgi apparatus through the enlargement of the secondary wall thickening, suggesting that new rosettes are continuously inserted into the plasma membrane through Golgi vesicles during secondary wall formation (Fig. 8) (Haigler and Brown, 1986). The rosettes may not be active within the Golgi apparatus and may be activated on the plasma membrane.

In addition to the increase in cellulose, there is a cessation of pectin deposition and an increase in hemicellulose deposition during secondary wall formation

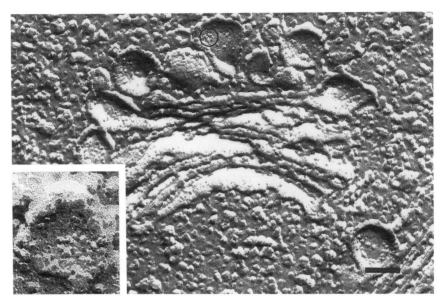

FIG. 8   A cross-fracture through a dictyosome shows vesicle formation at the *trans* face of the stack and a rosette (circled) on the PF face of a fractured vesicle which probably recently budded from the cisterna. Inset: Another example of a rosette (circled) within a vesicle. Bar, 100 μm. From Haigler and Brown (1986) and courtesy of C. H. Haigler.

(Northcote, 1972). In angiosperms, the major component of the increased hemi-cellulose is xylan, which is deposited simultaneously with cellulose (Jeffs and Northcote, 1966; Bolwell and Northcote, 1981; Ingold *et al.*, 1988). Northcote *et al.* (1989) found that an antiserum to β-1→4-oligoxylosides showed the presence of xylose not only in the secondary thickening, but also in the Golgi vesicles of TEs in the vascular tissue of bean root or in the differentiating *Zinnia* cells, although some xylan was located in the primary wall. In gymnosperms, the β-1→4-linked linear glucomannans are a major component of the hemicellulose fraction of secondary cell wall (Dalessandro *et al.*, 1986). Hogetsu (1990) found that fluorescein-conjugated wheat germ agglutinin was bound to the secondary wall of TEs in stems of many angiosperm plants, but not of gymnosperm and fern plants. The characterization of the binding sites suggested that the lectin bound to hemicelluloses, probably xylan.

   The synthesis of wall polysaccharides may be regulated in various ways: (1) amount or activity of a synthase, (2) amount of the precursor, (3) formation of a synthase complex, (4) transport of precursor to a synthase complex in a membrane, and (5) fusion to the plasma membrane of vesicles with polysaccharides (North-cote, 1989). Figure 9 shows the synthetic pathway of hemicellulose and pectin during secondary wall formation. Dalessandro and Northcote (1977a,b) measured activities of two epimerases which interconvert UDP-galactose or UDP-arabinose, and UDP-glucose or UDP-xylose in cambial cells, differentiating xylem cells, and differentiated xylem cells in trees, and showed little difference between the enzyme activities of different cells. On the other hand, the activities of two enzymes which catalyze the conversion from UDP-glucose to UDP-xylose increased during differentiation of xylem cells of sycamore (Dalessandro and Northcote, 1977a). These results show that the synthesis of hemicellulose precursors is activated during differentiation. In addition, the activity of xylan synthase, which works at the final step of hemicellulose synthesis, increased in differentiating and differentiated xylem cells compared with cambial cells of sycamore trees. The increase in activity of xylan synthase, in relation to secondary wall formation during TE differentiation, has been reported in *Phaseolus* callus. *Phaseolus* hypocotyl (Bolwell and Northcote, 1981), and cultured *Zinnia* cells (Suzuki *et al.*, 1991). These results indicated that the synthesis of xylan was also regulated at the level of the synthase activity. The activity of the synthase was inhibited by either actinomycin D or D-2-(4-methyl-2,6-dinitroanilino)-*N*-methylpropionamide, suggesting that the enzyme activity was controlled at the level of transcription and translation (Bolwell and Northcote, 1983). The synthase was located in the Golgi apparatus in bean (Bolwell and Northcote, 1983). The activity may be due to a single transglycosylase, since no lipid or proteinaceous intermediate acceptors were

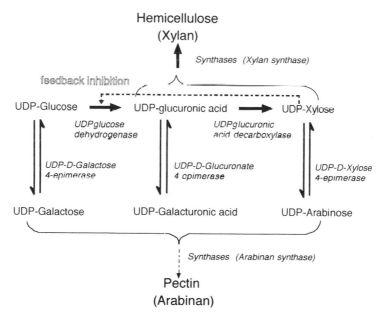

FIG. 9 Diagram to show the possible control of enzyme activities involved in polysaccharide synthesis during secondary wall formation. Modified from Northcote (1982).

found (Bolwell and Northcote, 1983). On the other hand, the decrease in pectin deposition occurs, coupled with the increase in xylan synthesis, during the differentiation. This may depend on a decrease of polygalacturonic acid synthase (Bolwell *et al.*, 1985b) and of arabinosyl transferase (Bolwell and Northcote, 1981). Unfortunately, the enzymes bound to the membrane have not been isolated yet, because the activity was lost during the separation from the membrane. There is an interesting approach that can be employed to characterize such membrane-bound enzymes that are difficult to isolate: Bolwell and Northcote (1984) raised, with total membrane proteins, an antibody that was monoclonal to endoplasmic and Golgi membranes and which inhibited an arabinosyl transferase localized in the membrane. However, there is no direct evidence to prove that this is a real antibody against an arabinosyl transferase.

Particulate membrane preparations isolated from differentiating xylem cells of *Pinus sylvestris* synthesized a β-1$\rightarrow$4-glucomannan from guanosine 5'-diphosphate (GDP)-mannose (Dalessandro *et al.*, 1986, 1988). This membrane preparation included epimerase activity which interconverts GDP-glucose and GDP-mannose, and glucomannan was synthesized by a supplement of only GDP-mannose. Because the addition of GDP-glucose to an *in vitro* system inhibited the incorporation of GDP-mannose into glucomannan, it was thought that the epimerase formed an enzyme complex with the synthase (glucosyltransferase) and may function in the conversion of the mannose to glucose at a definite ratio to maintain the synthesis of the glucomannan at an optimum rate (Dalessandro *et al.*, 1988). Increase in activity of the glucomannan synthase was found to be associated with TE differentiation in suspension-cultured cells of pine (Ramsden and Northcote, 1987a,b). There was another glucosyltransferase which synthesized β 1$\rightarrow$3 and 1$\rightarrow$4 mixed glucans from UDP-glucose in the membrane preparation of xylem cells (Dalessandro *et al.*, 1988). In addition to glycosyltransferase and epimerase, the enzyme complex on a membrane may include transporters, binding proteins to hold the acceptor molecules, and subsidiary proteins for regulation (Northcote, 1989). The enzyme complexes for hemicelluloses, similarly to that for cellulose, are partially constructed on the endoplasmic reticulum and then transferred to the Golgi apparatus, where they are modified and sorted. The hemicelluloses, therefore, are formed within vesicles and cisternae of the Golgi apparatus. The vesicles are transported to the plasma membrane, where fusion occurs, and the polysaccharides are released into cell wall. The orientation of the vesicle transport may be regulated by the cytoskeleton, e.g., by microtubules or actin filaments.

3. Lignin

Lignin deposition is a characteristic process of secondary wall formation (Wardrop, 1981). Lignification also occurs at wound sites to protect against microbial attack when plants are wounded or subjected to elicitors derived from fungal cell walls and culture fluids. Lignification, which is detected by phloroglucinol staining, was found to occur restrictively on the bands of secondary wall several hours later than the onset of visible secondary wall thickening in *Zinnia* cells (Fukuda and Komamine, 1982). The biosynthesis of lignin involves many steps, including shikimate and phenylpropanoid pathways, with the final step being polymerization of cinnamyl alcohols in cell walls. PAL, 4-coumarate:CoA ligase, and cinnamyl alcohol dehydrogenase in the phenylpropanoid pathway, and wall-bound peroxidase, which functions in polymerization of precursors, have been suggested to be markers of lignification during TE differentiation (Haddon and Northcote, 1976a; Kuboi and Yamada, 1978; Fukuda and Komamine, 1982; Masuda *et al.*, 1983; Church and Galston, 1988c; Walter *et al.*, 1988). Lignin deposition is thought to be correlated with the deposition of cellulose and hemicellulose during secondary wall formation. 2-Aminooxy-3-phenylpropionic acid (AOPP), which is a specific inhibitor of PAL, prevented lignification of *Vigna* and *Zinnia* cells without influencing organized deposition of cellulose and hemicellulose (Smart and Amrhein, 1985; Ingold *et al.*, 1990). However, when *Zinnia* differentiating cells were cultured in the presence of AOPP, elevated levels of xylosyl polysaccharides were released in the medium and the total carbohydrates of cell walls increased (Ingold *et al.*, 1990). This implied that the integration of cell wall polysaccharides into the wall was disturbed as a result of the absence of lignin. These results indicated that lignin synthesis was not coupled with either the initiation of the synthesis of cell wall polysaccharides or patterned deposition of the polysaccharides, but that lignin deposition had a role in the integration of cellulose and other polysaccharides into the secondary wall.

PAL, which produces cinnamic acid from phenylalanine, is a key enzyme for supplying lignin precursors; PAL has been shown to be correlated with TE differentiation in many plant materials (Rubery and Northcote, 1968; Rubery and Fosket, 1969; Haddon and Northcote, 1975; Durst, 1976; Kuboi and Yamada, 1978; Fukuda and Komamine, 1982; Lin and Northcote, 1990). Cinnamic acid, however, is an important precursor of other phenolic compounds, such as flavonoid pigments and coumarin phytoalexins. PAL is a tetrameric enzyme and the molecular weight of the PAL subunit is about 77,000 in bean (Hahlbrock and Scheel, 1989). In *Zinnia* cells, PAL showed two peaks of activity during differentiation (Fig. 10A) (Fukuda and Komamine, 1982). The first peak probably depends on a response to the injury caused by the isolation of cells from leaves and is not specific for differentiation. In contrast, the second peak is specific for differentiating cells and is consistent with the time of active lignin synthesis. The increase in PAL activity has been suggested to be a result of translational and transcriptional activation of the PAL gene, based on the experiments with inhibitors of translation and transcription (Jones and Northcote, 1981). Lin and Northcote (1990) demonstrated, using anti-potato PAL serum and cDNA for bean PAL as probes, that the levels of protein and mRNA of PAL peaked at the same time when PAL activity peaked in *Zinnia* cells (Fig. 10B). This indicated that the

PAL gene was temporally and preferentially expressed in association with lignification during TE differentiation.

PAL is encoded by a small family of genes in bean (Cramer *et al.*, 1989), parsley (Lois *et al.*, 1989), rice (Minami *et al.*, 1989), and *Arabidopsis* (Ohl *et al.*, 1990). Liang *et al* (1989a), with RNase protection with gene-specific probes, revealed that three PAL genes in bean were expressed differentially during development and in response to different environmental stimuli. The 5' upstream sequence of the *PAL 2* gene from bean was fused to the coding region of the reporter gene encoding β-glucuronidase (GUS) and transferred to tobacco by *Agrobacterium tumefaciens*-mediated leaf disk transformation (Liang *et al.*, 1989b; Bevan *et al.*, 1989). Histochemical analysis of GUS expression showed that the *PAL* promoter was active in differentiating xylem cells. However, high expression of GUS was also found in other cell types that accumulated phenylpropanoid derivatives in response to mechanical wounding and illumination, and during flower develop-ment. These data indicate that the *PAL 2* promoter transduces a complex set of developmental and environmental stimuli into an integrated spatial and temporal program of gene expression (Liang *et al.*, 1989b). Analysis of the regulatory properties of 5' deleted *PAL 1* promoters of *Arabidopsis* showed that the proximal region of the promoter to −290 was sufficient to establish the full tissue-specific pattern of expression and that the region proximal to −540 was responsive to

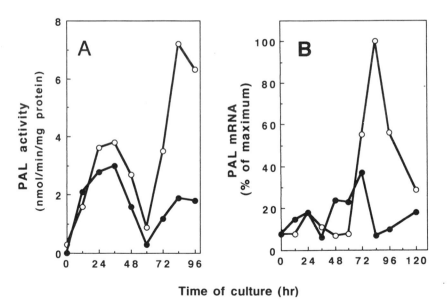

Time of culture (hr)

FIG. 10   Changes in PAL activity (A) and levels of PAL mRNA (B) in isolated *Zinnia* mesophyll cells in the TE-inductive (○) or the control (●) medium during culture. From Fukuda and Komamine (1982) and Lin and Northcote (1990).

environmental stimuli (Ohl *et al.*, 1990). Negative and positive elements were suspected to be located between −1816 and −823 and between −823 and −290, respectively. In addition to transcriptional regulation, there may be another type of regulational mechanism of PAL expression. Comparison between isoforms of PAL protein, labeled *in vivo* and *in vitro*, indicated that there are more isoforms of *in vivo*-labeled PAL (Bolwell *et al.*, 1985a). This suggested extensive post-translational modification of PAL subunits.

Kuboi and Yamada (1978) indicated that activities of shikimate dehydrogenase, cinnamate hydroxylase, caffeic acid-*O*-methyltransferase, and 5-hydroferulic acid-*O*-methyltransferase, as well as PAL, increased in association with TE differentiation in tobacco suspension cultures. Although the activity of *O*-methyltransferase was found to increase during the differentiation of bean callus (Haddon and Northcote, 1976a), the activity was not coupled with lignin synthesis during the differentiation of *Zinnia* cells (Fukuda and Komamine, 1982). More recently, a *S*-adenosylmethionine synthase gene (*sam-1*) of *Arabidopsis thaliana* was cloned (Peleman *et al.*, 1989). This enzyme synthesizes *S*-adenosylmethionine, which serves as a methyl group donor in numerous transmethylation reactions, including the *O*-methyltransferase reaction. Histochemical analysis demonstrated that *sam-1* was expressed in lignifying tissues.

The reduction of cinnamic acids to cinnamyl alcohols, which is a branch pathway specific for lignin synthesis, requires the activation of the acids to thioesters catalyzed by 4-coumarate:CoA ligase (4CL). 4CL is a monomeric enzyme that occurs in two isoforms in parsley, each of which is encoded by a single-copy gene (Hahlbrock and Scheel, 1989). The activity of this enzyme increased during secondary wall formation in *Zinnia* differentiating cells, and thus it may be a marker of secondary wall formation during TE differentiation (Church and Galston, 1988c). After CoA ligation, two reductive steps proceed which are catalyzed by cinnamoyl-CoA reductase and cinnamyl-alcohol dehydrogenase. Walter *et al.* (1988) isolated cDNA of the cinnamyl-alcohol dehydrogenase from suspension-cultured bean cells by immunoscreening with antiserum to the enzyme. Using the cDNA as a probe, the transcript was found to increase coupled with lignin deposition by addition of a fungal elicitor. However, it still remains to be resolved whether this gene is involved in lignin synthesis during TE differentiation.

The cinnamyl alcohols are delivered by vesicles derived from the Golgi apparatus or the endoplasmic reticulum to cell walls (Pickett-Heaps, 1968), where they are polymerized into lignin in a free radical reaction by peroxidases (Lewis and Yamamoto, 1990). The cinnamyl alcohols may form β-glucoside, which may be important for the transport of the alcohols. It has been shown biochemically (Mäder *et al.*, 1980) and cytochemically (Hepler *et al.*, 1972) that peroxidase isoenzymes bound to cell walls function in lignification. In *Zinnia* differentiating cells, there were three groups of peroxidases, i.e., peroxidases that are soluble in buffer, ionically bound to cell walls, or tightly bound to cell walls (Fukuda and Komamine, 1982). The activities of peroxidases ionically and tightly bound to cell walls were elevated markedly in the early and late stages of lignification, respectively, in cells cultured in a differentiation-induced medium but not in those in the control medium. Analysis of peroxidase isozymes revealed that two basic isoforms appeared in a differentiation-specific manner among several peroxidase isozymes that bound ionically to cell walls (Masuda *et al.*, 1983). One appeared when lignin synthesis was most active and the other appeared before lignin synthesis. The latter isoenzyme may be the same as the differentiation-specific peroxidase that was reported by Church and Galston (1988c). Ionically bound peroxidase isozymes in cultured *Zinnia* cells showed different substrate specificities and different requirements for metal ions (Y. Sato, unpublished data). Mäder *et al.* (1980) indicated for tobacco callus that two different acidic peroxidase isoenzyme groups in cell walls play different roles in lignification, that is, the production of H_2O_2 and polymerization of cinnamyl alcohols at the expense of H_2O_2. Although there is a difference between wall peroxidases of tobacco and *Zinnia*, i.e., they are acidic and basic, respectively, it is possible that different isoenzymes of wall-bound peroxidase with different substrate specificities are induced at different stages of differentiation and function differently in lignification. The wall peroxidase may establish other covalent linkages between the polypeptides whose tyrosine residues are oxidized by the enzyme and between polysaccharides by forming diferulic acid from ferulic acid (Northcote, 1989). A gene of an acid peroxidase isozyme of tobacco has been isolated (Lagrimini *et al.*, 1987). Lagrimini *et al.* (1990) made transgenic plants that had peroxidase activity that was 10-fold higher than in wild-type plants by introducing a chimeric gene composed of the cauliflower mosaic virus 35S promoter and cDNA of the acid peroxidase, resulting in a unique phenotype with chronic severe wilting, but not affecting xylem formation. At present, attempts to isolate a peroxidase gene involved in lignification during TE differentiation have not been successful. Further information on lignification can be obtained from the reviews by Higuchi (1985) and Lewis and Yamamoto (1990).

4. Cell Wall Proteins

In addition to enzymes involved in lignin synthesis, gene expression of structural proteins of cell walls during TE differentiation has been analyzed. Glycine-rich proteins (GRPs) are a class of such proteins that have a very repetitive primary structure, containing about 60% glycine which is predominantly arranged in $(Gly-X)_n$ repeats (Condit and Meagher, 1986, 1987; Keller *et al.*, 1988). GRPs were shown, using an antibody, to be located in secondary wall thickenings of xylem cells in *Phaseolus vulgaris* (Keller *et al.*, 1989a) and petunia (Condit *et al.*, 1990). Transgenic tobacco plants containing a chimeric gene of a *GRP* promoter fused to the GUS reporter gene expressed the GUS activity in the vascular tissue of roots, stems, leaves, and flowers (Keller *et al.*, 1989b). The gene was expressed during TE differentiation in both primary and secondary vascular tissues. The *GRP* promoter becomes active after the secondary wall thickening starts, while the

PAL 2 promoter described above is active in the initial stages of xylem differentiation (Liang *et al.*, 1989b). Thus, the *PAL 2* and *GRP* promoters may be induced by different components of a putative signal cascade controlling secondary wall formation during TE differentiation. Wounding induced the expression of the *GRP* promoter on the outside of the vascular cylinder in tobacco stem, similar to the expression of the *PAL 2* promoter (Keller *et al.*, 1989b).

Hydroxyproline-rich glycoproteins (HRGPs) are another class of wall-associated proteins that may function in secondary wall formation. Extensins, which are the best characterized of the HRGPs, contain a characteristic repeat of Ser-Pro$_4$ in which proline residues are hydroxylated and glycosylated (Cassab and Varner, 1988). Extensins contribute to the mechanical strength of cell walls (Cassab and Varner, 1988) and also to cell function by controlling microtubule organization (Akashi and Shibaoka, 1990). Extensins have been located in the cell wall of specific types of cells, including vascular elements of soybean cotyledons (Cassab and Varner, 1987). Stiefel *et al.* (1990), using *in situ* hybridization, studied the spatial pattern of expression for a maize gene encoding a HRGP. The expression of the gene was transient and high in regions initiating vascular elements and sclerenchyma in embryos, leaves, and roots. The most abundant signal appeared at a morphological stage in which TE differentiation was in progress. This pattern was similar to that observed in bean for GRP (Keller *et al.*, 1989b). The elucidation of the function of these wall proteins in secondary wall formation is one of the most important problems which should be examined.

B. Autolysis

The final important step of TE differentiation is cell autolysis, in which vacuole and other cytoplasmic organelles, including the nucleus, plastids, mitochondria, Golgi apparatus and the endoplasmic reticulum, are disrupted, and the cell dies and becomes functional in conduction (O'Brien and Thimann, 1967; Torrey *et al.*, 1971; Srivastava and Singh, 1972; Esau and Charvat, 1978; O'Brien, 1981). During this stage, the end wall between two TEs aligned longitudinally is hydrolyzed, creating a vessel or tracheary tube. Hydrolysis of the wall also occurs in lateral portions of the primary wall, leading to the maturation of TEs. Disruption of the vacuole occurs several hours after visible secondary wall formation in cultured differentiating *Zinnia* cells, and is followed by a rapid degradation of macromolecules in the cytoplasm, and then rapid loss of the cell contents (A. Minami and H. Fukuda, unpublished observation). This autolytic process is thought to be a programmed death and to involve various active events. However, there are few reports on biochemical or molecular genetical analysis of the process (Gahan, 1978; Thelen and Northcote, 1989).

The involvement of acid phosphatase, which is located in the vacuole, in the autolysis has been reported in *Vicia faba* (Gahan and Maple, 1966), *Coleus* (Jones and Villiers, 1972), various mosses (Hébant, 1973), and *Vigna* (De and Roy, 1984). De and Roy (1984) indicated that an anionic isoenzyme of acid phosphatase was expressed specifically during TE differentiation of *Vigna* callus. Membrane-bound hydrolytic enzymes, such as plasma membrane-bound K^+-activated acyl-phosphatase and endoplasmic reticulum-bound glucose-6-phosphatase, may also be involved in autolysis during differentiation (Gahan, 1978).

O'Brien (1981) suggested that the swelling of the primary wall regions that would be digested marked an early action of wall-degrading enzymes which had been inserted in the wall prior to cytoplasmic disruption.

Thelen and Northcote (1989) found that activities of RNase and single-strand (ss)-specific DNase increased dramatically in association with visible develop-ment of *Zinnia* TEs, and fell off rapidly after maturation was completed. This indicated that the activities were induced at the late stage of differentiation and were probably involved in autolysis. Among nucleases that are expressed specifically during differentiation (Fig. 11), there were three RNases (17, 22, and 25 kDa), and a nuclease (43 kDa) that shows both RNase and ss-DNase activities. The nucleases (43, 22, and 25 kDa) appeared 12 hr prior to the visible formation of TEs and increased conspicuously during the maturation phase of differentiation. A 17-kDa RNase was present transiently during the initial induction of other nucleolytic enzymes. These enzymes are considered to be good markers of the autolytic stage of differentiation, but the compartment and function of these enzymes still remain to be analyzed. A specific nuclease is known to appear in a developmentally programmed process in which nuclear structure should be de-

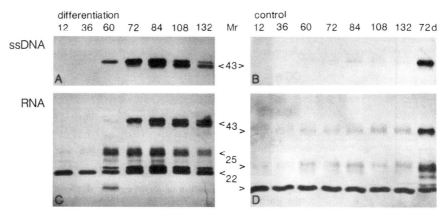

FIG. 11   Changes in single-strand-specific DNase (A, B) and RNase (C, D) in isolated *Zinnia* meso-phyll cells in the TE-inductive (A, C) or the control medium (B, D) during culture. The times (in hours) at which cell samples were taken during the course of culture are given at the top of each lane. The sample applied in the last lane on the right of gels B and D (72d) was from TEs at 72 hr. From Thelen and Northcote (1989).

stroyed. As to maternal inheritance, nuclease C was found to be activated and to function in preferential exclusion of male chloroplast DNA (Ogawa and Kuroiwa, 1985a,b). Thus, nucleases are expressed in a developmentally regulated manner and cause disruption of a specific nuclear structure to reach a new developmental stage. The 43-kDa nuclease has been isolated and characterized in part (Thelen and Northcote, 1989). We expect that further analysis of this enzyme will cast a new light on the problem of autolysis during TE differentiation.

Substances released from autolyzing TEs can promote new cytodifferentiation. Sheldrake and Northcote (1968a,b) suggested that auxin and cytokinin, produced as a consequence of cellular autolysis, may induce the formation of new TEs. Thus, autolysis is not only a final step of differentiation but also the first step in an adjacent cell.

C. Other Changes

Around 12 hr before secondary wall deposition starts, minor but differentiation-specific changes were observed on two-dimensional polypeptide maps of *Zinnia* cells (Fig. 12) (Fukuda and Komamine, 1983). Two newly synthesized polypeptides appeared in cells cultured in TE-inductive medium, but not in those in the control medium, and their synthesis continued at least until the time when a secondary wall begins to form. At the same time, two other polypeptides disappeared in a differentiation-specific manner. Recently, a few kinds of cDNA have been isolated that encode mRNAs which appear in a similar pattern to the newly synthesized polypeptides (T. Demura and H. Fukuda, unpublished data). The function of such translational products is unknown at present. However, because they appeared at the transition point from the early process to the late process, it is likely that they play important roles in the initiation or progress of the late events. In addition, these polypeptides and cDNA can be good markers for investigating the molecular mechanism of the transition from the early to the late process.

VI. Conclusions

To conclude, I will discuss various aspects of TE differentiation based on the results obtained with the extensively studied *Zinnia* system (Fukuda and Komamine, 1980a,b, 1981a,b, 1982, 1983, 1985; Burgess and Linstead, 1984; Falconer and Seagull, 1985a,b, 1986, 1988; Sugiyama *et al.*, 1986, 1990; Fukuda, 1987, 1989a,b; Kobayashi *et al.*, 1987, 1988; Church and Galston, 1988a,b,c, 1989; Roberts and Haigler, 1989, 1990; Fukuda and Kobayashi, 1989; Sugiyama and Komamine, 1987a,b, 1990; Thelen and Northcote, 1989; Lin and Northcote, 1990;

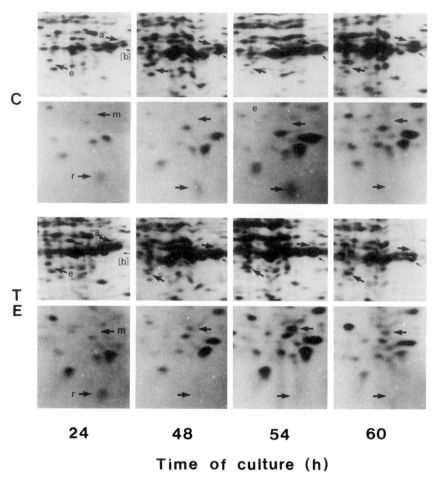

24   48   54   60

Time of culture (h)

FIG. 12   Changes in polypeptides in isolated *Zinnia* mesophyll cells in the TE-inductive (TE) or the control (C) medium during culture. The polypeptides were extracted from cells that were labeled with [^{35}S]methionine for 2 hr at the indicated hours. Note that polypeptides a and r were preferentially synthesized in control cells and e and m in TEs between 48 and 60 hr of culture. From Fukuda and Komamine (1983).

Suzuki *et al.*, 1991). Figure 13 is a summary of various events occurring in the process of redifferentiation into TEs from isolated mesophyll cells of *Zinnia elegans*. In the differentiation, initiation may occur by wounding and by the cooperative action of auxin and cytokinin, leading 30–60% of isolated mesophyll cells to differentiate into TEs. The process of differentiation after receiving the stimuli is divided into early and late processes, which may be defined as pre-restricted and restricted processes, respectively.

The early process is thought to be a complex process involving a variety of events which may be grouped into three, that is, house-keeping, dedifferentiation-related, and differentiation-specific events. Most of the events in this early process occur in almost all cultured mesophyll cells, including differentiating and non-differentiating cells; therefore they are not specific for differentiation, but are necessary for the sequence of differentiation. For example, active tubulin synthesis is observed at the early stage both in differentiation-induced culture and in control culture (Fukuda, 1987). However, the synthesis is essential for the new construction of microtubule arrays in the late process of differentiation, which controls patterned thickenings of secondary wall. Similarly, the synthesis of repair-type DNA that occurs in the early process is not specific for differentiating cells, but the inhibition of the synthesis causes the blockage of differentiation, suggesting that this event is also necessary for the sequence of differentiation (Sugiyama and Komamine, 1987a,b). These events may be involved in the "dedifferentiation process" in which isolated mesophyll cells lose their potential as photosynthetic cells and acquire the ability to grow and differentiate in the new environment. I would like to emphasize that "dedifferentiation" does not mean either callus formation or cell proliferation *in vitro*, but does mean a change in cell function.

A good example of dedifferentiation is a change in the organization of arrays of actin filaments in the early process of differentiation. Actin filaments of isolated mesophyll cells reorient from a reticulate array between the plasma membrane and each chloroplast to a three-dimensional network during the very early process of culture (Kobayashi *et al.*, 1987). This change causes the release of chloroplasts from the vicinity of the plasma membrane, and consequently, the loss of photosynthetic activity. At the same time, this forms a new intracellular organization which can function in cell development. Therefore, the reorganization of actin filaments is considered as a typical dedifferentiation-related event. The main role in TE differentiation of events in the early process may be to offer materials and cellular conditions that will be used for making structures and functions specific for TE differentiation in the late process. Unfortunately, differentiation-specific events have not as yet been found in the early process. However, it has been discovered that at least three different tubulin genes are expressed during TE differentiation in *Zinnia* cells (T. Yoshimura and H. Fukuda, unpublished data). This makes us expect that some of the tubulin isogenes may be expressed in the early process in a differentiation-specific manner.

The late process involves a variety of differentiation-specific events, most of which have been found in association with secondary wall thickening and autolysis. In this process, cells are engaged, probably irreversibly, in TE differentiation. As shown in Fig. 13B, most late events found in *Zinnia* cells, such as increases in the activities of PAL, xylosyltransferase, and RNases, occur after 60 hr of culture and function in secondary wall formation or autolysis. The synthesis of new species of proteins at 48 hr of culture is one of the earliest events which occurs specifically for TE differentiation in the late process (Fukuda and Komamine,

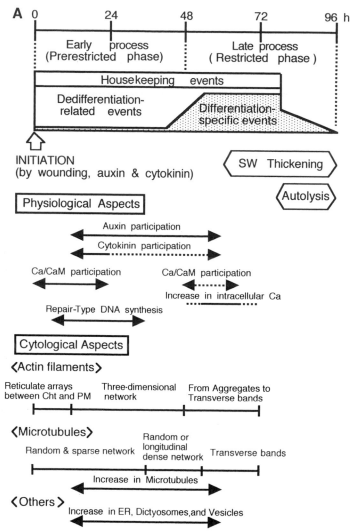

FIG. 13   (A, B) Sequential events in the process of TE differentiation from single mesophyll cells of *Zinnia elegans*. SW, Secondary wall; CaM, calmodulin; Cht, chloroplasts; PM, plasma membrane; IBW, ionically bound; TBW, tightly bound; 4CL, 4-coumarate:CoA ligase.

1983). Some cDNAs that encode mRNAs which appear in a differentiation-specific manner at 48 hr of culture have also been isolated (T. Demura and H. Fukuda, unpublished data). At present, the function of these differentiation-specific proteins is unknown, but, because they appear at the transition point to the irreversible process, they may play an important role in the initiation of secondary wall formation or autolysis. Analysis of the expression of these mRNAs may also

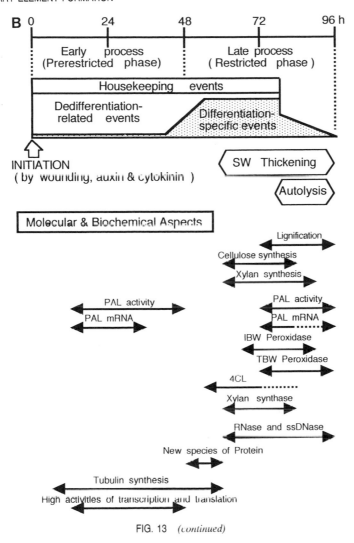

FIG. 13 *(continued)*

cast a new light on the elucidation of the regulation mechanism of gene expression in TE differentiation.

The presence of two different processes in TE differentiation is not specific for *Zinnia* cells, but is observed in other *in vitro* systems and even in the formation of primary xylem *in situ*. Gahan (1988) separated the process of TE differentiation *in situ* into five stages: competence of target cell, cell elongation, endoduplication, secondary wall formation, and autolysis. This subdivision shows the presence of a late process, including secondary wall formation and autolysis, and a process preceding the late process. The events in the late process have been considered to be fairly common in both *in vitro* and *in situ* differentiation. In the early process, there are also common events both *in vitro* and *in situ*, such as development of dictyosomes and the endoplasmic reticulum. However, we do not know to what extent the early process of differentiation in *in vitro* systems reflects the process *in situ*.

The information obtained about TE differentiation so far is still limited and fragmentary. Much more work should be done with the *Zinnia* system to elucidate the regulation mechanism of differentiation. In particular, the initiation of both the early process and the transition from the early to the late process should be studied extensively. In addition, detailed analysis of the content of the early process is needed.

Acknowledgments

I would like to thank Professors L. W. Roberts and P. Tandon for reading this manuscript. This work was supported in part by Grant-in-Aids from the Ministry of Education, Science and Culture of Japan and from the Inamori Foundation.

References

Ainley, W. M., Walker, J. C., Nagao, R. T., and Key, J. L. (1988). *J. Biol. Chem.* **263,** 10658–10666.
Akashi, T., and Shibaoka, H. (1990). *Planta* **182,** 363–369.
Albinger, G., and Beiderbeck, T. (1983). *Z. Pflanzenphysiol.* **112,** 443–448.
Alliotte, T., Tiré, C., Engler, G., Peleman, J., Caplan, A., van Montagu, M., and Inzé, D. (1989). *Plant Physiol.* **89,** 743–752.
Aloni, R. (1987). *Annu. Rev. Plant Physiol.* **38,** 179–204.
Aloni, R., and Jacobs, W. P. (1977). *Am. J. Bot.* **64,** 395–403.
Aloni, R., and Plotkin, T. (1985). *Planta* **163,** 126–132.
An, G., Costa, M. A., and Ha, S.-B. (1990). *Plant Cell* **2,** 225–233.
Avanzi, S., Maggini, F., and Innocenti, A. M. (1973). *Protoplasma* **76,** 197–210.
Barbier-Brygoo, H., Ephritikhine, G., Klambt, D., Ghislain, M., and Guern, J. (1989). *Proc. Natl. Acad. Sci. U.S.A.* **86,** 891–895.
Barnett, J. R. (1979). *Curr. Adv. Plant Sci.* **11,** 33.1–33.13.
Barnett, J. R., ed. (1981). "Xylem Cell Development." Castle House Publ. Ltd., Tunbridge Wells.
Basile, D. V., Wood, H. N., and Braun, A. C. (1973). *Proc. Natl. Acad. Sci. U.S.A.* **70,** 3055–3059.
Bevan, M., and Northcote, D. H. (1979a). *Planta* **147,** 77–81.
Bevan, M., and Northcote, D. H. (1979b). *J. Cell Sci.* **39,** 339–353.
Bevan, M., Shufflebottom, D., Edwards, K., Jefferson, R., and Schuch, W. (1989). *EMBO J.* **8,** 1899–1906.
Bishop, P. D., Pearce, G., Bryant, J. E., and Ryan, C. A. (1984). *J. Biol. Chem.* **259,** 13172–13177.
Boller, T., and Kende, I. (1980). *Nature (London)* **286,** 259–260.
Bolwell, G. P., and Northcote, D. H. (1981). *Planta* **152,** 225–233.
Bolwell, G. P., and Northcote, D. H. (1983). *Biochem. J.* **210,** 497–507.
Bolwell, G. P., and Northcote, D. H. (1984). *Planta* **162,** 139–146.

Bolwell, G. P., Bell, J. N., Cramer, C. L., Schuch, W., Lamb, C. J., and Dixon, R. A. (1985a). *Eur. J. Biochem.* **149,** 411–419.

Bolwell, G. P., Dalessandro, G., and Northcote, D. H. (1985b). *Phytochemistry* **24,** 699–702.

Bornman, C. H., and Ellis, R. P. (1971). *J. S. Afr. Bot.* **37,** 281–289.

Burgess, J., and Linstead, P. (1984). *Planta* **160,** 481–489.

Cassab, G. I., and Varner, J. E. (1987). *J. Cell Biol.* **105,** 2581–2588.

Cassab, G. I., and Varner, J. E. (1988). *Annu. Rev. Plant Physiol.* **39,** 321–353.

Chen, J., and Varner, J. E. (1985). *EMBO J.* **4,** 2145–2151.

Church, D. L., and Galston, A. W. (1988a). *Plant Physiol.* **88,** 92–96.

Church, D. L., and Galston, A. W. (1988b). *Phytochemistry* **27,** 2435–2439.

Church, D. L., and Galston, A. W. (1988c). *Plant Physiol.* **88,** 679–684.

Church, D. L., and Galston, A. W. (1989). *Plant Cell Physiol.* **30,** 73–78.

Cleveland, D. W. (1987). *J. Cell Biol.* **104,** 381–383.

Clutter, M. E. (1960). *Science* **132,** 548–549.

Comer, A. E. (1978). *Plant Physiol.* **62,** 354–359.

Condit, C. M., and Meagher, R. B. (1986). *Nature (London)* **323,** 178–181.

Condit, C. M., and Meagher, R. B. (1987). *Mol. Cell. Biol.* **7,** 4272–4279.

Condit, C. M., McLean, B. G., and Meagher, R. B. (1990). *Plant Physiol.* **93,** 596–602.

Cramer, C. L., Edwards, K., Dron, M., Liang, X., Dildine, S. L., Bolwell, G. P., Dixon, R. A., Lamb, C. L., and Schuch, W. (1989). *Plant Mol. Biol.* **12,** 367–383.

Cronshaw, J. (1965). *In* "Cellular Ultrastructure of Woody Plants" (W. A. Cote, Jr., ed.), pp. 99–124. Syracuse Univ. Press, Syracuse, New York.

Dalessandro, G. (1973). *Plant Cell Physiol.* **14,** 1167–1176.

Dalessandro, G., and Northcote, D. H. (1977a). *Biochem. J.* **162,** 267–279.

Dalessandro, G., and Northcote, D. H. (1977b). *Biochem. J.* **162,** 281–288.

Dalessandro, G., and Roberts, L. W. (1971). *Am. J. Bot.* **58,** 378–385.

Dalessandro, G., Piro, G., and Northcote, D. H. (1986). *Planta* **169,** 564–574.

Dalessandro, G., Piro, G., and Northcote, D. H. (1988). *Planta* **175,** 60–70.

Datta, P. C., Mukherjee, S., Chakrabarti, S., and Saha, S. (1979). *Indian J. Exp. Biol.* **17,** 46–49.

Dauwalder, M., Roux, S. J., and Hardison, L. (1986). *Planta* **168,** 461–470.

Davies, E., and Schuster, A. (1981). *Proc. Natl. Acad. Sci. U.S.A.* **78,** 2422–2426.

De, K. K., and Roy, S. C. (1984). *Theor. Appl. Genet.* **68,** 285–287.

Dodds, J. H. (1980). *Z. Pflanzenphysiol.* **99,** 283–285.

Dodds, J. H. (1981a). *In* "Xylem Cell Development" (J. R. Barnett, ed.), pp. 153–167. Castle House Publ. Ltd., Tunbridge Wells.

Dodds, J. H. (1981b). *Plant, Cell Environ.* **4,** 145–146.

Dodds, J. H., and Phillips, R. (1977) *Planta* **135,** 213–216.

Doley, D., and Leyton, L. (1970). *New Phytol.* **69,** 87–102.

Durante, M., Cremonini, R., Brunori, A., Avanzi, S., and Innocenti, A. M. (1977). *Protoplasma* **93,** 289–303.

Durante, M., Frediani, M., Mariani, L., Citti, L., Geri, C., and Cremonini, R. (1990). *Protoplasma* **158,** 149–154.

Durst, F. (1976). *Planta* **132,** 221–227.

Dyer, C. M., Henstrand, J. M., Handa, A. K., and Herrmann, K. M. (1989). *Proc. Natl. Acad. Sci. U.S.A.* **82,** 6731–6735.

Esau, K., and Charvat, I. (1978). *Ann. Bot. (London)* [N.S.] **42,** 665–677.

Falconer, M. M., and Seagull, R. W. (1985a). *Protoplasma* **125,** 190–198.

Falconer, M. M., and Seagull, R. W. (1985b). *Protoplasma* **128,** 157–166.

Falconer, M. M., and Seagull, R. W. (1986). *Protoplasma* **133,** 140–148.

Falconer, M. M., and Seagull, R. W. (1988). *Protoplasma* **144,** 10–16.

Farmer, E. E., Pearce, G., and Ryan, C. A. (1989). *Proc. Natl. Acad. Sci. U.S.A.* **86,** 1539–1542.

Fosket, D. E. (1968). *Proc. Natl. Acad. Sci. U.S.A.* **59,** 1089–1096.

Fosket, D. E. (1970). *Plant Physiol.* **46,** 64–68.
Fosket, D. E., and Roberts, L. W. (1964). *Am. J. Bot.* **51,** 19–25.
Fosket, D. E., and Torrey, J. G. (1969). *Plant Physiol.* **44,** 871–880.
Freundlich, H. F. (1909). *Jahrb. Wiss. Bot.* **46,** 137–206.
Fukuda, H. (1987). *Plant Cell Physiol.* **28,** 517–528.
Fukuda, H. (1989a). *Plant Cell Physiol.* **30,** 243–252.
Fukuda, H. (1989b). *Bot. Mag.* **102,** 491–501.
Fukuda, H., and Iwata, N. (1986). *Plant Cell Physiol.* **27,** 273–283.
Fukuda, H., and Kobayashi, H. (1989). *Dev. Growth Differ.* **31,** 9–16.
Fukuda, H., and Komamine, A. (1980a). *Plant Physiol.* **65,** 57–60.
Fukuda, H., and Komamine, A. (1980b). *Plant Physiol.* **65,** 61–64.
Fukuda, H., and Komamine, A. (1981a). *Physiol. Plant.* **52,** 423–430.
Fukuda, H., and Komamine, A. (1981b). *Plant Cell Physiol.* **22,** 41–49.
Fukuda, H., and Komamine, A. (1982). *Planta* **155,** 423–430.
Fukuda, H., and Komamine, A. (1983). *Plant Cell Physiol.* **24,** 603–614.
Fukuda, H., and Komamine, A. (1985). *In* "Cell Culture and Somatic Cell Genetics of Plants" (I. K. Vasil, ed.), Vol. 2, pp. 149–212. Academic Press, Orlando, Florida.
Gahan, P. B. (1978). *Ann. Bot. (London)* [N.S.] **42,** 755–758.
Gahan, P. B. (1988). *In* "Vascular Differentiation and Plant Growth Regulators" (L. W. Roberts, P. B. Gahan, and R. Aloni, eds.), pp. 1–21. Springer-Verlag, Berlin.
Gahan, P. B., and Maple, A. J. (1966). *J. Exp. Bot.* **17,** 151–155.
Gahan, P. B., Rana, M. A., and Phillips, R. (1983). *Cell Biochem. Funct.* **1,** 109–111.
Graham, J. S., Hall, G., Pearce, G., and Ryan, C. A. (1986). *Planta* **169,** 399–405.
Gronewald, J. W., and Hanson, J. B. (1980). *Plant Sci. Lett.* **18,** 143–150.
Gunning, B. E. S., and Hardham, A. R. (1982). *Annu. Rev. Plant Physiol.* **33,** 651–698.
Gunning, B. E. S., Hughes, J. E., and Hardham, A. R. (1978a). *Planta* **143,** 121–144.
Gunning, B. E. S., Hardham, A. R., and Hughes, J. E. (1978b). *Planta* **143,** 145–160.
Gunning, B. E. S., Hardham, A. R., and Hughes, J. E. (1978c). *Planta* **143,** 161–179.
Haddon, L. E., and Northcote, D. H. (1975). *J. Cell Sci.* **17,** 11–26.
Haddon, L. E., and Northcote, D. H. (1976a). *Planta* **128,** 255–262.
Haddon, L. E., and Northcote, D. H. (1976b). *J. Cell Sci.* **20,** 47–55.
Hagen, G., Uhrhammer, N., and Guilfoyle, T. J. (1988). *J. Biol. Chem.* **263,** 6442–6446.
Hahlbrock, K., and Scheel, D. (1989). *Annu. Rev. Plant Physiol. Plant Mol. Biol.* **40,** 347–369.
Haigler, C. H., and Brown, R. M., Jr. (1986). *Protoplasma* **134,** 111–120.
Hammersley, D. R. H., and McCully, M. E. (1980). *Plant Sci. Lett.* **19,** 151–156.
Hardham, A. R., and Gunning, B. E. S. (1978). *J. Cell Biol.* **77,** 14–34.
Hardham, A. R., and Gunning, B. E. S. (1979). *J. Cell Sci.* **37,** 411–442.
Hardham, A. R., and Gunning, B. E. S. (1980). *Protoplasma* **102,** 31–51.
Hardham, A. R., and McCully, M. E. (1982a). *Protoplasma* **112,** 143–151.
Hardham, A. R., and McCully, M. E. (1982b). *Protoplasma* **112,** 152–166.
Havránek, P., and Movák, F. J. (1973). *Z. Pflanzenphysiol.* **68,** 308–318.
Hawkins, S. W., and Phillips, R. (1983). *Plant Sci. Lett.* **32,** 221–224.
Hébant, C. (1973). *Protoplasma* **77,** 231–241.
Hepler, P. K. (1981). *In* "Cytomorphogenesis in Plants" (O. Kiermayer, ed.), pp. 327–347. Springer-Verlag, Berlin and New York.
Hepler, P. K., and Fosket, D. E. (1971). *Protoplasma* **72,** 213–236.
Hepler, P. K., and Newcomb, E. H. (1964). *J. Cell Biol.* **20,** 529–533.
Hepler, P. K., and Wayne, R. O. (1985). *Annu. Rev. Plant Physiol.* **36,** 397–439.
Hepler, P. K., Rice, R. M., and Terranova, W. A. (1972). *Can. J. Bot.* **50,** 977–983.
Hershko, A., and Ciechanover, A. (1982). *Annu. Rev. Biochem.* **51,** 335–364.
Herth, W. (1985). *Planta* **164,** 12–21.

Hesse, T., Feldwisch, J., Balshusemann, D., Bauw, G., Puype, M., Vandekerckhove, J., Lobler, M., Klambt, D., Schell, J., and Palme, K. (1989). *EMBO J.* **8**, 2453–2461.

Hewitt, E. J. (1963). *In* "Plant Physiology: A Treatise" (F. C. Steward, ed.), Vol. 3. Academic Press, New York.

Higuchi, T. (1985). *In* "Biosynthesis and Biodegradation of Wood Components" (T. Higuchi, ed.), pp. 141–160. Academic Press, Orlando, Florida.

Hogetsu, T. (1990). *Protoplasma* **156**, 67–73.

Hoyle, H. D., and Raff, E. C. (1990). *J. Cell Biol.* **111**, 1009–1026.

Ingold, E., Sugiyama, M., and Komamine, A. (1988). *Plant Cell Physiol.* **29**, 295–303.

Ingold, E., Sugiyama, M., and Komamine, A. (1990). *Physiol. Plant.* **78**, 67–74.

Inohara, N., Shimomura, S., Fukui, T., and Futai, M. (1989). *Proc. Natl. Acad. Sci. U.S.A.* **86**, 3564–3568.

Iwasaki, T., and Shibaoka, H. (1991). *Plant Cell Physiol.* **32**, 1007–1014.

Iwasaki, T., Fukuda, H., and Shibaoka, H. (1986). *Plant Cell Physiol.* **27**, 717–724.

Iwasaki, T., Fukuda, H., and Shibaoka, H. (1988). *Protoplasma* **143**, 130–138.

Jacobs, W. P. (1952). *Am. J. Bot.* **39**, 301–309.

Jacobs, W. P. (1954). *Am. Nat.* **90**, 163–169.

Jeffs, R. A., and Northcote, D. H. (1966). *Biochem. J.* **101**, 146–152.

Jeffs, R. A., and Northcote, D. H. (1967). *J. Cell Sci.* **2**, 77–88.

Jones, D. H., and Northcote, D. H. (1981). *Eur. J. Biochem.* **116**, 117–125.

Jones, D. T., and Villiers, T. A. (1972). *J. Exp. Bot.* **23**, 375–380.

Kakimoto, T., and Shibaoka, H. (1987). *Protoplasma* **140**, 151–156.

Keil, M., Sánchez-Serrano, J., Schell, J., and Willmitzer, L. (1990). *Plant Cell* **2**, 61–70.

Keller, B., Sauer, N., and Lamb, C. J. (1988). *EMBO J.* **7**, 3625–3633.

Keller, B., Schmid, J., and Lamb, C. J. (1989a). *EMBO J.* **8**, 1309–1314.

Keller, B., Templeton, M. D., and Lamb, C. J. (1989b). *Proc. Natl. Acad. Sci. U.S.A.* **86**, 1529–1533.

Kikuchi, M., Imazeki, H., and Sakai, S. (1989). *Plant Cell Physiol.* **30**, 765–773.

Kobayashi, H., Fukuda, H., and Shibaoka, H. (1987). *Protoplasma* **138**, 69–71.

Kobayashi, H., Fukuda, H., and Shibaoka, H. (1988). *Protoplasma* **143**, 29–37.

Kohlenbach, H. W., and Schmidt, B. (1975). *Z. Pflanzenphysiol.* **75**, 369–374.

Kohlenbach, H. W., and Schöpke, C. (1981). *Naturwissenschaften* **68**, 576–577.

Kohlenbach, H. W., Korber, M., and Li, L. (1982). *Z. Pflanzenphysiol.* **107**, 367–371.

Koritsas, V. M. (1988). *J. Exp. Bot.* **39**, 375–386.

Kuboi, T., and Yamada, Y. (1978). *Biochim. Biophys. Acta* **542**, 181–190.

Lagrimini, L. M., Burkhart, W., Moyer, M., and Rothstein, S. (1987). *Proc. Natl. Acad. Sci. U.S.A.* **84**, 7542–7546.

Lagrimini, L. M., Bradford, S., and Rothstein, S. (1990). *Plant Cell* **2**, 7–18.

Lawton, M. A., and Lamb, C. J. (1987). *Mol. Cell. Biol.* **7**, 335–341.

Lewis, N. G., and Yamamoto, E. (1990). *Annu. Rev. Plant Physiol. Plant Mol. Biol.* **41**, 455–496.

Liang, X., Dron, M., Cramer, C. L., Dixon, R. A., and Lamb, C. J. (1989a). *J. Biol. Chem.* **264**, 14486–14492.

Liang, X., Dron, M., Schmid, J., Dixon, R. A., and Lamb, C. J. (1989b). *Proc. Natl. Acad. Sci. U.S.A.* **86**, 9284–9288.

Lin, Q., and Northcote, D. H. (1990). *Planta* **182**, 591–598.

Little, C. H. A., and Savidge, R. A. (1987). *Plant Growth Regul.* **6**, 137–169.

Lois, R., Dietrich, A., Hahlbrock, K., and Schulz, W. (1989). *EMBO J.* **8**, 1641–1648.

Ludwig, S. R., Oppenheimer, D. G., Silflow, C. D., and Snustad, D. P. (1988). *Plant Mol. Biol.* **10**, 311–321.

Mäder, M., Ungemach, J., and Schloss, P. (1980). *Planta* **147**, 467–470.

Malawer, C. L., and Phillips, R. (1979). *Plant Sci. Lett.* **15**, 47–55.

Masuda, H., Fukuda, H., and Komamine, A. (1983). *Z. Pflanzenphysiol.* **112**, 417–426.

May, G. S. (1989). *J. Cell Biol.* **109**, 2267–2274.

McClure, B. A., and Guilfoyle, T. (1987). *Plant Mol. Biol.* **9**, 611–623.

McClure, B. A., and Guilfoyle, T. (1989). *Science* **243**, 91–93.

McClure, B. A., Hagen, G., Brown, C. S., Gee, M. A., and Guilfoyle, T. (1989). *Plant Cell* **1**, 229–239.

Miller, A. R., and Roberts, L. W. (1982). *Ann. Bot. (London)* [N.S.] **50**, 111–116.

Miller, A. R., and Roberts, L. W. (1984). *J. Exp. Bot.* **35**, 691–698.

Miller, A. R., Pengelly, W. L., and Roberts, L. W. (1984). *Plant Physiol.* **75**, 1165–1166.

Miller, A. R., Crawford, D. L., and Roberts, L. W. (1985). *J. Exp. Bot.* **36**, 110–118.

Minami, E., Ozeki, Y., Matsuoka, M., Koizuka, N., and Tanaka, Y. (1989). *Eur. J. Biochem.* **185**, 19–25.

Minocha, S. C. (1984). *J. Exp. Bot.* **35**, 1003–1015.

Minocha, S. C., and Halperin, W. (1974). *Planta* **116**, 319–331.

Mizuno, K., and Komamine, A. (1978). *Planta* **138**, 59–62.

Mizuno, K., Komamine, A., and Shimokoriyama, M. (1971). *Plant Cell Physiol.* **12**, 823–830.

Nakajima, N., Nakagawa, N., and Imazeki, H. (1988). *Plant Cell Physiol.* **29**, 989–998.

Nelson, C., Walker-Simmons, M., Makus, K., Zuroske, G., Graham, J., and Ryan, C. A. (1983). *ACS Symp. Ser.* **208**, 103–122.

Northcote, D. H. (1972). *Annu. Rev. Plant Physiol.* **30**, 425–484.

Northcote, D. H. (1982). *In* "Encyclopedia of Plant Physiology, New Series" (D. Boulter and B. Parthier, eds.), Vol. 14A, pp. 637–655. Springer-Verlag, Berlin and New York.

Northcote, D. H. (1989). "Plant Cell Wall Polymers," pp. 1–15. Am. Chem. Soc., Washington, D.C.

Northcote, D. H., Davey, R., and Lay, J. (1989). *Planta* **178**, 353–366.

O'Brien, T. P. (1974). *In* "Dynamic Aspects of Plant Ultrastructure" (A. W. Robards, ed.), pp. 414–440. McGraw-Hill, New York.

O'Brien, T. P. (1981). *In* "Xylem Cell Development" (J. R. Barnett, ed.), pp. 14–46. Castle House Publ. Ltd., Tunbridge Wells.

O'Brien, T. P., and Thimann, K. V. (1967). *Protoplasma* **63**, 443–478.

Ogawa, K., and Kuroiwa, T. (1985a). *Plant Cell Physiol.* **26**, 481–491.

Ogawa, K., and Kuroiwa, T. (1985b). *Plant Cell Physiol.* **26**, 493–503.

Ohl, S., Hedrick, S. A., Chory, J., and Lamb, C. J. (1990). *Plant Cell* **2**, 837–848.

Oppenheimer, D. G., Haas, N., Silflow, C. D., and Snustad, D. P. (1988). *Gene* **63**, 87–102.

Parthasarathy, M. V., Perdue, T. D., Witztum, A., and Alvernaz, J. (1985). *Am. J. Bot.* **72**, 1318–1323.

Pearce, D., Miller, A. R., Roberts, L. W., and Pharis, R. P. (1987). *Plant Physiol.* **84**, 1121–1125.

Peleman, J., Boerjan, W., Engler, G., Seurinck, J., Botterman, J., Alliotte, T., van Montagu, M., and Inzé, D. (1989). *Plant Cell* **1**, 81–93.

Phillips, R. (1980). *Int. Rev. Cytol. Suppl.* **11A**, 55–70.

Phillips, R. (1981a). *Planta* **153**, 262–266.

Phillips, R. (1981b). *Ann. Bot. (London)* [N.S.] **47**, 785–792.

Phillips, R. (1987). *Ann. Bot. (London)* [N.S.] **59**, 245–250.

Phillips, R., and Dodds, J. H. (1977). *Planta* **135**, 207–212.

Phillips, R., and Hawkins, S. W. (1985). *J. Exp. Bot.* **36**, 119–128.

Phillips, R., and Torrey, J. G. (1973). *Dev. Biol.* **31**, 336–347.

Pickard, B. G. (1973). *Bot. Rev.* **39**, 172–201.

Pickett-Heaps, J. D. (1967). *Dev. Biol.* **15**, 206–236.

Pickett-Heaps, J. D. (1968). *Protoplasma* **65**, 181–205.

Ramsden, L., and Northcote, D. H. (1987a). *J. Cell Sci.* **88**, 467–474.

Ramsden, L., and Northcote, D. H. (1987b). *Phytochemistry* **26**, 2679–2683.

Rana, M. A., and Gahan, P. B. (1982). *Ann. Bot. (London)* [N.S.] **50**, 757–762.

Rana, M. A., and Gahan, P. B. (1983). *Planta* **157**, 307–316.

Ray, T. B. (1984). *Plant Physiol.* **75**, 827–831.

Reuther, G., and Werckmeister, P. (1973). *Z. Pflanzenphysiol.* **70**, 276–282.

Reynolds, T. L. (1986). *J. Plant Physiol.* **125,** 179–184.

Rier, J. P., and Beslow, D. T. (1967). *Bot. Gaz. (Chicago)* **128,** 73–77.

Robbertse, P. J., and McCully, M. E. (1979). *Planta* **145,** 167–173.

Roberts, A. W., and Haigler, C. H. (1989). *Protoplasma* **152,** 37–45.

Roberts, A. W., and Haigler, C. H. (1990). *Planta* **180,** 502–509.

Roberts, L. W. (1976). "Cytodifferentiation in Plants: Xylogenesis as a Model System." Cambridge Univ. Press, London and New York.

Roberts, L. W. (1988). *In* "Vascular Differentiation and Plant Growth Regulators" (L. W. Roberts, P. B. Gahan, and R. Aloni, eds.), pp. 1–21. Springer-Verlag, Berlin.

Roberts, L. W., and Baba, S. (1987). *Environ. Exp. Bot.* **27,** 289–295.

Roberts, L. W., and Miller, A. R. (1982). *What's New Plant Physiol.* **13,** 13–16.

Roberts, L. W., Gahan, P. B., and Aloni, R., eds. (1988). "Vascular Differentiation and Plant Growth Regulators." Springer-Verlag, Berlin.

Ronchi, V. N., and Gregorini, G. (1970). *Bot. Ital.* **104,** 443–455.

Rost, T. L. (1984). *J. Plant Growth Regul.* **3,** 51–63.

Rubery, P. H., and Fosket, D. E. (1969). *Planta* **87,** 54–62.

Rubery, P. H., and Northcote, D. H. (1968). *Nature (London)* **219,** 1230–1234.

Ryan, C. A. (1978). *Trends Biochem. Sci.* **5,** 148–150.

Ryder, T. B., Hendrick, S. A., Bell, J. N., Liang, X., Clouse, S. D., and Lamb, C. J. (1987). *Mol. Gen. Genet.* **210,** 219–233.

Sachs, T. (1969). *Ann. Bot. (London)* [N.S.] **33,** 263–275.

Sachs, T. (1981). *Adv. Bot. Res.* **9,** 151–262.

Sachs, T. (1986). *In* "Plant Growth Substances" (M. Bopp, ed.), pp. 231–235. Springer-Verlag, Berlin and New York.

Saltveit, M. E., Jr., and Dilley, D. R. (1978). *Plant Physiol.* **61,** 447–450.

Sánchez-Serrano, J., Schmidt, R., Schell, J., and Willmitzer, L. (1986). *Mol. Gen. Genet.* **203,** 15–20.

Savidge, R. A., and Wareing, P. F. (1981). *In* "Xylem Cell Development" (J. R. Barnett, ed.), pp. 192–235. Castle House Publ. Ltd., Tunbridge Wells.

Seagull, R. W., Falconer, M. M., and Weerdenburg, C. A. (1987). *J. Cell Biol.* **104,** 995–1004.

Sheldrake, A. R., and Northcote, D. H. (1968a). *Planta* **80,** 227–236.

Sheldrake, A. R., and Northcote, D. H. (1968b). *New Phytol.* **67,** 1–13.

Shimomura, S., Sotobayashi, T., Futai, M., and Fukui, T. (1986). *J. Biochem. (Tokyo)* **99,** 1513–1524.

Shininger, T. L. (1975). *Dev. Biol.* **45,** 137–150.

Shininger, T. L. (1978). *In Vitro* **14,** 31–50.

Shininger, T. L. (1979). *Annu. Rev. Plant Physiol.* **30,** 313–337.

Shinshi, H., Mohnen, D., and Meins, F., Jr. (1987). *Proc. Natl. Acad. Sci. U.S.A.* **84,** 89–93.

Simon, S. (1908). *Ber. Dtsch. Bot. Ges.* **26,** 364–396.

Sinnott, E. W., and Bloch, R. (1944). *Proc. Natl. Acad. Sci. U.S.A.* **30,** 388–392.

Sinnott, E. W., and Bloch, R. (1945). *Am. J. Bot.* **32,** 151–156.

Smart, C. C., and Amrhein, N. (1985). *Protoplasma* **124,** 87–95.

Srivastava, L. M., and Singh, A. P. (1972). *Can. J. Bot.* **50,** 1795–1804.

Stanford, A., Bevan, M., and Northcote, D. H. (1989). *Mol. Gen. Genet.* **215,** 200–208.

Stiefel, V., Avila, L. R., Raz, R., Vallés, M. P., Gómez, J., Pagès, M., Izquiedo, J. A. M., Ludevid, M. D., Langdale, J. A., Nelson, T., and Puigdomènech, P. (1990). *Plant Cell* **2,** 785–793.

Sugiyama, M., and Komamine, A. (1987a). *Plant Cell Physiol.* **28,** 541–544.

Sugiyama, M., and Komamine, A. (1987b). *Oxford Surv. Plant Mol. Cell Biol.* **4,** 343–346.

Sugiyama, M., and Komamine, A. (1990). *Cell Differ. Dev.* **31,** 77–87.

Sugiyama, M., Fukuda, H., and Komamine, A. (1986). *Plant Cell Physiol.* **27,** 601–606.

Sugiyama, M., Fukuda, H., and Komamine, A. (1990). *Plant Cell Physiol.* **31,** 61–67.

Sussex, I. M., and Clutter, M. E. (1968). *In Vitro* **3,** 3–12.

Sussex, I. M., Clutter, M. E., and Goldsmith, M. H. M. (1972). *Am. J. Bot.* **59,** 797–804.

Suzuki, K., Ingold, E., Sugiyama, M., and Komamine, A. (1991). *Plant Cell Physiol.* **32,** 303–306.

Takahashi, Y., Kuroda, H., Tanaka, T., Machida, Y., Takabe, I., and Nagata, T. (1989). *Proc. Natl. Acad. Sci. U.S.A.* **86,** 9279–9283.

Thelen, M. P., and Northcote, D. H. (1989). *Planta* **179,** 181–195.

Theologis, A., and Laties, G. G. (1981). *Plant Physiol.* **68,** 53–58.

Theologis, A., Huynh, T. V., and Davis, R. W. (1985). *J. Mol. Biol.* **183,** 53–68.

Torrey, J. G. (1975). *Physiol. Plant.* **35,** 158–165.

Torrey, J. G., and Fosket, D. E. (1970). *Am. J. Bot.* **57,** 1072–1080.

Torrey, J. G., Fosket, D. E., and Hepler, P. K. (1971). *Am. Sci.* **59,** 338–352.

Tucker, W. Q. J., Wilson, J. W., and Gresshoff, P. M. (1986). *Ann. Bot. (London)* [N.S.] **57,** 675–679.

van der Zaal, E. J., Memelink, J., Mennes, A. M., Quint, A., and Libbenga, K. R. (1987). *Plant Mol. Biol.* **10,** 145–157.

Verma, D. P. S., and van Huystee, R. B. (1970). *Can. J. Bot.* **48,** 429–431.

Vöchting, H. (1892). Über Transplantation am Pflanzenkörper. Untersuchungen zur Physiologie and Pathologie." H. Laupp, Tübingen.

Walter, M. H., Grima-Pettenati, J., Grand, C., Boudet, A. M., and Lamb, C. J. (1988). *Proc. Natl. Acad. Sci. U.S.A.* **85,** 5546–5550.

Wardrop, A. B. (1965). *In* "Cellular Ultrastructure of Woody Plants" (W. A. Cote, ed.), pp. 61–97. Syracuse Univ. Press, Syracuse, New York.

Wardrop, A. B. (1981). *In* "Xylem Cell Development" (J. R. Barnett, ed.), pp. 168–191. Castle House Publ. Ltd., Tunbridge Wells.

Wetmore, R. H., and Rier, J. P. (1963). *Am. J. Bot.* **50,** 418–430.

Wetmore, R. H., and Sorokin, S. (1955). *J. Arnold Arbor., Harv. Univ.* **36,** 305–317.

Wooding, F. B. P., and Northcote, D. H. (1964). *J. Cell Biol.* **23,** 327–337.

Wright, K., and Northcote, D. H. (1973). *J. Cell Sci.* **12,** 37–53.

Yokota, T., Nakamura, Y., Takahashi, N., Nonaka, M., Sekimoto, H., Oshio, H., and Takatsuto, S. (1991). *In* "Gibberellin" (N. Takahashi, B. O. Phinney, and J. MacMillan, eds.), pp. 339–349. Springer-Verlag, Berlin.

INDEX